A CENTURY OF MATHEMATICS

Benjamin Franklin Finkel, Editor of the MONTHLY, 1894–1912

A CENTURY OF MATHEMATICS

Through the Eyes of the *Monthly*

edited by John Ewing, Indiana University

Library of Congress Catalog Number 93-08 1168
ISBN 0-88385-459-7
Copyright © 1994
The Mathematical Association of America (Incorporated)
PRINTED IN THE UNITED STATES OF AMERICA

Contents

4. Battles and Wars: 1941–1950 131

5. Mathematics Gets Serious: 1951–1960 209

6. Mathematics Expands: 1961–1970 237

Preface

This book celebrates the Centennial of the American Mathematical Monthly, which was first published in January of 1894. It is not merely a book about the Monthly, however; it is a book about American Mathematics.

This is *not* The Best of the Monthly. Many of the best (mathematics) articles in the Monthly are long—others are dated. Choosing a small number would have been impossible; choosing a large number would have produced a giant volume that everyone would admire, and no one would read (or buy). Mathematics articles from 100 or even 50 years ago do not make good reading. The articles were chosen to tell a story, not because they are the best articles ever published.

This is *not* a Monthly Sampler. To fairly represent 100 years and over 70,000 pages of a journal in one volume, we would have to include articles on many subjects, in many styles, by many authors. The result would have been an encyclopedia: We wanted to celebrate the Monthly, not to memorialize it.

Our aim was rather to give the reader an opportunity to skim the contents of 100 years of the Monthly without actually opening 100 volumes. We wanted to condense a century of history into a few hundred pages by allowing the reader to see that history unfold in the pages of the Monthly. Along the way, readers can glimpse the American mathematical community at the turn of the century, naive and struggling for respectability. We see the beginning of the Association and the divisions between the mathematical communities of teachers and researchers. One reads about the struggle to prevent colleges from eliminating mathematics requirements in the 20's, the controversy about Einstein and relativity, the debates about formalism in logic, the immigration of mathematicians from Europe, and the frantic effort to organize as the war began. At the end of the war, one hears about new divisions between pure and applied mathematics, heroic efforts to deal with large numbers of new students in the universities, the advent of computers and computer science, and the under-representation of women and minorities.

Many of the selections are only excerpts rather than whole articles; this was necessary to keep the volume small. The mathematics that *is* included is often here because it is short and charming rather than because it was the most important mathematics of its time. We have included one *minor* mathematical theme—Mersenne primes and perfect numbers. This is *not* always lofty mathematics, but it is one of the few topics that can be found throughout the 100 volumes of the Monthly. Reading about the search for Mersenne primes and perfect numbers allows the reader to see the evolution of mathematics over a century—to read about the amazing feat of proving a number with 37 digits was prime (in 1911), as well as finding primes with hundreds of thousands of digits (in 1992).

Viewing history through the eyes of a journal is like learning about someone's life from their diary. The story is biased and incomplete. By careful reading, one can distill the truth.

This view of American mathematics from the Monthly is not meant to be objective. It reflects the unique place of the Monthly in American Mathematics, straddling two cultures that developed during the twentieth century—one concentrating on teaching and one on research. The Monthly was born to address the first and it grew into a favorite child of the second. The great issues of research in mathematics are often seen only indirectly, especially in the early years. Concerns about teaching and the profession are always more evident. Nonetheless, because the Monthly was part of the professional life of most mathematicians, its pages reflect the mathematical issues that everyone discussed.

During the past century, the Monthly was the only journal that contained articles by some of the greatest mathematicians in the world, as well as articles by students and faculty from small midwestern colleges where those great names were barely known. This book gives a glimpse of both worlds. It tells a story rather than the details of history. This is the story of a century of mathematics in America.

John Ewing, 1993

Acknowledgements

How does one sift through one hundred years of the Monthly? Certainly not alone.

I could not have completed this book without the help of the Associate Editors of the Monthly, each of whom spent many hours browsing through a decade or more of Monthlies to provide the raw material for this volume. Together they provided several thousand pages of fascinating material from which to choose. The framework of this book was created by those editors: *Peter Borwein*, *Richard Bumby*, *Dennis DeTurck*, *Woody Dudley*, *John Duncan*, *Joseph Gallian*, *Steve Galovich*, *Richard Guy*, *Darrell Haile*, *Paul Halmos*, *Richard Nowakowski*, *Stan Wagon*, *and Herb Wilf*.

Fitting that framework together was possible only with the help of *Misty Cummings*, the current editorial assistant to the Monthly, who cut and glued and proofread for many hundreds of hours.

The foundation of this book, however, was built during the past 100 years by the authors and the editors of the Monthly. Their craftsmanship produced a journal that occupies a special place in American mathematics—and *that* special place is what makes reading 100 years of the Monthly interesting.

Finally, while he may not have been the architect of the present Monthly, Benjamin Franklin Finkel chose the site on which to build it. He provided the vision and the dedication to nurture the Monthly through its first two decades of life. Without him the Monthly would not exist. This book is dedicated to Benjamin Franklin Finkel, a remarkable man who loved mathematics, and who made many others love it as well.

1. THE EARLY YEARS, 1894–1920

Florian Cajori, President of the MAA 1917

The Early Years, 1894–1920

America was naive. In 1894, most people in the 44 United States believed in their Manifest Destiny. The country slipped into a four year depression following a financial panic the year before. Thomas Edison gave his first public showing of the kinetoscope. Within the next ten years, Americans would go to war in Cuba and the Philippines, would discover oil in Texas, and would live through the assassination of a President. Movies raged throughout the country. And the Supreme court, in Plessy v. Ferguson, would approve segregation under the "separate but equal" doctrine.

American mathematics was young and simple, and the Monthly was unsophisticated. The Monthly carried articles about Infinity and Differentials, struggling with distinctions between lower and higher analysis. Mathematical articles often concerned plane or analytic geometry. Number theory was almost exclusively elementary. The most popular problem was to prove $-1 \times -1 = 1$ (using as little as possible). A few articles were plainly wrong.

American mathematicians drew their inspiration from Europe. Klein visited the Chicago World's Columbian Exposition in 1893 and mathematicians remembered for the next 20 years. Almost every mathematician of ability was educated (at least partially) in Europe. This included two well known women, Charlotte Angus Scott, who was educated in London, and Mary Winston Newson, who was a student of Klein. The early Monthly carried a series of biographical sketches, most of them about famous European mathematicians. Important books from Germany and France dominated the reviews column. News of developments in Europe were reported regularly to readers.

But the naiveté slowly dissolved as America produced her own great mathematicians. The New York Mathematical Society became the American Mathematical Society. One of its early presidents, E. H. Moore at Chicago, produced nearly 30 Ph.D.'s, including L. E. Dickson, Oswald Veblen, G. D. Birkoff, and T. H. Hildebrandt. Chicago, Harvard, and Princeton became leading centers of mathematics. America became a leader in at least one area of research—Finite Group Theory—largely because of Moore, Dickson, and G. A. Miller (who holds the record for contributions to the Monthly). American mathematicians were invited to major events, and to International Congresses. They developed a culture of their own.

There was a second mathematical culture, however, which was larger and more visible than the first. Most mathematicians were teachers, who perhaps dabbled in research but who viewed themselves as teachers of mathematics, rather than mathematicians who teach. The cultures were not always disjoint. They came together, for example, in E. H. Moore who had a profound interest in mathematics education throughout his career. (Mathematics education was the subject of his retiring presidential address in 1902.) But most teachers of mathematics were unaware of major mathematical developments.

Benjamin Franklin Finkel was surely a part of this second culture. He was a school teacher who had almost no editorial experience. After moving to the Kidder Institute in Missouri, he decided to publish a new journal, which he called the American Mathematical Monthly. Later in life, he recalled that he "had the ambition to publish a journal devoted solely to mathematics and suitable to the needs of teachers of mathematics ... " because of the deplorable condition of the high schools. The distinction between the two culures was made clear on the first page of the first issue. "Most of our existing Journals deal almost exclusively with subjects beyond the reach of the average student or teacher of Mathematics,"

Finkel wrote. "The American Mathematical Monthly will also endeavor to reach the average mathematician ... "

In one sense, Finkel's plan was a failure. Few high school teachers subscribed. But he had the genius to enlist the help of college and university faculty to provide both funding and written material for the early Monthly. Those faculty embraced the Monthly with enthusiasm. E. H. Moore, who became president of the American Mathematical Society in 1901, showed a special interest in the Monthly and encouraged others as well. Moore enlisted the help of H. E. Slaught, who helped Finkel to reshape the journal. The Monthly was soon directed towards a new audience (college and university teachers) while maintaining its goal of making mathematics accessible to all.

By 1914, the Monthly had outgrown its financial arrangements, and it was Slaught who turned to the American Mathematical Society to adopt the Monthly as an official journal. But American mathematics was growing as fast as the Monthly, and the Society was already plagued by factional disputes between the Eastern establishment (the Ivy league schools) and the Midwest (led by Chicago). Slaught's request became a controversy. Should an organization dedicated to promoting mathematical research support a journal like the Monthly? Many, especially in the East (led by Osgood), thought it should not, and the Society voted narrowly to give the Monthly a pat on the back rather than money.

The controversy concerning the Monthly and the American Mathematical Society showed the turbulence at the boundary of the two cultures. The founding of the MAA in 1916, and the formulation of its mission during its early years, shaped American mathematics (and its two cultures) for the rest of the twentieth century.

By the close of the decade, there were two mathematical organizations in America, the AMS and the MAA—but with many connections. (The first President of the MAA, Earle Raymond Hedrick, was also President of the AMS a dozen years later, and was editor of both the Monthly *and* the Bulletin.) Topology was newly born and thriving in the work of R. L. Moore, J. W. Alexander, and Solomon Lefschetz. American mathematics grew stronger while Europeans recovered from the war and an influenza epidemic that killed nearly 20 million people worldwide. Ramanujan died in 1920. And the International Mathematical Union was founded; it carefully excluded all mathematicians from the Central Powers. While European mathematicians bickered, American mathematicians learned to live with their two cultures, and they began a decade of prosperity.

THE
AMERICAN
MATHEMATICAL MONTHLY.

| VOL. 1 | JANUARY, 1894. | No. 1. |

INTRODUCTION.

It has seemed to the Editors that there is not only room but a real need for a mathematical Journal of the character and scope of this MONTHLY. At the present time there is no Mathematical Journal published in the United States sufficiently elementary to appeal to any but a very limited constituency, and that comes to its readers at regular intervals. Most of our existing Journals deal almost exclusively with subjects beyond the reach of the average student or teacher of Mathematics or at least with subjects with which they are not familiar, and little, if any space, is devoted to the solution of problems. While not neglecting the higher fields of mathematical investigation, THE AMERICAN MATHEMATICAL MONTHLY will also endeavor to reach the average mathematician by devoting regular departments to the important branches of Mathematical Science.

It is recognized that those improvements in the Science are most fruitful, which lead to improvements in the elementary treatises, and yet it must be admitted that little has been accomplished by previous mathematical journals in this line, as the crudities and solecisms handed down from one text-book to another bear witness.

While realizing that the solution of problems is one of the lowest forms of Mathematical research, and that, in general, it has no scientific value, yet its educational value can not be over estimated. It is the ladder by which the mind ascends into the higher fields of original research and investigation. Many dormant minds have been aroused into activity through the mastery of a single problem.

THE AMERICAN MATHEMATICAL MONTHLY will, therefore, devote a due portion of its space to the solution of problems, whether they be the easy problems in Arithmetic, or the difficult problems in the Calculus, Mechanics, Probability, or Modern Higher Mathematics. Papers and other interesting features will be presented, including portraits of prominent mathematicians and their biographies, a column of Queries and Information in which readers may have information furnished and their doubts cleared up by the aid of the contributors and editors, a column of Notes, and Book reviews.

No pains will be spared on the part of the Editors to make this the most interesting and most popular journal published in America. In order to do this, we must have the earnest co-operation of our readers. Teachers, students and all lovers of mathematics are, therefore, cordially invited to contribute problems, solutions and papers on interesting and important subjects in mathematics. We will be pleased to note your success, and all information of interest in regard to our contributors will be cheerfully received and noted.

5

All problems, solutions, and articles intended for publication in the February Number, should be received on or before February 1st, 1894. Solutions to problems in this Number will appear in March Number, but should be mailed to Editors before February 15th.

It will be our aim to have the MONTHLY reach its subscribers about the middle of each month. If you do not receive your Number about that time, inform us immediately.

B. F. Finkel, $\Big\}$ Editors.
J. M. Colaw,

1(1894), 1–2

47. Proposed by LEONARD E. DICKSON, A. M., Fellow in Mathematics, University of Chicago, Chicago, Illinois.
   Prove that $(-1)(-1) = +1$.

SOLUTIONS

   I. *Assuming* the distributive law to hold, $(-1)\{(+1) + (-1)\}$, or 0, $= (-1)(+1) + (-1)(-1)$. Assuming the commutative law, $(-1)(+1) = (+1)(-1) = -1$. $\therefore -1 + (-1)(-1) = 0$, or $(-1)(-1) = +1$.
   This proof was suggested by a longer one due to Professor D. A. Hull of Upper Canada College
[*L. E. Dickson.*]

   II. $(-1)(-1)$ means that -1 is to be taken subtractively one time.
$\therefore 0 - (-1) = +1$. $\therefore (-1)(-1) = +1$.
[*G. B. M. Zerr.*]

   III. $-1 \times a = -a$. $-1 \times (a - 1) = -(a - 1) = -a + 1$.
$\therefore -1 \times [(a - 1) - a] = -a + 1 - (-a) = -a + 1 + a = 1$
[*P. H. Philbrick.*]

2(1895), 272–273

46. Proposed by H. C. WHITAKER, A. M., Ph.D., Professor of Mathematics, Manual Training School, Philadelphia, Pennsylvania.

"There was an old woman tossed up in a basket
   Ninety times as high as the moon."
Mother Goose.

Neglecting the resistance of the air, how long did it take the old lady to go up?

3(1896), 281

51. Proposed by H. C. WHITAKER, A.M., Ph.D., Professor of Mathematics, Manual Training School, Philadelphia, Pennsylvania.

"Swift of foot was Hiawatha.
He could shoot an arrow from him
And run forward with such fleetness
That the arrow fell behind him!
Strong of arm was Hiawatha;
He could shoot ten arrows upward
Shoot them with such strength
   and swiftness
That the tenth had left the bowstring
Ere the first to earth had fallen."
Longfellow.

Assuming Hiawatha to have been able to shoot an arrow every second and to have aimed when not shooting vertically so that the arrow might have the longest range; what was Hiawatha's time in a hundred yard?

3(1896), 330

Mathematical Infinity and the Differential

Franklin A. Becher

Mathematics, as defined by the great mathematician, Benjamin Pierce, is the science which draws necessary conclusions. In its broadest sense, it deals with conceptions from which necessary conclusions are drawn. A mathematical conception is any conception which, by means of a finite number of specified elements, is precisely and completely defined and determined. To denote the dependence of a mathematical conception on its elements, the word "manifoldness," introduced by Riemann, has been recently adopted. Manifoldness may be looked upon as the genus, and function, as the species. This conception reaches down to the very foundation of mathematical concepts and principles. It is the central idea from which the whole field and range of the mathematical sciences may be surveyed. Time, space, and numbers are included in the notion, manifoldness.

Manifoldness may be defined according to Dr. Cantor as being in general every *muchness* or complexity which may be conceived as a unit, or a number of objects, conceptions, or elements which are united in one law or system.

. . .

Number, in all its forms, whether finite or infinite, rational or irrational, constant or variable, continuous or discontinuous, is included as one of the elements of manifoldness.

We will now consider number with special reference to its limits, infinity and zero, by the introduction of the conception of variability, of continuity, and of the differential.

By means of an unlimited continuous series of rational numbers, $\alpha_1, \alpha_2, \alpha_3, \ldots, \alpha_n$, whose terms have the property that there be given to every number δ, however small, a place n, from which the difference of all succeeding numbers remains smaller than δ, we define a definite number which is called a limit of this series. The creation of this conception admits of a comparison of rational numbers with respect to their magnitude. If all the numbers of the series differ after the place α_n by less than the number δ, then the limit is a number which lies between $\alpha_n - \delta$ and $\alpha_n + \delta$, which, because δ may be chosen as small as we please, can be expressed by a rational number as near as we please.

The totality of all numbers of an interval, for example, 0 to 1, consists not only of all numbers between 0 and 1, but of the totality of all numbers which may be interpolated between the limiting values of the defined series of numbers. This totality we dominate the aggregate or inclusive of the continuous series of numbers.

It is apparent that the conception of a limit of variability and of continuity have their root in irrationality. The two conceptions attached to a limit are in their

nature entirely different. In the first instance, a limit may be defined as a limit of a
variable, a limitless increasing or decreasing; in the second instance a limit means
that which exceeds all limits of measurable number, either because it possesses no
magnitude or because the amount or extent would not be exhausted by means of
all the series of all numbers though they were being perfected. In the first case, we
deal with variable numbers; in the second case with the conception of the absolute
value of the numbers derived from the formation of zero and conception of infinity.

Zero and infinity are the limits of the natural series of numbers. They are
derived in the same manner as the rational series of numbers. Infinity is the result
of unlimited addition of unity or other positive numbers, the unlimited multiplication of whole numbers except unity. Zero is derived from the subtraction of two
equal numbers. These are the fundamental conceptions of zero and infinity as
derived in the lower analysis. It is evident from the different ways in which each of
these symbols are derived that they have different meanings attached to them. We
may note here that every problem carries inherently with it its solution. The
meaning of every symbol depends upon its origin, derivation and relation. In
different problems they may have different meanings. Symbols of quantity, like
words, have different definitions, and these are to be determined according to the
nature of the problem and their relation to other symbols.

In the higher analysis, the conceptions of infinity and zero present themselves
more systematically in the developement of infinite series, infinite products,
infinite continued fractions, etc.

. . .

3(1897), 229–232

THE SINKING-FUND OF THE UNITED STATES.

By G. B. M. ZERR.

The public debt of the United States is being paid by the sinking-fund in the following manner.
During each fiscal year a sum is paid equal to one per cent. of the principal of the current debt,
plus a sum equal to the interest on the part of the debt already paid at the rate of interest the
debt bears. If such a sinking-fund had been operated under the same law from the beginning,
how long would it require to pay the public debt, if the rate of interest the debt draws is four per
cent. per annum?

A very excellent solution of the above problem is given in the *Mathematical Magazine* for
September, 1904, by Theodore L. DeLand, who employs the Calculus of Finite Differences.
As the great debt of the United States will fall due in a few years, and, as its payment, then, will
have to be met by a long-time loan at a different rate, which will change the present
sinking-fund, we believe that a simple algebraic solution of this national problem will be
interesting to the readers of the MONTHLY.

Let p = principal of the public debt at the beginning; $r = .01$, the rate per annum on the current
principal; $R = .04$, the rate of interest the debt draws per annum; n = number of years required
to pay the debt.

Then rp = first payment;

$p(1 - r)$ = unpaid part of debt after first payment;

$rp(1 - r) + Rrp = pr(1 - r + R)$ = second payment;

Encyklopaedie der Mathematischen Wissenschaften

Dr. George Bruce Halsted

Mit Unterstuetzung der Akademicen der Wissenschaften zu Muenchen und Wien und der Gesellschaft der Wissenschaften zu Goettingen, herausgegeben von *H. Burkhardt* and *W. F. Meyer*. Band I. Heft 1. Leipzig, Teubner. 1898. Pages 1–112.

This is an undertaking of extraordinary importance and promise. Its aim is to give a conservative presentation of the assured results of the mathematical sciences in their present form, while, by careful and copious references to the literature, giving full indications regarding the historic development of mathematical methods since the beginning of the nineteenth century. The work begins with 27 pages on the foundations of arithmetic by Hermann Schubert of Hamburg. Schubert's reputation was made by his remarkable book on enumerative geometry. He has since applied modern ideas in an elementary arithmetic, and is known in America as a contributor to the *Monist*. Unfortunately, Schubert has made in public some strange slips. In an article "On the nature of mathematical knowledge," in the *Monist*, Vol. 6, page 295, he says: "Let me recall the controversy which has been waged in this century regarding the eleventh axiom of Euclid, that only one line can be drawn through a point parallel to another straight line. The discussion merely touched the question whether the axiom was capable of demonstration solely by means of the other propositions or whether it was not a special property, apprehensible only by sense-experience, of that space of three dimensions in which the organic world has been produced and which therefore is of all others alone within the reach of our powers of representation. The truth of the last supposition affects in no respect the correctness of the axiom but simply assigns to it, in an epistemological regard, a different status from what it would have if it were demonstrable, as was one time thought, without the aid of the senses, and solely by the other propositions of mathematics."

If Schubert had written this seventy-five years ago it might have been pardonable. Just at the beginning of this century Gauss was trying to prove this Euclidean parallel-postulate. Even up to 1824 he was in Schubert's state of mind, for he then writes Taurinus: "Ich habe daher wohl zuweilen *in Scherz* den Wunsch geaeussert, dass die Euclidische Geometrie nicht die Wahre waere." But the joke had even then gone out of the matter if Gauss had but known it, for in 1823 Bolyai Janos had written to his father, "from nothing I have created a wholly new world." Of the geometry of this world as given also by Lobachevski, Clifford wrote: "It is quite simple, merely Euclid without the vicious assumption." But this assumption is only vicious if supposed to be "apprehensible by sense-experience" or "demonstrable by the aid of the senses." That "the organic world has been produced" in Euclidean space can never be demonstrated in any way whatsoever. On the other hand, the mechanics of actual bodies might be shown by merely approximate methods to be non-Euclidean. Therefore Schubert's contribution on the foundations of arithmetic may fairly be read critically. He begins with counting, and defines number as the result of counting. This is in accord with the theory that their laws alone define mathematical operations, and the operations define the various kinds of number as their symbolic outcome. There is no word of the primitive number-idea, which is essentially prior to counting and necessary to explain the cause and aim of counting. This primitive number-idea is a creation of the human mind, for it only pertains to certain other creations of the human mind which I call artificial individuals. The world we consciously perceive is a mental phenomenon. Yet certain separable or distinct things or primitive individuals we cannot well help believing to subsist somehow 'in nature' as well as in conscious perception. Now by taking together certain of these permanently distinct things or natural individuals the human mind makes an artificial individual, a conceptual unity.

· · ·

The second section of the Encyklopaedie is "Kombinatorik" by E. Netto. This is a part of mathematics which never fulfilled the hopes of the school which was lost in it during the early part of this century. Of the most comprehensive monographs the last two are in 1826 and 1837. For us it has gone over into determinants, and more than half of Netto's article is devoted to determinants. This article is particularly valuable from a bibliographic and historic point of view.

The third section is "Irrationalzahlen und Konvergenz nnenlicher Prozesse," by A. Pringsheim. It begins on page 47, and goes past the end of the Heft. This is a modern subject, of intense living interest. How entirely modern it is might not be suspected by readers of such sentences in Cajori's excellent history of mathematics as those on page 70; "The first incommensurable ratio known seems to have been that of the side of a square to its diagonal, as $1 : \sqrt{2}$. Theodorus of Cyrene added to this the fact that the sides of squares represented in length by $\sqrt{3}$, $\sqrt{5}$, etc., up to $\sqrt{17}$, and Theaetetus, that the sides of any square, represented by a surd, are incommensurable with the linear unit."

· · ·

6(1899), 7–11

COMMITTEE REPORT

"The defects in the mathematical training of the student of engineering appear to be largely in the knowledge and grasp of fundamental principles, and the constant effort of the teacher should be to ground the student thoroughly in these fundamentals, which are too often lost sight of in a mass of details.

18(1911), 24

A Popular Account of Some New Fields of Thought in Mathematics

Dr. G. A. Miller*

At the beginning of the nineteenth century elementary arithmetic was a Freshman subject in our best colleges. In 1802 the standard of admission to Harvard College was raised so as to include a knowledge of arithmetic to the 'Rule of Three.' A boy could enter the oldest college in America prior to 1803 without a knowledge of a multiplication table.* From that time on the entrance requirements in mathematics were rapidly increased, but it was not until after the founding of Johns Hopkins University that the spirit of mathematical investigation took deep root in this country.

The lectures of Sylvester and Cayley at Johns Hopkins University, the founding of the *American Journal of Mathematics* and the young men who received their training abroad coöperated to spread the spirit of mathematical investigation throughout our land. This has led to the formation of the American Mathematical Society eight years ago as well as to the starting of a new research journal, *The Transactions of the American Mathematical Society*, at the beginning of this year. While these were some of the results of mathematical activity, they, in a still stronger sense, tend to augment this activity.

In Europe such men as Descartes, Newton, Leibniz, Lagrange and Euler laid the foundation for the development of mathematics in many directions. These men, as well as a few of the most prominent names in the early part of the nineteenth century, were not specialists in mathematics. They were familiar with all the fields of mathematical activity in their day and some of them were well known for their contributions in other fields of knowledge. The last three-quarters of a century and especially the last two or three decades have witnessed a marvelous change in the mathematical activity of Europe. Mathematical periodicals have sprung up on all sides. A number of mathematical societies have been organized and many of the leading mathematicians have confined their investigations to comparatively small fields of mathematics.

The rapid increase of the mathematical literature created an imperative need of bibliographical reviews. This need was met in part by the establishment at Berlin, in 1869, of a year-book devoted exclusively to the reviews of mathematical articles, *Jahrbuch über die Fortschritte der Mathematik*. The 28th volume of this work reached our library a short time ago. It contains over 900 pages, and gives a review of over 2000 memoirs and books. With a view towards further increasing the facilities to keep in

---

*Read at the regular Winter Term Meeting of the Alpha Chapter of Sigma Xi, Cornell University.
*Cajori, *The History and Teaching of Mathematics*, 1890, page 60.

11

touch with the growing literature, the Amsterdam Mathematical Society commenced the publication of a semiannual review, *Revue Semestrielle*, in 1893. In the last number of this, 236 periodicals are quoted, each of which contains, at times, mathematical articles that are of sufficient merit to be noted. Each of the four countries, France, Germany, Italy, and America publishes over thirty such periodicals.

One of the characteristic features of our times is the prominence of the spirit of coöperation. The mathematical periodicals and the mathematical societies are evidences of this spirit. In quite recent years international mathematical congresses have given further expression of the wide-spread desire to coöperate with even the most remote workers in the same fields. The first of these congresses was held in Zurich in 1897, and the second is to be held during the coming summer in connection with the Paris Exposition. The same spirit led in 1894 to the starting of a periodical, *L'intermédiaire des Mathématiciens*, which is devoted exclusively to the publishing of queries and answers in regard to different mathematical subjects.

This desire for extensive coöperation is tending towards unifying mathematics and towards laying especial stress on those subjects which have the widest application in the different mathematical disciplines. This explains why the theory of functions of a complex variable and the theory of groups are occupying such prominent places in recent mathematical thought.*

Before entering upon a description of some of the fields included in these subjects and the interesting problems which they present, it may be well to state explicitly that our remarks on mathematics will have very little reference to its application to other sciences. To the pure mathematician a result that has extensive application in mathematics is just as important and useful as one which applies to the other sciences. Mathematics is a science which deserves to be developed for its own sake. The thought that some of its results may find application in other sciences is, however, a continual inspiration, and those who investigate such applications sometimes add materially to the development of mathematics.

· · ·

7(1900), 91–99

---

231. Proposed by A. J. KEMPNER, University of Illinois.

Is the series whose terms are the reciprocals of all positive integers not containing a given combination of figures, for example not containing the combination 37, convergent or divergent? Numbers such as $\dfrac{1}{37}, \dfrac{1}{370}, \dfrac{1}{5371}$ shall be omitted, numbers such as $\dfrac{1}{73}, \dfrac{1}{307}, \dfrac{1}{5317}$ shall be admitted as terms of the series. (Compare AMERICAN MATHEMATICAL MONTHLY, Volume XXI, page 123.)

22(1915), 131–132

---

*Cf. Klein, *Chicago Mathematical Papers*, 1893, page 134.

Multiply Perfect Odd Numbers with Three Prime Factors

R. D. Carmichael

The object of this note is to prove the following

PROPOSITION. *There exist no multiply perfect odd numbers containing only three primes.*

Let the numbers here considered be of the form $p_1^{a_1} p_2^{a_2} p_3^{a_3}$, where p_1, p_2, p_3 are distinct odd primes, and $p_1 < p_2 < p_3$. And also let the multiplicity be m, where $m > 1$. Now, by definition, the multiplicity times the number equals the sum of the factors. Hence,

$$(1) \qquad m p_1^{a_1} p_2^{a_2} p_3^{a_3} = \frac{p_1^{a_1+1} - 1}{p_1 - 1} \cdot \frac{p_2^{a_2+1} - 1}{p_2 - 1} \cdot \frac{p_3^{a_3+1} - 1}{p_3 - 1} \cdot$$

$$(2) \qquad m = \frac{p_1^{a_1+1} - 1}{p_1^{a_1}(p_1 - 1)} \cdot \frac{p_2^{a_2+1} - 1}{p_2^{a_2}(p_2 - 1)} \cdot \frac{p_3^{a_3+1} - 1}{p_3^{a_3}(p_3 - 1)} \cdot$$

$$(3) \qquad m < \frac{p_1}{p_1 - 1} \cdot \frac{p_2}{p_2 - 1} \cdot \frac{p_3}{p_3 - 1} \cdot$$

The right member of (3) is greatest when the primes are smallest. By substituting 3, 5, 7 for p_1, p_2, p_3, we find that $m \lesseqgtr 2$. Now, with $m = 2$ in (3), it is easily shown that we must have $p_1 = 3, p_2 = 5, p_3 < 16$, whence $p_3 = 7, 11$, or 13. Then (1) becomes

$$(4) \qquad 2^4 \cdot 3^{a_1} \cdot 5^{a_2} \cdot p_3^{a_3}(p_3 - 1) = (3^{a_1+1} - 1)(5^{a_2+1} - 1)(p_3^{a_3+1} - 1) \cdot$$

When $p_3 = 7$, equation (4) becomes

$$(5) \qquad 2^5 \cdot 3^{a_1+1} \cdot 5^{a_2} \cdot 7^{a_3} = (3^{a_1+1} - 1)(5^{a_2+1} - 1)(7^{a_3+1} - 1) \cdot$$

If $a_3 + 1$ is even, $7^{a_3+1} - 1$ is divisible by $7^2 - 1$, which contains 2^4. But, in any event, $5^{a_2+1} - 1$ contains 2^2 and $3^{a_2+1} - 1$ contains 2. The right member will then contain 2^7, which is impossible. Hence, $a_3 + 1$ is odd. Since odd powers of 7 end in 7 or 3, $7^{a_3+1} - 1$ is not divisible by 5. This requires $a_1 + 1$ to be even; as odd powers of 3 end in 3 or 7, and $3^{a_1+1} - 1$ is therefore not then divisible by 5. Since $a_1 + 1$ is even, $3^{a_1+1} - 1$ is divisible by $3^2 - 1 = 2^3$. The right member of (5) then contains 2^6. This is impossible. Hence, there are no numbers of the type here considered when $p_3 = 7$.

Next, for $p_3 = 11$, (4) becomes

$$(6) \qquad 2^5 \cdot 3^{a_1} \cdot 5^{a_2+1} \cdot 11^{a_3} = (3^{a_1+1} - 1)(5^{a_2+1} - 1)(11^{a_3+1} - 1) \cdot$$

We can here, as above, show that $a_3 + 1$ is odd. Likewise, that $a_1 + 1$ is odd. Then $3^{a_1+1} - 1$ does not contain 5. Hence, the right member contains the factor 5 only in $11^{a_3+1} - 1$. But this factor must occur at least twice, as $a_2 \neq 0$. By writing

13

$(10 + 1)^{a_3+1} - 1$ and expanding, we may easily show that it contains 5 only once unless $a_3 + 1$ is divisible by 5. Then, let $a_3 + 1 = 5n$. Now, $11^{5n} - 1$ is divisible by $11^5 - 1$, which contains a prime greater than 11. Hence, $p_3 = 11$ yields no numbers of the type here considered.

Finally, for $p_3 = 13$, (4) becomes

$$(7) \qquad 2^6 \cdot 3^{a_1+1} \cdot 5^{a_2} \cdot 13^{a_3} = (3^{a_1+1} - 1)(5^{a_2+1} - 1)(13^{a_3+1} - 1) \cdot$$

If $a_3 + 1$ is even, $13^{a_3+1} - 1$ is divisible by $13^2 - 1$. This introduces the inadmissible factor 7. Hence, $a_3 + 1$ is odd. The odd powers of 13 end in 3 or 7. Hence $13^{a_3+1} - 1$ is not now divisible by 5. If $a_1 + 1$ is odd, $3^{a_1+1} - 1$ is not divisible by 5. But to satisfy the equation, it must contain 5. Hence, $a_1 + 1$ is even, and $3^{a_1+1} - 1$ then contains the factor $3^2 - 1 = 2^3$. $5^{a_2+1} - 1$ always contains 2^2, and $13^{a_3+1} - 1$ always contains 2^2. Hence, the right member contains 2^7, which is impossible. Therefore, this case yields no numbers of the type here considered.

13(1906), 35–36

A PERFECT MAGIC SQUARE.

By PROFESSOR F. ANDEREGG.

In the accompanying magic square, the sum of the numbers in any line, column, or diagonal is the same (2056). Instead of continuous lines, any four numbers can be taken from the first and third groups of four numbers, and any four from the second and fourth groups in the first line, and the corresponding numbers in the last line. Similar combinations can be made for the second and fifteenth lines, the third and fourteenth, etc. Similar combinations can also be made for the columns.

Instead of a complete diagonal two incomplete diagonals can be taken on opposite sides of either complete diagonal so that the two together contain sixteen numbers.

The sum of the sixteen numbers in any four two-squares symmetrically situated with respect to the center of the sixteen-square is 2056. The sum of the sixty-four numbers in any eight-square taken at random, or of any four four-squares symmetrically situated with respect to the center of the sixteen-square is 8224.

1	2	3	4	248	247	246	245	73	74	75	76	192	191	190	189
17	18	19	20	232	231	230	229	89	90	91	92	176	175	174	173
33	34	35	36	216	215	214	213	105	106	107	108	160	159	158	157
49	50	51	52	200	199	198	197	121	122	123	124	144	143	142	141
128	127	126	125	137	138	139	140	56	55	54	53	193	194	195	196
112	111	110	109	153	154	155	156	40	39	38	37	209	210	211	212
96	95	94	93	169	170	171	172	24	23	22	21	225	226	227	228
80	79	78	77	185	186	187	188	8	7	6	5	241	242	243	244
181	182	183	184	68	67	66	65	253	254	255	256	12	11	10	9
165	166	167	168	84	83	82	81	237	238	239	240	28	27	26	25
149	150	151	152	100	99	98	97	221	222	223	224	44	43	42	41
133	134	135	136	116	115	114	113	205	206	207	208	60	59	58	57
204	203	202	201	61	62	63	64	132	131	130	129	117	118	119	120
220	219	218	217	45	46	47	48	148	147	146	145	101	102	103	104
236	235	234	233	29	30	31	32	164	163	162	161	85	86	87	88
252	251	250	249	13	14	15	16	180	179	178	177	69	70	71	72

OBERLIN COLLEGE, *April 26, 1905.*

12(1905) 195–196

The Teaching of Mathematics in the Colleges

H. E. Slaught

The first decade of the twentieth century is witnessing a wide spread wave of interest in the better teaching of secondary mathematics. This interest includes both the subject-matter and the form of presentation and arrangement in the curriculum and also the preparation of teachers for secondary schools.

Among the important activities pertaining to the subject-matter and its presentation may be mentioned the large number of associations of teachers, banded together for the purpose of improving the teaching of mathematics, in cities, in states, and in large sections of the country, such as the New England Association, the Middle States and Maryland Association, and the Central Association. In all of these associations, by papers, discussions, syllabi, committee reports, and experience meetings, every phase of secondary mathematical work has undergone the most careful reconsideration.

With respect to the preparation of teachers of secondary subjects, a great change has also taken place in recent years. Whereas, formerly there was no prevalent demand for college graduates as high school teachers, it is now true that it is very difficult for a teacher without a college degree to secure an appointment in any first-class high school. But further than this, specific training in the principles and practice of education is coming to be demanded. A notable instance of this is the state requirement in California, where no teacher can be certificated without a year's work in education in the University of California or in some institution where work of a high grade is offered.

While doubtless this standard of preparation has been set by the better equipped universities and by the few schools or colleges of education, yet the demand for such training is now being reflected back from the schools and compelling colleges and universities all over the country to turn their attention to the departments of education. The writer, in connection with the Board of Recommendations of the University of Chicago, had occasion recently to ask information from about two hundred institutions with respect to the training and appointment of teachers. The replies indicated that about fifty colleges and universities either already have, or are about to establish, schools or colleges or departments of education, or are taking steps to strengthen this work where it already exists. It would seem, therefore, with normal schools and certain schools of education devoting themselves to the training of teachers for the elementary schools, and with the colleges and universities awakening to their responsibility for the better preparation of teachers for the secondary schools, that education below the college grade is in a fair way to receive due attention, and with respect to mathematics that it will come in for its share of attention both as to subject-matter and as to the preparation of teachers.

But what of the teaching of college mathematics and the preparation of teachers for work in colleges? Where are the associations devoted to the improvement of teaching mathematics to Freshmen and Sophomores? What universities are offering courses in education especially intended for the preparation of teachers for colleges? Is it to be inferred that special training is regarded as necessary for a secondary teacher of mathematics, but that any one who knows the subject can teach a Freshman? An instructive address* bearing upon this subject was delivered before the Association of Doctors of Philosophy of the University of Chicago last June by Professor Charles Hubbard Judd, the new director of the School of Education at Chicago. The Association had been considering for two years the Relation of the Doctorate to the Teaching Profession, and a symposium† including the opinions of many of the Doctors, together with many important addresses on the subject, have been printed and distributed. Professor Judd's topic was: "The Department of Education in American Universities," and the following scattered sentences will indicate its bearing:

I have been invited to discuss further the subject which you had under consideration last year, namely, the problem of turning Doctors of Philosophy into efficient teachers. There are some of you, I see by the printed reports of your earlier meeting, who regard the ability to teach as a natural gift. There are apparently many of you who are persuaded that courses in pedagogy cannot contribute materially to the improvement of a graduate of a university. In the face of such settled views, backed up by the success that many of you have obtained in the teaching profession, it seems to be a bold and from some points of view a useless undertaking to come before you as I must with the assertion that ability to teach is not a natural gift and that every Doctor of Philosophy would be improved by a careful consideration in a scientific and historical way of the problems of education. ...

Until very recently university and college organizations have been based on the opinion expressed by some of you that teaching cannot be made a subject of special study and instruction. Gradually, however, a change is being worked out before our eyes. In spite of opposition and indifference, courses in education are being organized even in the most conservative institutions. ...

The first fact which I wish to point out is that there is a great deal of poor teaching within our universities and colleges, and this fact can be traced to a neglect upon the part of academic men and women of the form in which they arrange their material. ...

It is widely recognized that such a neglect of form as I have indicated appears in much of our university lecturing. The assumption of the ordinary university lecturer is that if he presents a certain body of material so organized that it seems to him to be fairly coherent and logical in its character, it is a matter of small moment whether it appeals to the students because of its literary form or whether it is easily intelligible to them because of its careful adjustment to their present stage of developement. ...

The elective system has in part overcome this attitude and there is a very much more general conviction on the part of college instructors now than there was a generation ago that the demands of the student that the material shall be presented in clear and coherent form should be met. An instructor who is in competition with the other members of his own department for students in his course is likely to recognize the importance of preparing his material as clearly as possible for presentation to his class. ...

It is obvious that everyone who is successful in the art of teaching must have complied with the demands indicated in the foregoing discussion; that is, he must have organized his material in such a way that it has significance not only for his own mind but also for the minds of others. The teacher who does not sympathize with his pupils fails commonly because he does not recognize the type of fact which we have just been discussing.

---

*Printed in the November, 1909, *School Review*. A few reprints may still be had.

†This has been reprinted and a few copies may still be had by applying to the Secretary of the Association at the University of Chicago.

It should not be inferred from the opening sentence above quoted that all of the five hundred Doctors of the University of Chicago look with distrust upon the proposition that a university man who expects to teach in a college or university should know something of the principles, history and practice of education in general and much of these things as they pertain to his own department in particular, in order that he may, at least, avoid the historic blunders of his predecessors. The following quotation from the symposium bears upon this point:

As matters now stand, whatever the *ideal* held by the university world in regard to the high calling of the doctor in the field of pure research, the *fact* is that the great majority of the doctors now turned out in this country are nominated and pushed by their respective departments, for teaching positions, which they must fill successfully or be counted as failures by all who measure the ratio of results accomplished to tasks undertaken.

Success in teaching is well-nigh *indispensable* (save perhaps in a great university, where failure is less conspicuous because of the large numbers and multifarious interests). If the doctor is to gain encouragement and opportunity to go on with his research, he must first establish a reputation for himself in the community where he goes, for soundness of judgment, clearness of presentation, and power to inspire and lead students; then, having made sure of his ground, he can have pretty much his own way in planning and carrying out his work in the interest of his research. But, on the contrary, if in his blind devotion to his ideal he fails at the outset to teach successfully, to inspire and to lead students, he thereby cuts himself off from his very best resources for ultimately accomplishing his highest aims.

These being the facts as realized in everyday experiences, the conclusion is forced upon us that the universities must prepare for teaching those doctors whom it is proposed to recommend as teachers, and preparations must include breadth of culture, eliminations of angularities, development of pedagogical sense, and some acquaintance with the great educational movements of the past and present. If it be urged that these, some or all of them, are either matters of personal quality or are foreign to the great purpose of the graduate school as director of explorations in unknown fields, then the other conclusion is forced upon us that only those doctors should be recommended as teachers who possess by nature the prerequisities or have developed them *outside* or *in spite of* the graduate school, and that the present indiscriminate practice of recommending any doctor as a teacher, however narrow or however lacking in the elements of pedagogic sense and power to teach, be superseded by a careful discrimination and selection of those who are prepared to teach and a refusal to nominate or recommend those who are not.

Such a discrimination will lead, as it should, to a more careful and personal consideration of every individual's candidacy for the doctorate, and will compel a readjustment of curriculum and a recognition as a distinct field of that broad, thorough, and scientific training, commensurate with any standard now set up for the doctorate, which is demanded by the man or woman as a preparation for the highest attainment in teaching, leaving undisturbed the narrowest and possibly the deepest channels of pure research to be followed by those who either have not the time or have not the inclination, or perchance have not the personal qualities needed in preparation for teaching.

This leads again and finally to the conclusion either that the number of those who are encouraged to go on to the doctorate should be greatly diminished, or else that the basis of the doctorate should be greatly broadened so as to provide the highest standards and the strongest preparation for the noble art of teaching, as well as to produce fine investigators.

Whatever may have been the attitude of the universities up to this time in regard to training men as teachers for college positions, it is clear that from this time forward the demand is likely to increase for men thus trained, not only because the colleges themselves are awakening to the necessity of better teaching, but also because the irresistible wave of progress in this respect in the secondary schools is bound to reach up into the colleges and shame them into action. Fortunately, however, the agitation is already begun to some extent with respect to the colleges, under the challenge of such men as Abraham Flexner and with the support of many who are loyal to the cause of better teaching.

But while the question of better training for teachers of mathematics is sure to get consideration along with the like questions for all other departments in the general upward movement, it is the special responsibility of the mathematicians themselves to consider the subject-matter of the collegiate curriculum, with respect to its better arrangement, better form of presentation, closer contact with concrete applications, closer correlation with related departments, such as Physics, Astronomy, Chemistry, Geology, and better adjustment to both the earlier and later work in the department. In this connection also belongs the consideration of all these questions concerning mathematics as related to students in the literary courses, in the general science courses, in the specialized science courses, in the small college and in the great university.

It is well known that changes have been going on in respect to the status of various branches of mathematics in the college curriculum; for instance, as to the content of College Algebra, the scope of Trigonometry and of Analytic Geometry, the proper adjustment of theory and applications in the Calculus, and the whole question as to whether all of these branches should not be considered as one subject—Mathematics—rather than as separate topics carefully partitioned off from each other both theoretically and pedagogically.

The editors of the MONTHLY have decided to open its columns to contributions on the teaching of collegiate mathematics, in the hope that some impetus may be given to the better training of teachers, the better arrangement and coordination of material and the better form and methods of presentation. Already some important papers have been promised and others will be announced in the near future.

Moreover, no less than six sub-committees, under Klein's International Commission on the Teaching of Mathematics, are now working in this country on topics connected with mathematics of a collegiate grade, and it is supposed that the substance of their reports will be made known in America and will become the basis of discussion at the various meetings of mathematical societies during the Autumn and Winter. From these sources also much should be expected both in pointing out existing conditions and in arousing interest in questions of possible improvements.

16(1909), 173–177

NOTE ON PRIME NUMBERS.

By DERRICK N. LEHMER, University of California.

It is a well known theorem that it is possible to find an arbitrarily great number of consecutive composite numbers. This appears from the values which the expression $n! + r$ takes for $r = 2, 3, \ldots, n$. This theorem furnishes an interesting proof of the theorem that the number of primes less than or equal to x is not determined by a function of x which is a polynomial in x of finite degree. For if $f(x)$ were such a function of degree n, then for $x = (n + 2)! + r$, $f(x)$ must keep the same value for $r = 2, 3, 4, \ldots, n + 2$. If this value is k, then $f(x) - k = 0$ is an equation of degree n with $n + 1$ roots, which is impossible.

19(1912), 50

The Tenth Perfect Number

R. E. Powers

A number which is equal to the sum of all its divisors is called a "perfect number." Thus, the divisors of 6 are 1, 2, and 3, the sum of which is equal to 6; the divisors of 28 are 1, 2, 4, 7, and 14, whose sum is 28. Euclid (IX, 36) proved that if $2^p - 1$ is prime, then $2^{p-1}(2^p - 1)$ is a perfect number, and no other perfect numbers are known. In 1644 Mersenne, in the preface to his *Cogitata Physico-Mathematica*, stated, in effect, that the only values of p not greater than 257 which make $2^p - 1$ prime are 2, 3, 5, 7, 13, 17, 19, 31, 67, 127, and 257. Regarding these "Mersenne's Numbers" ($2^p - 1$), W. W. Rouse Ball, in his *Mathematical Recreations and Essays* (4th Edition, pages 262, 263, 269), says:

"I assume that the number 67 is a misprint for 61. With this correction, we have no reason to doubt the truth of the statement, but it has not been definitely established... . Seelhoff showed that $2^p - 1$ is prime when $p = 61,...$ and Cole gave the factors when $p = 67...$. One of the unsolved riddles of higher arithmetic... is the discovery of the method by which Mersenne or his contemporaries determined values of p which make a number of the form $2^p - 1$ a prime... . The riddle is still, after nearly 250 years, unsolved."

No exception to Mersenne's assertion (corrected by the substitution of 61 for 67) is known at the present time. Below we show, however, that $2^{89} - 1$ is a prime number, contrary to his statement.

Following is a list of Mersenne's Numbers thus far proved to be prime, with the corresponding perfect numbers:

p	$2^p - 1$	Perfect Numbers
2	3	6
3	7	28
5	31	496
7	127	8,128
13	8,191	33,550,336
17	131,071	8,589,869,056
19	524,287	137,438,691,328
31	2,147,483,647	(19 digits)
61	2,305,843,009,213,693,951	(37 digits)

To these must now be added the prime number $2^{89} - 1$, so that the tenth perfect number is $2^{88}(2^{89} - 1)$, or

191,561,942,608,236,107,294,793,378,084,303,638,130,997,321,548,169,216

(it is known that $2^p - 1$ is composite for all other values of p not greater than 100).

In his *Théorie des Nombres*, page 376, Lucas says: "Nous pensons avoir démontré par de très longs calculs qu'il n'existe pas de nombres parfaits pour $p = 67$ et $p = 89$." While this result has since been verified for $p = 67$, the opinion has been expressed that also the case $p = 89$ needed an independent examination. The result here shown that $2^{89} - 1$ is a prime is therefore in conflict with Lucas' computation. The same writer, in an article entitled "Théorie des Fonctions Numériques Simplement Périodiques," Section XXIX, in the *American Journal of Mathematics*, Volume 1 (1878), proved the following remarkable theorem (the theorem appears on page 316 of the volume):

"If $P = 2^{4q+1} - 1$, and we form the series of residues (modulo P)

$$4, 14, 194, 37634, \ldots,$$

each of which is equal to the square of the preceding, diminished by two units: the number P is composite if none of the $4q + 1$ first residues is equal to O; P is prime if the first residue O lies between the $2q$th and the $(4q + 1)$th term."

Applying the above theorem to the number $2^{89} - 1$, and denoting the terms of the series by $L_1, L_2, L_3, \ldots$, we found the folowing residues (modulo $2^{89} - 1$):

m	L_m
1	4
2	14
3	194
10	$-115{,}113{,}975{,}804{,}653{,}882{,}052{,}836{,}464$
20	$36{,}000{,}517{,}785{,}442{,}762{,}303{,}479{,}300$
30	$-204{,}144{,}540{,}641{,}167{,}292{,}618{,}604{,}303$
40	$-126{,}791{,}709{,}316{,}676{,}382{,}795{,}042{,}761$
50	$-90{,}990{,}560{,}635{,}837{,}660{,}454{,}542{,}648$
60	$-206{,}308{,}592{,}424{,}355{,}282{,}693{,}419{,}690$
70	$99{,}498{,}791{,}857{,}820{,}493{,}810{,}407{,}653$
80	$269{,}783{,}273{,}665{,}984{,}523{,}074{,}966{,}550$
86	$-309{,}403{,}333{,}482{,}440{,}150{,}628{,}882{,}422$
87	$-35{,}184{,}372{,}088{,}832$
88	0

Since the first (and only) residue 0 occurs at the 88th term of the above series, it follows, from the foregoing theorem, that $2^{89} - 1$, or

$$618{,}970{,}019{,}642{,}690{,}137{,}449{,}562{,}111$$

is a PRIME NUMBER.

As M. Lucas points out, his method used above is free from any uncertainty as to the accuracy of the conclusion that the number under consideration is prime, in case our attempt to arrive at the residue 0 meets with success, since an error in calculating any term of the series would have the effect of preventing the appearance of the residue 0. We would add that, denoting the number $2^{89} - 1$ by N, we have verified that

$$3^{N-1} - 1 \text{ is divisible by } N,$$

which is in accordance with Fermat's well-known theorem.

18(1911), 195–197

Foreword on Behalf of the Editors

E. R. Hedrick

Collegiate Mathematics.—The lifework of hundreds of men is the teaching of collegiate mathematics; hundreds of other men are preparing for this same lifework. Thousands of students are now working and will work under the guidance of these men in collegiate courses in mathematics.

Suggestions of moment arise in the minds of many, yet there exist few means for transmitting an idea in this field to others. Any organized effort toward the dissemination of methods that are new, or of facts that are interesting, in this field is totally lacking.

These remarks apply, of course, only to the United States. Abroad, in every country in which mathematics holds a place commensurate with its position here, definite journals exist whose purpose is almost wholly to foster these interests. THE AMERICAN MATHEMATICAL MONTHLY, beginning with this issue, proposes to afford an opportunity for any discussions that seem valuable upon collegiate mathematics, and the editors invite contributions concerning the methods of instruction as well as those that treat special topics or theorems.

Problems of Instruction.—Every teacher in this field has asked himself what topics he should best teach, why these topics are taught, how they should be approached, what purpose should be recognized as the main aim of collegiate mathematics. Ordinarily he has either found no answer whatever and has settled down to the basis of traditional procedure, or else he has satisfied himself perforce with partial answers that he could himself discover. Each man has been a power unto himself, but no man has received or given aid, at least to the extent of his ability.

Some conspicuous exceptions to this general state of affairs are mentioned below. What has been done is, however, sporadic, discontinuous, and of little influence—certainly no such effort has been general, and the effects of what has been attempted are far from general.

The most general and most sustained work in this direction has recently culminated in the report of a committee of twenty mathematicians and engineers, under the chairmanship of Professor E. V. Huntington of Harvard University. This committee, was originally appointed at a joint meeting* of mathematicians and engineers, organized in 1907 by the Chicago Section of the American Mathematical Society; but it was authorized to report to the Society for the Promotion of Engineering Education, which body had already carried on a consistent series of discussions through a period of years on the teaching of mathematics to students of engineering; and these engineers, rather than the mathematicians, have fur-

---

*A few printed copies of the papers read at this meeting may still be had from the secretary of the Chicago Section, H. E. Slaught, 58th Street and Ellis Ave., Chicago, Ill.

nished the medium, through their journal, for the publication of these discussions and of this final report.†

Though it may be contended that the aims of this Society are at least special, and touch only one side of our work, and though the spirit of many of the discussions may seem narrow to many, it is at least true that the attempt to suggest means for the improvement of mathematical teaching has been carried on consistently, courageously, and earnestly.

The efforts of this society, whose special aim is the improvement of engineering education, emphasize the variety of conditions that must be met by collegiate courses in mathematics at this time. The reasonable demands of the rapidly increasing schools of engineering, together with the normally increasing enrollments in academic classes, have vastly increased the problems that confront us.

Origin of New Problems.—The problems of to-day differ from those of a generation ago on account of three great movements that lie wholly outside the field of mathematics. Of these, the first is the spread and almost universal acceptance of the elective principle in colleges and in academic classes in institutions of every kind. To be sure, the tendency to-day seems to be to limit the scope of the elective system, but the days of the old fixed curriculum, with mathematics as a required study for all students, has passed, and forever.

Another movement of no less extent is the rapid increase in the number of those entering engineering and other semi-technical professions. In schools of engineering, a certain amount of mathematics is required of all students, though the amount varies. But this requirement that all students take collegiate courses in mathematics is not made, as was the older academic requirement, for the sake of general mental discipline and the training of the mind as a whole. While this motive may persist in part, the recent discussions by engineers, mentioned above, leave no room for doubt that the inherent reason, in this instance, is the belief in the direct usefulness to the engineer of the facts, the principles, and the processes to be learned in these courses.

The last, in point of time, of the three principal movements mentioned above is the present rapid change in educational theory, which is based to a large extent on experimental psychology, and which bears the same relation to the older pedagogy that chemistry bears to alchemy. Not all its problems are yet settled, many of them are now under intense discussion, but the manner of that discussion is the manner of science rather than the manner of controversy. One of its contributions that is widely accepted is that there are no such *fixed* faculties of the mind as those we had believed in rather implicitly; and that the training of such supposed faculties as the memory, for example, is impossible in its widest sense because there seems to be no one such faculty. The memory for number, for example, seems to be wholly unrelated to the memory for poetry, or the memory for forms.

· · ·

Selection of Topics.—Teachers of mathematics will not lightly abandon the value of the disciplinary training afforded by the subject, but the ability and scientific attitude of mind of teachers of collegiate mathematics, as compared with teachers in more elementary schools, will lead them to search for the truth about

---

†This final report is in the form of a syllabus of courses in mathematics for students of engineering, and is for sale by the Society for the Promotion of Engineering Education, 43 E. Nineteenth St., New York.

these questions and will make them cautious in basing their faith entirely upon an aim that is made questionable by these modern investigations in education.

One immediate result is the realization that the same amount of mental training, whatever that may be, seems likely to result from any one of several possible topics, if each of them is studied with the same thoroughness and attention. Does this make possible the selection of topics on grounds entirely different from mental discipline and, if the topics to taught are selected for other reasons, will the resultant training be just as efficient?

This consideration, and the changed conditions which seem to raise the same demand, would point toward a very thorough and critical study of the motives, other than disciplinary training, for the study of mathematical courses, and of topics within those courses. Without discarding discipline as one of the desirable results of a mathematical training, a knowledge of other motives for the study of any one topic will serve only to enhance its value.

Comparative merits of different topics and of different methods of instruction can be realized most effectively if these other motives are considered, and if the disciplinary factors are temporarily ignored, for the claims of disciplinary value, as between two topics, are at best conflicting and uncertain, and the wholesome effect of any topic that is studied intensively will be recognized by all.

Interest.—The qualification that a study of any topic, in order to be effective, must be intensive and earnest will be insisted upon by all. It would seem therefore that profitable discussion may well take place concerning the means by which such work can be obtained. To arouse the student to his maximum effort, to stimulate his greatest efficiency, to gain their fullest measure of disciplinary good, there may be many plans of merit; but it is certain that all of them will have in common the holding of the student's attention, without which no effective study is thinkable. Any process for holding the attention in a sustained and serious fashion is described by saying that the student has become "interested" in the study or that his "interest" has been secured. But the word interest has been misused. The doctrine of interest has led to a mistaken doctrine of "amusement," than which no interpretation of the doctrine of interest could be more absurdly irrelevant. It is scarcely to be wondered at that the whole doctrine, sound as its real basis may be, has been misunderstood, abused, ridiculed, and that it appears to many as the antithesis of that very doctrine of discipline of which it is the sole effective support.

The Province of the Monthly.—This article is not itself a pedagogical document. What is here stated is perfectly well known in pedagogical circles, and we trust that we have not through ignorance, as we have certainly not through intention, overstepped the bounds of that which is generally recognized as finally established in the theory of education.

It is not the province of the MONTHLY to enter the field of general pedagogy, nor will the MONTHLY entertain discussions that are concerned with research in general pedagogy, or that deal with new theories of pedagogy, however important these contributions may seem. Journals for the publication of such contributions abound, and any article of worth in the general field is already more than liberally provided with possible means of publicity. In such matters, we should not venture to pose as expert critics of new matter, nor to pass upon the merits of articles. We must accept, as from a foreign field, the results of general pedagogy and of experimental psychology, and in order to be secure, we must remain always a little behind, rather than in advance of the times, on pedagogical theory.

What we do desire is to inspire, not a discussion along these lines, but rather a discussion of definite *mathematical* problems. It is our hope that this article will spur mathematicians to the realization of the new conditions that confront us, to the variety of problems that demand discussion, to the vital importance which their solution holds to a great body of intelligent people. For the discussion from this standpoint of mathematical questions, properly speaking, the MONTHLY will open its columns without reserve, except for those usual discretionary and advisory powers that are traditional in any editorship.

The Future.—This is not the last work that the MONTHLY will have to say upon this question. We shall indulge in rather little general pedagogical preaching, for we claim no professional attitude toward that subject. What seemed necessary has been said above. For this issue this is enough. We shall, however, not place any limitation upon the scope of our suggestions to readers and to possible contributors; and we shall hope, by these suggestions, to point out very explicit problems of a mathematical nature which might well be treated. We shall aim to awaken an interest in their discussion, even among those who now regard their own profession —the teaching of collegiate mathematics—with distrust as a possible field for that type of human thinking that is known as scientific research. In this distrust, we, the editors of the MONTHLY, emphatically announce that we do not share.

20(1913), 1–5

On January 19, 1913, occurred in Denver the death of Robert Gauss, who had been long connected with the *Denver Republican*. On the same day died also his brother, Charles H. Gauss, of St. Louis. They were sons of Eugene Gauss, and grandsons of the mathematician, CARL FRIEDRICH GAUSS.

20(1913), 71

According to cable reports from London, the Council of Trinity College, Cambridge, has removed Professor BERTRAND RUSSELL from his lectureship in logic and principles of mathematics on account of his having been convicted under the defense of the realm act for publishing a leaflet defending the "Conscientious Objector" to service in the British army. Professor Russell is well known in this country through his mathematical writings.

23(1916), 317

At the University of Maine, Dr. NORBERT WIENER has resigned to enter the military service, and Associate Professor H. R. WILLARD has been appointed statistician under Mr. HOOVER and reported at Washington in September.

24(1917), 399

The name of Professor FELIX KLEIN, of the University of Göttingen, together with those of six other German educators, has been cancelled from the roll of honorary members of the National Education Association in response to a persistent demand from active members of the association, from members of the Council of National Defense, and from others.

25(1918), 331

THE AMERICAN
MATHEMATICAL MONTHLY

OFFICIAL JOURNAL OF

THE MATHEMATICAL ASSOCIATION
OF AMERICA

VOLUME XXIII	JANUARY, 1915	NUMBER 1

THE MATHEMATICAL ASSOCIATION OF AMERICA.

In accordance with the call for an organization meeting of a new national mathematical association, signed by four hundred and fifty persons representing every state in the Union, the District of Columbia, and Canada, such a meeting was held at Columbus, Ohio, on Thursday, December 30, 1915, in room 101 of Page Hall, on the Campus of Ohio State University. The meeting extended through two sessions, the latter being held on Friday morning, December 31, at which time the constitution was finally adopted.

The following one hundred and four persons were in attendance, a large number of whom took active part in the proceedings:

. . .

When the meeting was called to order, E. R. HEDRICK, University of Missouri, was elected temporary Chairman, and W. D. CAIRNS, Oberlin College, temporary Secretary. Upon the request of the Chairman, some introductory remarks were made by H. E. SLAUGHT, as the representative of the Board of Editors of the AMERICAN MATHEMATICAL MONTHLY who had been responsible for proposing the call for the meeting. Referring to the history of this movement, as outlined in the October, 1915, issue of the MONTHLY, he emphasized the fact that this journal had stood consistently, since its reorganization, for advancing the interests of mathematics in the collegiate and advanced secondary fields, and expressed the hope that the new organization might carry forward these aims with still greater effectiveness, coöperating, on the one hand, with the various well-organized secondary associations, and, on the other hand, with the American Mathematical Society in its chosen field of scientific research, but being careful to encroach upon neither of these fields.

The meeting was then resolved into a committee of the whole for the consideration, section by section, of a constitution and by-laws, tentative drafts of which had been prepared in advance. After three hours of patient and pains-taking deliberation, all mooted questions were settled except the name of the new organization. This was left to a committee to choose from the eighteen different variations which had been proposed and to report the following morning. A committee was also delegated to assist the temporary secretary in smoothing out any verbal inconsistencies or inaccuracies in the constitution and by-laws. The name finally chosen by the committee was adopted without a dissenting vote, as embodying more favorable points and fewer objections than any other that had been suggested.

23(1916), 1–6

LETTER TO EDITOR

The following passage of a personal letter to the editor, from a distinguished professor in a Scottish University, will be of general interest.

"But what I want most to write about today is to thank you for, and to offer you a word of appreciation of, your bulletin of the 'Undergraduate Clubs'. We have nothing like this over here, and the whole thing strikes me as admirable. It seems to me that it might with immense advantage be extended (as perhaps it already is on your side) to other sciences as well as mathematics. Our better students either drop their work altogether on graduation, or either insist on attempting, or perhaps are tempted (for instance by the Carnegie Scholarships) to attempt 'original investigation' before they are fit for it. There is no temptation and little opportunity to prolong their own studies, to engage in wider reading or to make acquaintance with the historical aspect of their science. Often a comparatively raw youth or girl comes to me and wants to 'do original research'. I say 'Why, you have read nothing but a text-book or two; you have read nothing worth speaking of. Why not read for a year or two; make yourself master of what has been done in this field or that, and widen your horizon. Epitomize the literature of some theme, or of some historical period.' But the reply is always the same. '*I want to do "original research"*; and the Carnegie Trust will give me a good scholarship for doing so, and for nothing else in the world.' Now your plan seems to me to precisely meet the case. You encourage your young people to do *work*—not necessarily 'original work'—which is more than any reasonable man can expect of them; but at least work which involves just so much originality or at least independence as can fairly be expected. And it is work by which they learn something; while heaps of so-called original work, as I see it done, teaches nothing, for it is too often confined to some tiny problem, and only means watching the spot of a galvanometer, or making endless and all but identical titrations or measurements."

25(1918), 462–463

A Tentative Platform of the Association

E. R. Hedrick, President

No man can speak with authority concerning the future of this new Association which was created by those who met at Columbus last December. Its future lies with those who constitute its membership. Any statement must be rather a history of past events than a prediction for the future.

What were the causes which led so many to wish for, to exert themselves and to struggle for a new society in the mathematical field? What motives lay behind the movement which culminated in this organization? These are questions which are distinctly answerable. I shall try to show for those who did form the Association what were their purposes and what is now their aim. If these purposes or aims are wrong or insufficient, they will perish and newer and better policies will supplant them. The great fact which we cannot overlook is that we now have a large and representative body of men and women interested in mathematics joined together in this association to foster whatever they believe to be worthy and beneficial.

The chief motive may well be said to be that of service to the whole body of teachers of mathematics in American colleges. If I am right, the Association will not stop at anything which will serve this body of men.

Perhaps there is one exception. The majority of those responsible for the new organization are themselves members of the American Mathematical Society. This older organization is itself bound by its constitution to promote the interests of mathematics in this country. That there should be any conflict between the two organizations would defeat the ends of both, and would not give the maximum service which can be rendered to American mathematicians. The American Mathematical Society has chosen, through the action of its Council,[1] to restrict its activities to the field of pure research in mathematics, and to the promotion of those phases of mathematics which are commonly associated with that word. Those responsible for the new organization are by no means at variance with this determination, and it is their aim to carry out in good faith the separation of fields of activity provided for by the action just mentioned. This one limitation to the activities of the Association should therefore be mentioned prominently.

Another restriction which is imposed, not by any agreement but by the dictates of good judgment, is that matters dealing with secondary and elementary schools should be left to the organizations already in existence devoted to that field. The new organization will not undertake to discuss or to print papers specially dealing with the details of secondary instruction, though it may well undertake to discuss and define questions concerning the preparation of students who enter colleges, particularly with respect to training in mathematics.

---

[1]*Bulletin of the American Mathematical Society*, volume 21, page 482 (July, 1915); this MONTHLY, volume 22, page 252 (October, 1915).

This question of college entrance is one of such vital importance that some leadership of national standing is desirable to crystallize and to formulate the views of mathematicians of all grades of schools. Such questions cannot be said to be the primary function of any one class of organization, and it is thought that secondary school teachers will be the first to welcome a strong national leadership in this matter. It may be well to add that the Council of the American Mathematical Society decided specifically about a year ago not to undertake work along this line, in its relation to the attack upon mathematics in the secondary schools now being made in various quarters.

In general, however, the activities of the Association will be centered strongly in the collegiate field, and it is expected that the great majority of the work fostered by the Association will be on questions directly affecting collegiate courses in mathematics. That the range of topics which may be concerned is rather large, and that the considerations which may be presented are varied and complex is reasonably forecast by the papers which have appeared in the MONTHLY since its reorganization three years ago.

There will doubtless be many articles of historical interest. These will deal with topics which may lie anywhere in the entire range of mathematics. They may be said to be allied with the collegiate courses on the history of mathematics. Thus, the interesting paper by Professor Karpinski presented before the Columbus meeting was entitled "The Story of Algebra." It might be thought that this paper was therefore of secondary character. But the merest inspection of it will suffice to demonstrate that it lies beyond the secondary field and that its association is strictly with the history of mathematics. Other papers, such as Professor Cajori's remarkable series of articles on the History of Logarithms, and the various interesting papers on Number Systems which have appeared in the MONTHLY, are further indication of the intention to deal with matters of this type.

That elementary college courses are still open to serious reconsideration is evidenced by the appearance of several important papers in the MONTHLY during the last two years which deal with subjects taught in the freshman year. Other weighty contributions of this character are in type awaiting their turn.

Perhaps more deserving of mention is the fact that advanced college subjects should properly fall within the field of this association, and that discussions which affect such topics as projective geometry, second courses in calculus, the elementary theory of functions, and other courses commonly given to undergraduates, are properly subjects for discussion.

One more idea would seem to me to clarify the situation very materially. There have appeared in the MONTHLY from time to time articles which cannot be said to be of research character from the standpoint of the common acceptation of that word, but which nevertheless represent a great deal of labor of a purely investigational sort which would seem quite worthy of being called research in a broader interpretation of that word. This again is well illustrated by the historical papers mentioned above, all of which certainly constitute a very dignified form of research in this broader sense, though they may not satisfy the stricter interpretation placed ordinarily on the word research. The same thing can be said of a number of other papers which have appeared in the MONTHLY which deal essentially with college subjects. It is held that a dignified discussion which involves investigation from a scientific point of view is worthy of the name research in its broader application. While the new association will recognize fully the prior right of the American Mathematical Society in all questions which would ordinarily be termed research under the common interpretation, the attitude of this Association will be to encourage and to dignify all investigations of character which have here been called research in the broader sense.

If I have tried to say what seems to me to be the policy of those responsible for the organization of the ASSOCIATION, I should perhaps add a word concerning the questions distinctly avoided thus far, which may be said to be not within our present intentions. One such which certainly deserves mention is the general notion of pedagogy in its more restricted interpretation. All of those questions which are termed pedagogical in the strict sense of that word have been held, and are held, by those responsible for the organization of the Association to belong to the field of education and to be wholly outside the field of the present association. Just as research will be held to be within its province only if the word is given a broad interpretation, it may be said also that the discussions which this association will foster may be termed pedagogical only if that word is used in a much broader sense than is common. I may define this broader sense to include those questions affecting instruction in which a *professional knowledge of the subject-matter* is a necessary element toward the formation of any dignified conclusion. That there is no doubt about the existence of such questions is amply proved by the files of the MONTHLY during the last two years.

This statement does not pretend to be exhaustive or infallible. The intention is to give as clear an idea as may be in a short space of characteristic topics which this Association will discuss. That the policy of the Association may be changed in the future and that the statements of this article are by no means binding upon the Association will be quite evident upon even a casual examination of the Constitution.

23(1916), 31–33

At the recent election Professor D. A. ROTHROCK, of Indiana University, was elected a member of the State Legislature. He is relieved of duties at the University during the legislative session beginning in January.

26(1919), 86

Professor MAX NOETHER, of the University of Erlangen, who is now seventy-five year of age, was relieved from lecturing after April 1, 1918.

26(1919), 86

Dr. G. M. GREEN, instructor in mathematics at Harvard University, and one of the most promising young geometers of the country, died at Cambridge on January 25, in the twenty-ninth year of his age. He graduated from College of the City of New York, B.Sc. 1911, Columbia University, M.A. 1912, Ph.D. 1913. His thesis was entitled *Projective differential geometry of triple systems of surfaces* (Lancaster, Pa., 1913, 28 pp.). His mathematical papers appeared in: *Transactions of the American Mathematical Society*, 1914–17; *Bulletin of the American Mathematical Society*, 1917–18; *Annals of Mathematics*, 1918; and *Proceedings of the National Academy of Sciences*, 1915–18.

26(1919), 86

Top: B. F. Sine, J. M. Bandy; Bottom: Shas. E. Cross, Associate Editor J. M. Colaw, Benj.
F. Yanney, from the MONTHLY 6(1899), 52.

The Significance of Mathematics*

E. R. Hedrick

Several circumstances combine to render peculiarly fitting a consideration at this time of the significance of mathematics. Of late we have heard much from real or alleged educators, tending to show a lack of appreciation on their part, if not on the part of the public, of the vital role which mathematics plays in the affairs of humanity. These attacks were beginning to receive some hearing in the educational world, on account of their reiteration and their vehemence, if not through intrinsic merit.

A counter influence of tremendous public force, whose import is as yet seen only by those most nearly interested, has now arisen through the existence of war and the necessities of war. To the layman, lately told by pedagogical orators that mathematics lacks useful applications, the evident need of mathematical training on every hand now comes as a distinct surprise.

The attacks on mathematics, and the lay conception of the entire subject, center naturally around elementary and secondary instruction. We ourselves, college teachers of mathematics, have commonly talked of current practice and of reforms largely with respect to secondary education. The third influence which contributes toward the present situation and which may strongly affect its future development is the formation and the existence of this great Association, which affords for the first time in the history of America an adequate forum for the discussion of the problems of collegiate instruction in mathematics.

As retiring president of the Association, I know of no more fitting topic than that which I have chosen. It vitally concerns us; it is bound up with the functions of this Association; and the times in which we live seem to point forcibly towards its consideration. I shall attempt to outline to you my own views on the true significance of mathematics, and to sketch what I for one would be glad to see this Association promote.

In speaking of the significance of mathematics, I understand that we mean not at all the baser material advantage to the individual student, not at all a narrow utilitarianism, but rather a comprehensive grasp of the usefulness of mathematics to society as a whole, to science, to engineering, to the nation. Any narrower view would be unworthy of us; any narrower demand by educators means a degraded view of the purposes of education in a democracy.

Especially under the stress of war, public attention may be secured for the real claim of mathematics as a public necessity, not only to be employed by a few

---

*Retiring Presidential Address delivered at the second summer meeting of the Mathematical Association of America, at Cleveland, Ohio, September 6, 1917.

specialists, but also to influence and to determine the conduct and the efficiency of thousands.

. . .

In algebra, as taught in colleges, among the topics always considered are fractional exponents, logarithms, and arithmetic and geometric progressions. To many, fractional exponents remain a pure formalism, learned by rote and unappreciated, connected neither with the other topics just mentioned nor with any realities of life. That fractional exponents occur in expressions for air-resistance (as in airplanes), in water resistance (as in measuring stream-flow), in electricity (as in induction), would surprise most students who pass our courses. That these exponents are determinable and are determined by logarithms would surprise students and some teachers, even if the essential equivalence of exponents and logarithms is adequately emphasized. The idea of a compound interest law, namely, that one quantity may proceed in arithmetic progression as another related quantity proceeds in geometric progression, is ordinarily not brought out, nor is the fact that this same situation leads to a logarithmic law.

The omission of these and similar vital connections, both of mathematics to the exterior world and of one topic in mathematics to another, is directly responsible for the failure of algebra to reach the hearts of our students, and for the failure of the students to gain real insight into the significance of the subjects they so dully learn.

. . .

That the calculus is regarded as dry and uninteresting by many students, and that its value is occasionally doubted, is the strongest proof possible that its significance is not grasped. Here the connection with realities is so easy and so abundant that it is actually a skillful feat to conceal the fact. Yet it is done. I know personally of courses in the calculus (and so may you) in which the pressure to obtain and to enforce memory of formal algebraic rules has resulted in absolute neglect of the idea that a derivative represents a rate of change! I know students whose whole conception of integration is the formalistic solution of integrals of set expressions by devices whose complexity you well know. That an integral is indeed the limit of a summation, and that results of science may be reached through such summation is often nearly ignored and not at all appreciated. That the ideas of the Calculus should fall so low as to consist mainly in formal differentiation and integration of set expressions must indeed astound anyone to whom the wonderful significance of the subject is at all known. Moreover, it must convince any liberally minded educator who takes our own courses as a true representation of mathematical values that even the calculus is of no importance for real life or for society.

. . .

To the same end, may I now emphasize what seems to me a great, if not the greatest, function of this Association. In America, up to recent years, the beauty and interest centering in pure mathematics has so absorbed all mathematical talent that we have almost if not quite neglected that other phase of mathematics in which the significance of all we do is so self-evident: applied mathematics. This Association has, through its journal and through its meetings, already demonstrated its willingness and its ability to foster mathematics of this type. On this side of mathematics, not only discussion of the mathematics taught or to be taught, but even research papers of high grade have had in the past no adequate means of exposition. The wonderful work of Gibbs was for this reason long buried in the obscure Connecticut Academy, and mathematical advancement along the important lines that he laid down was delayed or wholly prevented. The great work of G. W. Hill, which included profound work on infinite determinants, was for the same reason unknown and unappreciated by many mathematicians in this country until near his death, and work by others along the lines he mapped out was discouraged and delayed. Thus American mathematics has suffered not only in reputation, through the suppression of what are perhaps the greatest American achievements in mathematics, but also in that encouragement necessary to the establishment of a strong school. The same may be said of the essentially mathematical researches of other men still living, whom I hesitate to name, whose work is scattered through journals on general science, journals on astronomy, journals on life insurance, journals on engineering, and so fourth.

· · ·

24(1917) 401–406

George Bruce Halstead, from the MONTHLY 1(1894), 336.

MATHEMATICS CLUB OF CONNECTICUT COLLEGE

There were ten members of the club in 1918–19 and due to influenza and diphtheria quarantines as well as to war conditions only four formal meetings were held.

27(1920), 28

Members of the Kansas Section of the MAA, meeting on March 18, 1916

To the Council

The National Committee on Mathematical Requirements was requested by your body at your meeting in December, 1918, and by the Council of the American Mathematical Society also in December, 1918, to undertake a reconsideration of the definitions of college entrance requirements as formulated by a Committee of the American Mathematical Society in 1903.

The National Committee is concerned with the study of a desirable reorganization of courses in mathematics in secondary schools and colleges and with proposals for the improvement in the teaching of this subject generally. The subject of college entrance requirements is only one phase of the work for which this committee was organized.

. . .

The following principles have accordingly been tentatively adopted to govern our proposed reconsideration of college entrance requirements in mathematics:

1. The scholarship implied in the present customary requirement of 2 1/2 units should not be lowered.

2. Drill in algebraic manipulation should be limited to those processes and to the degree of complexity necessary for a thorough understanding of principles and required by the probable applications either in common life or in subsequent mathematics.

3. More emphasis should be given to such immediately useful elementary topics as:

(*a*) The understanding and use of the formula.

(*b*) The interpretation of graphic representation and the use of graphic methods.

4. More emphasis should be placed on the acquisition of insight and power and less time devoted to acquiring mere facility in the solution of formal exercises.

5. While specific minimum requirements in separate subjects like algebra, geometry, etc., are still necessary, adequate provision should be made, perhaps through the so-called comprehensive examination, for the pupils in those schools who are developing "general courses" in mathematics where these subjects are not taught in separate courses.

6. College entrance requirements should be stated not only in terms of subjects and topics but also in terms of specific mathematical abilities to be developed.

7. Means should be found to introduce a proper and desirable amount of flexibility into the requirements in order to encourage progress in secondary education through the experimental introduction of new topics and methods. On the basis of these principles we hope to formulate our final report for submission to the Council not later than its next summer meeting.

Respectfully submitted, for the Committee,

J. W. YOUNG,
Chairman

27(1920) 102–104

George B. McClellan Zerr, from the MONTHLY 18(1911)

2. A MATURING ASSOCIATION, 1921–1930

Dunham Jackson, President of the MAA 1926

A Maturing Association

In 1921, Congress finally declared peace with Germany, Austria, and Hungary. Americans looked inward for inspiration and prosperity. The year before, more than 2,700 "anarchists" had been arrested in the great Red Scare. In 1921, Sacco and Vanzetti were found guilty of murder and anarchy, and Congress passed a series of laws that sharply curtailed immigration. The Ku Klux Klan initiated a revival that spread violence across the country.

It was also a time of expansion, hope, and prosperity. Just the year before, the 19th Amendment was passed, granting nationwide suffrage to women. In the next decade, America would experience a roller coaster ride of financial extravagance and failure. The first talking pictures would attract audiences across the country. Charles Lindbergh and Amelia Earhart would become heroes. It was the age of George Gershwin and Thomas Wolfe and William Faulkner. It was also the age of the Harley-Smoot Tariff act, which withdrew America from world trade and ensured a lasting depression.

The International Congress of Mathematics was held again in 1924 in Toronto (in place of New York, which was the original choice). Once again, mathematicians from the Central Powers were excluded, but it was the young Americans who led the fight to include *all* mathematicians in the next meeting. While the American public looked inward for strength, American mathematicians began to look outward with new confidence.

The leading figures in American mathematics were Bliss, Veblen and G. D. Birkhoff, who each served in turn as president of the AMS. Each produced a steady stream of outstanding young Ph.D.'s. There was new mathematics (Marston Morse began laying the foundations of Morse theory at this time) and old (Algebraic Functions became a mature subject). The close association between mathematics and astronomy was renewed in a new setting, as Einstein's revolutionary work on general relativity captivated the imagination of differential geometers. At the 1928 Congress, two Americans (Birkhoff and Veblen) were invited to give plenary addresses. Birkhoff spoke on aesthetics; Veblen on relativity.

As the prestige of American mathematics increased, the newly formed Mathematical Association of America found its place in the community. A series of distinguised presidents, including E. R. Hedrick, Florian Cajori, H. E. Slaught, and David Eugene Smith, established goals and an agenda for the new Association. Titles such as *Mathematics and Music* and *The Human Significance of Mathematics* (both presidential addresses) show how the Association expanded its interests to every aspect of mathematics and its popular image. The Carus Monograph series was begun (with *The Calculus of Variations* by Bliss); regular meetings were held jointly with the Society and the American Association for the Advancement of Science; the Association was incorporated.

Nonetheless, much of the Association's focus remained on education. In 1923, they published a 650 page report on *The Reorganization of Mathematics in Secondary Education*. It was an ambitious report from an energetic committee led by John Wesley Young, and after nearly seven years of committee work was greeted with enthusiasm and expectation by mathematics educators across the country.

Throughout the period, debates raged about whether Calculus should be taught in the first year of college (Osgood thought it should) and whether Freshman mathematics courses should be designed to suit the needs of special groups (J. W. Young thought they should not).

The Monthly reflected every aspect of the maturing Association and yet it
maintained much of its original character as well, continuing to publish articles
and problems that were aimed at Finkel's "average mathematician." As America
eased into a fitful depression, American mathematicians spoke out against an
increasing public indifference to mathematics and laid the groundwork for the next
great expansion of American mathematics—from abroad.

Emmy Noether (1882–1935)

Mathematicians and Music*

R. C. Archibald

"Mathematics and Music, the most sharply contrasted fields of intellectual activity which one can discover, and yet bound together, supporting one another as if they would demonstrate the hidden bond which draws together all activities of our mind, and which also in the revelations of artistic genius leads us to surmise unconscious expressions of a mysteriously active intelligence." In such wise wrote one supremely competent to represent both musicians and mathematicians, the author of that monumental work, *On the Sensations of Tone as a Physiological Basis for the Theory of Music.*

"Bound together?" Yes! in regularity of vibrations, in relations of tones to one another in melodies and harmonies, in tone-color, in rhythm, in the many varieties of musical form, in Fourier's series arising in discussion of vibrating strings and development of arbitrary functions, and in modern discussions of acoustics.

This suggests that the famous affirmation of Leibniz, "Music is a hidden exercise in arithmetic, of a mind unconscious of dealing with numbers," must be far from true if taken literally. But, in a very general conception of art and science, its verity may well be granted; for, in creating as in listening to music, there is no realization possible except by immediate and spontaneous appreciation of a multitude of relations of sound.

. . .

The manner in which music, as an art, has played a part in the lives of some mathematicians is recorded in widely scattered sources. A few instances are as follows.

Maupertuis was a player on the flageolet and German guitar and won applause in the concert room for performance on the former. At different times William Herschel served as violinist, hautboyist, organist, conductor, and composer (one of his symphonies was published) before he gave himself up wholly to astronomy. Jacobi had a thorough appreciation of music. Grassmann was a piano player and composer, some of his three-part arrangements of Pomeranian folk-songs having been published; he was also a good singer and conducted a men's chorus for many years. János Bolyai's gifts as a violinist were exceptional and he is known to have been victorious in 13 consecutive duels where, in accordance with his stipulation, he had been allowed to play a violin solo after every two duels. As a flute player De Morgan excelled. The late G. B. Mathews knew music as thoroughly as most professional musicians; his copies of Gauss and Bach were placed together on the same shelf. It was with good music that Poincaré best liked to occupy his periods of leisure. The famous concerts of chamber music held at the home of Emile Lemoine during half a century exerted a great influence on the musical life of Paris. And in

---

*Presidential Address delivered before the Mathematical Association of America, September 6, 1923.

America we have only to recall colleagues in the mathematics departments of the Universities of California, Chicago and Iowa, and of Cornell University, who are, to use Shakespeare's phrase, "cunning in music and mathematics."

While Friedrich T. Schubert, the Russian astronomer and mathematician, played the piano, flute, and violin in an equally masterly fashion, his great-grand-daughter Sophie Kovalevsky was devoid of musical talent; but she is said to have expressed her willingness to part with her talent for mathematics could she thereby become able to sing. Abel had no interest in music as such, but only for the mathematical problems it suggested. His close attention to a performer at a piano was once explained by the fact that he sought to find a relation between the number of times that each key was struck by each finger of the player. Lagrange welcomed music at a reception because he could by the fourth measure become oblivious to his surroundings and thus work out mathematical problems; for him the most beautiful musical work was that to which he owed the happiest mathematical inspirations. Dirichlet seemed to be sensible to the charms of music in a similar manner.

Such are a few instances, which could be considerably multiplied, of the relation of mathematicians to the art of music

"that gentlier on the spirit lies
Than tir'd eyelids upon tir'd eyes."

They suggest the accuracy of at least a part of the following observations of Möbius in his book on mathematical abilities: "Musical mathematicians are frequent ... but there are wholly unmusical mathematicians and many more musicians without any mathematical capability." That there are musicians with some mathematical ability will be granted when we recall, not only that Henderson, the prominent New York music critic and the author of many works on musical topics, has written a little book on navigation, but also that the late Sergei Tanaïeff, pupil of Rubenstein and Tchaikovsky, successor of the latter as professor of composition and instrumentation at the Moscow Conservatory of Music, and one of the most prominent of modern Russian composers, found algebraic symbolism and formulæ of fundamental importance in his lectures and work on counterpoint.

A question which has interested more than one group of inquirers is: Can one establish any relationship between mathematical and musical abilities? Within the past year two Jena professors, Haecker and Ziehen, published the results of an elaborate inquiry as to the inheritance of musical abilities in *musical* families. As a by-product of the inquiry they arrived at the result that in only about 2 per cent of the cases considered was there any appreciable correlation between talent for music and talent for mathematics; they found also that the percentage of males lacking in talent for music but showing a talent for mathematics was comparatively high, about 13 per cent. At the Eugenics Record Office of Cold Spring Harbor, Long Island, there has been collected a considerable body of data upon which a study of the correlation of mathematical and musical abilities could be based. It will be interesting to see if the conclusions of Haecker and Ziehen are here checked, and also if some results are found as to the extent to which musical abilities are present in a group of mathematicians.

. . .

31(1924), 1–5

Is the Universe Finite?[1]

Archibald Henderson

1. On one occasion, after finishing a highly theoretical piece of research, the English mathematician H. J. S. Smith is reported to have made the delightful remark:[2] "Thank God, there is something which can never by any possibility have a practical application!" It was perhaps with some such feeling as this that Bolyai and Lobatchewsky developed the non-Euclidean geometry associated with their names, although the latter did dabble a little in parallaxes; and perhaps too, little thought of practical considerations animated the minds of Klein and Newcomb in the analysis of elliptical and spherical space. Little did Ricci and Levi-Civita, I daresay, as they were evolving it, dream of the extraordinary rôle in a new mechanics their theory of the absolute differential calculus was so soon destined to play.

. . .

2. Certain physicists and astronomers during the past thirty years have raised the query as to whether the Newtonian mechanics, while holding with remarkable accuracy for our planetary system, also holds exactly for bodies at immeasurably great distances apart. In his paper, "Concerning Newton's law of gravitation" (*Astronomische Nachrichten*, 137, 1895), Seeliger affirms that we must make one or the other assumption: (1) If the common mass of the universe is immeasurably large, then the Newtonian law is not valid as mathematically strict expression for the controlling powers; or (2) If the Newtonian law is absolutely exact, then the common mass of the universe must be finite. Moreover, either an absolutely empty space or one filled with infinitely tenuous matter would not be in conformity with Newtonian mechanics. For instance, if the universe were infinite and there were an attenuated swarm of fixed stars of approximately the same kind and density, however far we might penetrate the interstellar spaces, then matter would have a finite mean density; and in accordance with Newton's law of gravitation and a theorem due to Gauss, a body at the surface of a very large spherical portion of the universe would be attracted by a force proportional to the product of the radius of the sphere and the mean density of matter. As the radius of the sphere increases without limit, the intensity of the gravitational field at the boundary of the universe would be infinite. This is manifestly impossible, as it would give rise to velocities of a magnitude unobserved by astronomers.

---

[1]This paper, on behalf of the Mathematical Association of America, was delivered on January 1, 1925, before the American Mathematical Society, the Mathematical Association of America, and Sections A, B, and D of the American Association for the Advancement of Science, at Washington, D.C.

[2]Smith's words, probably spoken in regard to some method in the theory of numbers, are thus quoted by D. E. Smith, *History of Mathematics*, I, 467: "It is the peculiar beauty of this method, gentlemen, and one which endears it to the really scientific mind, that under no circumstances can it be of the smallest possible utility."

On the other hand, if the mean density of matter were infinitesimally small, the cosmos must present the picture of an island of finite extent surrounded on all sides by infinite empty space. Such a view is repugnant to our minds, since the light of the stars and isolated stars themselves would drift away into the infinite space devoid of matter, and this ephemeral cosmos would gradually melt away and disappear. Astronomical observation and physical research on the whole do not support the view that the energy of the cosmos is continually being dissipated.

Seeliger advanced two hypotheses to meet the dilemma presented by Newtonian mechanics: either the possibility of matter of negative density in order to produce a null mean density; or to substitute in the numerator of Newton's law of attraction the quantity $e^{-\lambda r}$, where e is the base of natural logarithms, which would require λ to have the value 0.00000038, in order to account for the discrepancy in the advance per century in the perihelion of Mercury, taken as 40″. Professor Hall suggested a modification of Newton's law by which, for great distances, the force of attraction between two masses diminishes more rapidly than would result from the inverse square law, the increment to the exponent 2 being chosen as 0.00000016, in order to explain Mercury's movement. Unfortunately none of these *ad hoc* hypotheses, which have neither empirical nor theoretical foundation, will serve; for while setting right the outstanding anomaly in the node of Venus, as de Sitter has pointed out, they at the same time introduce greater discrepancies in other elements.

3. Is the universe infinite? If a voyager of the skies travel deep into the inter-stellar spaces, past the great blue helium stars of Orion, past Betelgeuse and Antares, beyond the white variable Cepheids, the gaseous red and yellow giant-stars, the faintest of the super-nebulæ, "lying like silver snails in the garden of the stars" but whirling in fiery spirals in the dim void of remoter space—will he ever reach any limit to the universe? Astronomers are not yet agreed that the amount of matter in the material universe is finite. It is significant that the density of matter falls off quite rapidly the deeper we penetrate into the stellar universe. For example, Hale says that there is probably an actual thinning out of the stars towards the boundary of the stellar universe. However, the problem with which we are here concerned is a very different one, *viz.*, has space a curvature?

· · ·

On metaphysical grounds alone Kant affirmed that space is infinite and sown with similar stars in all parts. Descartes, confronted with the question, "What lies beyond?" always maintained that a finite universe was impossible. In his *Our Place among Infinities*, the astronomer, Proctor, says: "The teachings of science bring us into the presence of the unquestionable infinities of time and of space, and the presumable infinities of matter and of operation—hence therefore into the presence of infinity of energy." An awful image of an infinite void is procured us by the Book of Job: "He stretcheth out the north over the empty space, and hangeth the earth upon nothing."

In a paper entitled "Cosmological observations concerning general relativity," published in the *Report of the Berlin Academy of Sciences*, February 8, 1917, Einstein advanced the view that the universe is finite, but unbounded. Various and cogent reasons led him to this conclusion. He was in search of a theory of the universe in conformity with the principle of general relativity, by which, in contradistinction to Newtonian mechanics, no preference is given to any reference system, and the laws of nature remain unchanged irrespective of the frame of reference to which they may be assigned. Ignoring the local concentrations of matter, represented by bodies and systems of bodies, Einstein assumed that the matter of the universe is distributed with uniform density—expecting thereby to arrive at some approximate conception of the metrical character of space as a whole. In order to differentiate between the coördinates of space and the coordinate of time, he made the reasonable assumption that the stellar system is approximately at rest, since the motion of matter is very small as compared with the velocity of light. Einstein was now confronted with two alternatives: either to assume that the universe is infinite and Euclidean at infinity; or else to adopt the view of Mach that inertia depends upon a mutual action of matter. Although the former is consonant with our conventional view that Galilean behavior tends to set in as we recede from a massive body, it was rejected by Einstein on the ground that it involves the far-reaching limitation, lacking in a physical basis, namely that B_{iklm} shall vanish at infinity, twenty independent conditions, while only ten curvature components G_{ik} enter into the laws of the gravitational field. It is not difficult, however, to show that, in the important case of the radially symmetrical field, these remaining ten conditions, on being complied with, lead to a solution indistinguishable from the familiar Schwarzschild form—which when r becomes infinite gives the Galilean values for the metrical tensor g_{ik}.

In his choice, Einstein was controlled by his desire to retain Mach's principle of the "relativity of inertia"—according to which the inertia of a body is entirely due to all the remaining matter in the universe. Brief consideration suffices to show that the only set of boundary values at infinity for the g_{ik} which would be invariant for all transformations is that all g_{ik} be zero. This postulate has been termed by de Sitter the "mathematical postulate of relativity of inertia." In Einstein's original theory of 1915, the g_{ik} are determined by the covariant field equations inside matter

$$G_{ik} = -\kappa T_{ik} + \tfrac{1}{2}\kappa g_{ik}T,$$

or

$$G_{ik} - \tfrac{1}{2}g_{ik}G = -\kappa T_{ik},$$

$$G = \kappa T,$$

where T_{ik} is the energy tensor of matter, T is the invariant of the energy tensor, and $g^{ik}T_{ik} = T$.

In the modification of the Newtonian potential used by Neumann, which amounts to replacing the Laplace-Poisson equation by

$$\nabla^2\Omega - \lambda\Omega = -4\pi K\rho,$$

it had been shown by de Sitter that thereby was secured a distribution of matter of a non-vanishing though very small mean density, maintaining its equilibrium without an extra pressure. Guided by this alteration, Einstein modified his original field equations to the form

$$G_{ik} - \lambda g_{ik} = -\kappa\left(T_{ik} - \tfrac{1}{2}g_{ik}T\right),$$

which may be written

$$G_{ik} - \tfrac{1}{2}g_{ik}(G - 2\lambda) = -\kappa T_{ik},$$

since

$$G - 4\lambda = \kappa T.$$

$\cdots$

32(1925), 213–223

HIBBERT JOURNAL, London, volume 18, no. 3, April, 1920: "Euclid, Newton, and Einstein" by C. D. Broad, 425–458 [Last paragraph: "I have now fulfilled my promise to the best of my ability. We have seen what exactly Einstein's theory is and how it is related to Euclidean geometry and to Newtonian mechanics. The connection with the former is not really very intimate, and Einstein himself makes very little play with it. The connection with the latter is all-important. Einstein's discovery synthesizes Newton's two great principles—the laws of motion and the law of gravitation. It removes the obscurity that has always hung over the former, by working out the relativity of motion to the bitter end, whilst it generalizes and slightly corrects the latter and accounts for its peculiar position among all the other laws of nature. Such work can only be done by a man of the highest scientific genius, and we have no right and no need to enhance his greatness by decrying the immortal achievements of his predecessors. It is enough that we can, without the slightest flattery or hyperbole, class Einstein with Newton, and say of the former what is written on the tomb of the latter:—'Sibi gratulentur homines tale tantumque exstitisse humani generis decus.'"]

27(1920), 315

At the University of Texas, Assistant Professor R. L. MOORE, of the University of Pennsylvania, has been appointed associate professor of pure mathematics, and Dr. JESSIE M. JACOBS [1920, 92] instructor in pure mathematics.

27(1920), 337

SRINIVASA RAMANUJAN, brilliant Indian mathematician of "astonishing individuality and power" (Hardy), and with "powers as remarkable in their way as those of any living mathematician" (Hardy), died at Chetput, Madras Presidency on April 26, 1920 at the early age of thirty-one. He was a fellow of Trinity College, Cambridge, and a research fellow of the University of Madras, and was elected the first Indian fellow of the Royal Society, London, when only twenty-eight years of age.

27(1920), 338

American Contributions to Mathematical Symbolism

Florian Cajori

1. Maya Number-System. Probably seven or eight centuries prior to the introduction of the zero in the Hindu-Arabic numerals, the Maya of Central America had a fully developed number system on the scale of 20 (except in one step). This system, as it appears in Maya codices, had a symbol for zero, and an extended application in the wonderful system of Maya chronology.[1]

2. Peruvian Knot Records. The use of knots in cords for reckoning, and recording numbers, early practised by the Chinese, had a most remarkable development among the Inca of Peru, from the eleventh to the sixteenth century of our era. Upon a twisted woolen cord (quipu) other smaller cords of different colors were tied. The color, length and number of knots, and distance of one from the other, all had their significance.[2] Quipu-like string records have been found in North America among the Indians of the Northwest.[3]

3. Dollar Mark. An extended study of manuscripts has led to the conclusion that our dollar mark descended, during the last quarter of the eighteenth century, from the Spanish-American abbreviation "p^s" for "pesos."[4]

4. Sporadic Notations for Radicals. In an anonymous publication, *The Columbian-Arithmetician. By an American*, Haverhill, Mass., 1811, there is added to the usual exponential notation 4^2, 2^m the following bold innovation (p. 13): $^2 4$ to mean $\sqrt{4}$, $^3 8$ to mean $\sqrt[3]{8}$, $^m 8$ to mean $\sqrt[m]{8}$. This symbolism found no favor.

5. B. Peirce's Signs for our π, e and i. These are shown[1] in Figs. 1 and 2; they were used by his sons, J. M. Peirce and C. S. Peirce.

6. Equivalence in Geometry. This was expressed by the sign ⧢ which occurs in C. Davies' *Elements of Geometry and Trigonometry*,[2] 1851. For a quarter of a century following 1885, the sign enjoyed considerable popularity; the two signs = and ⧢ were used in Geometry to express congruence and equivalence.

---

[1] See S. G. Morley, *An Introduction to the Study of the Maya Hieroglyphs*. Government Printing Office, Washington, 1915.

[2] L. Leland Locke, *The Ancient Quipu or Peruvian Knot Record*. 1923.

[3] J. D. Leechman and M. R. Harrington, *String Records of the Northwest*, Indian Notes and Monographs, 1921.

[4] F. Cajori in *Popular Science Monthly*, vol. 81, 1912, p. 521; *Science*, N.S., vol. 38, 1913, p. 848.

[1] B. Peirce in Runkle's *Mathematical Monthly*, vol. 1, No. 5, 1859, pp. 167, 168, "Note on two new symbols."

EDITOR'S NOTE:—Professor W. E. Story of Clark University used these symbols in this sense in 1907 in a course of lectures on the calculus of finite differences. NORMAN ANNING.

[2] Charles Davies, *Elements of Geometry and Trigonometry*, from the works of A. M. Legendre, New York, 1851, p. 87.

Note on Two New Symbols

by Benjamin Pierce,
Professor of Mathematics in Harvard College, Cambridge, Mass.

The symbols which are now used to denote the Naperian base and the ratio of the circumference of a circle to its diameter are, for many reasons, inconvenient; and the close relation between these two quantities ought to be indicated in their notation. I would propose the following characters, which I have used with success in my lectures:—

     Ω to denote ratio of circumference to diameter,

     Ω to denote Naperian base.

It will be seen that the former symbol is a modification of the letter c (*circumference*), and the latter of b (*base*).

     The connection of these quantities is shown by the equation,

$$\Omega^{\Omega} = (-1)^{-\sqrt{-1}}.$$

Fɪɢ. 1. B. Peirce's signs for π and e in the *Mathematical Monthly*, 1859.

$$\sqrt{G}^{\,\partial} = \sqrt[J]{J}$$

Fɪɢ. 2. From J. M. Peirce's *Tables*, 1871.

7. Approaching the Limit was designated by $\doteq$. The sign is due to J. E. Oliver of Cornell who at first used it in the sense "is nearly equal to." In 1880, W. E. Byerly[3] gave it the meaning "approaches as a limit." This same symbol $\doteq$ was used in 1875 by A. Steinhauser[4] of Vienna in the sense "nahezu gleich," but to the best of our knowledge Oliver is in no way indebted to Steinhauser.

8. A Symbolism in Vector Analysis was invented by Josiah Willard Gibbs of Yale, a pioneer in this field. In 1881 he marked a scalar or "direct" product by $\alpha \cdot \beta$, a vector or "skew" product by $\alpha \times \beta$, where small Greek letters represent vectors. He let a_0 stand for the magnitude of vector α. He marked triple products, $\alpha \times \beta \cdot \gamma$, $(\alpha \cdot \beta)\gamma$, $\alpha[\beta \times \gamma]$.

9. Symbolic Logic was developed by C. S. Peirce. "He understood how to profit by the work of his predecessors, Boole and De Morgan, and built upon their

---

[3]W. E. Byerly, *Elements of the Differential Calculus*, Boston, 1880, p. 7. Oliver himself used the symbol in print in the *Annals of Mathematics*, Charlottesville, Va., vol. 4, 1888, pp. 187, 188. It is used also in Oliver, Wait and Jones' *Algebra*, Ithaca, 1887, pp. 129, 161.

[4]A. Steinhauser, *Lehrbuch der Mathematik*. Algebra. Vienna, 1875, p. 292.

foundations, and he anticipated the most important procedures of his successors even when he did not work them out himself." C. S. Peirce introduced a considerable number of new symbols, for instance, $\prec$ for "inclusion in" or "being as small as"; x, y signifies commutative multiplication; in multiplication, "identical with—" is 1; etc. Other symbols were proposed for symbolic logic by Mrs. Christine Ladd Franklin and O. H. Mitchell.

10. General Analysis. Of notations introduced in America in the present century, I mention only the symbolism used by E. H. Moore in his general analysis. He takes some of his logical signs from Peano's *Formulario mathematico*, 1906, and uses them approximately in the sense of Peano. Among Moore's other signs are $\neq$ for logical diversity, $\equiv$ for definitional identity, $\ni$ for "such that," x^P for "x has the property P." Moore's aim is different from that of Peano, Whitehead and Russell whose object was to proceed with absolute certainty in difficult, abstract studies of the foundations of mathematics, and who for that purpose used elaborate notations. Moore aims to meet the needs of the working mathematicians who consider extreme logical complications relatively unimportant, but desire simplicity and flexibility of notation.

None of the ten notations cited as originating in America has thus far found general acceptance in Europe, except, perhaps, the dollar mark.

32(1925), 414–416

The MONTHLY draws the attention of mathematicians to the excessive charges for publications of the Cambridge University Press made by the present American agent, The Macmillan Company. A single illustration will suffice. The third edition of Whittaker and Watson's *A Course of Modern Analysis* was published at 40 shillings (*1921*, 31); the American agent's price is $12.50. Hence by ordering from a London bookseller, and paying the duty, a saving of at least $3.00 on the purchase of this single volume could be effected.—There is no duty for books ordered for college and public libraries. The American Branch of the Oxford University Press appears to count more definitely on the ignorance of purchasers in the United States. To illustrate: H. Hilton's *Plane Algebraic Curves*, 1920, was published at 28 shillings (about $5.60 at the present rate of exchange); the price of the American Branch is $12.60!

28(1921), 218–219

The *Bulletin of the American Mathematical Society* has published the following note (compare *1920*, 340): "Professor J. H. TANNER, of Cornell University, and Mrs. TANNER have given to the trustees of that institution fifty thousand dollars to establish a mathematical institute under the following stipulations. The money is to be allowed to accumulate without diversion for seventy-five years. At the end of that time one professor shall be appointed, whose duty it shall be to begin the formulation of plans for the proposed institute. At the end of each of the four succeeding periods of five years one or more additional professorships shall be established, the incumbents to collaborate in the same plans. The stipends of these professors shall be paid from the fund, but no other demands shall be made upon it until one hundred years from the date of the deed of gift (June, 1920), from which time the income of the entire sum shall be devoted to the maintenance of the institute; half of the expenditure of each year is to be applied to research in the mathematical sciences."

28(1921), 151

THE CHAUVENET PRIZE

In March 1925, PRESIDENT COOLIDGE proposed that the Association establish a prize for special merit in mathematical exposition. The proposition was sanctioned by mail vote of the Trustees and a committee consisting of Professors A. J. KEMPNER, Chairman, LOUISE D. CUMMINGS and D. R. CURTISS was appointed to formulate the details. This committee presented a report at the Ithaca meeting which in somewhat modified form was adopted by the Trustees. The substance of the report is as follows:

The committee believe that the proposed prize will exert a desirable influence on the production of high-grade exposition articles. They adopted the name suggested by President Coolidge, namely, "The Chauvenet Prize for Mathematical Exposition." For a study of the life and influence of WILLIAM CHAUVENET, 1820–1870, Professor of mathematics in the U. S. Navy, 1847–1859, President of the Academic Board of the Navy, 1847–1850, Professor of mathematics and natural philosophy at Washington University, St. Louis, Mo., 1859–1869, Author of many works and treatises, they refer to an article by Professor W. H. ROEVER, in *Washington University Studies*, Vol. XII, *Scientific series*, No. 2, 1925.

The Chauvenet Prize is to be awarded every five years for the best article of an expository character dealing with some mathematical topic, written by a member of the Mathematical Association of America and published in English in a journal during the five calendar years preceding the award. This prize will not be awarded for books, even though a large portion of mathematical books are mainly or completely expository in character, such as textbooks. They bring their own reward in the form of royalties.

The amount of the prize was fixed at one hundred dollars, an amount which the committee deemed sufficiently large to be attractive apart from the honor of the award. The cash for the first award has been provided by a friend of the Association. Thereafter, it will be supplied from the Association treasury, one fifth of the amount being set aside each year for the purpose.

The first award is to be made at the annual meeting in December 1925, covering the five-year period ending with the calendar year 1924.

It is provided that the award shall be determined at each quinquennial period by a scrutinizing committee of three to be appointed by the president of the Association and that this committee should be restricted as little as possible, aside from the specifications mentioned in the foregoing paragraphs. President Coolidge appointed the scrutinizing committee for the first award as follows:

E. B. VAN VLECK, Wisconsin, *Chairman*,

ANNA J. P. WHEELER, Bryn Mawr,

W. C. GRAUSTEIN, Harvard.

We have already recorded (*1921*, 150) the presentation to Professor F. N. COLE of an address accompanied by a purse containing about four hundred and seventy-five dollars ($475.00). Professor Cole has presented this sum to the American Mathematical Society. The Council of the Society named it The Cole Fund and appointed a committee to report on the most desirable method of expending the income from the Fund.

The Human Significance of Mathematics[1]

Dunham Jackson

As students and teachers of mathematics, we are confronted all the time by the question: What is it all about? What are we doing it for? Are we accomplishing something of which we can give an intelligible account? Or are we merely doing what our predecessors did because somebody else had done it before them? Are we, like Christopher Robin,[2]

> " . . . . following an elephant
> That's following an elephant
> That's following an elephant
> That isn't really there"?

I do not believe we are in this unprofitable state. If I did, I should not have taken this occasion to say so. But it is no confession of weakness to recognize that the problem is always with us. In its nature it is one for each generation that comes along. We are not asking whether it is important that mathematical knowledge should exist and be on record. That is admitted without question. There are tons of mathematical books in our libraries, and we accept the fact of their existence without uneasiness. The question is why we and our pupils should make it a chief concern of our lives to study those books, and look beyond the printed pages for the meaning that is behind them, and try to do the things that the authors have left undone.

I do not pretend to offer a final and compelling answer, resolving all doubts and removing all difficulties. If I had such an answer, I should not have been so unmindful of my obligation to society as to keep it a secret all this time. But as years pass I do find certain convictions increasingly confirmed in my own mind.

The question must be discussed with reference to the circumstances of the people on whose behalf it is asked. A reason that was good under the conditions of twenty-five years ago may not be a good one now, and if it is, that is a new observation, and not a mere reiteration of an old one.

On the other hand, regarding the problem as personal rather than universal, we can limit its scope, and so escape being carried too far afield. I have no intention of arguing the value of mathematical instruction in the public schools, or the importance of the Pythagorean theorem to the man in the street. I am thinking specifically of the members of the Mathematical Association of America, and of the

---

[1] Address of the retiring president of the Mathematical Association of America, at Nashville, Tenn., December 29, 1927.

[2] The writer offers his apologies to Christopher Robin for a slight adaptation of the latter's phraseology.

organizations affiliated with it, a few thousand individuals at most in a population of a hundred millions. Why should we, few as we are, spend our time doing mathematics, when there is so much else to be done? What shall we tell our students, when they ask us what we are driving at?

With our own peace of mind is involved the effectiveness of our instruction. We are told constantly that motivation is a prime requisite in teaching. Whatever that may imply for the technique of elementary instruction, the college student is a person of some maturity, and does not require motivation hour by hour and page by page. If it is available, so much the better. A young instructor in perplexity came to my colleague Professor Brooke, and inquired: "What do you do when students ask you what this or that particular topic in the calculus is good for in engineering?" To which Professor Brooke replied: "I tell them." But generally speaking, our students have some continuity of purpose, to the extent that they have purpose at all, and the most effective motivation is confidence that the instructor knows what he is about.

There was a time when you could make a good beginning by a reference to a liberal education. We have to recognize now that the thing formerly called by that name no longer exists. There is no body of knowledge or type of intellectual experience which is the common possession of persons of academic training, a bond of fraternity and medium of understanding between one educated man and another. It is possible to be influential in the intellectual life of the community without ever having opened a Latin grammar or a trigonometry. It is possible, as I happen to know from a specific instance, to graduate from an American university and think that Columbus discovered America in 1642. The traditional education survives as an experience of individuals, but not as a badge of a coherent and dominant group in society.

I do not admit for a moment that this is a sign of retrogression or decay. It is an incident of progress into the unknown. There is so much more to be learned that there has to be a division of labor in the learning of it; and there are thousands of people who have never been educated in the old-fashioned way, but whose intellectual attainments the alumnus of the old-fashioned education is compelled to respect. We have achieved democracy, or it has been achieved for us, if not in perfection, at any rate to a degree never contemplated by those who invoked it. And now that we have it, we have the problem of discovering what to do with it. Incidentally, we can not pretend that its destinies are committed to the hands of a class characterized by an ability to solve quadratic equations.

Another quaint phrase which we have heard from the lips of old people is "academic leisure." According to the theory of academic leisure, a professor taught his classes for an hour or two, and then had nothing more to do for the rest of the day. He retired to his desk to study undisturbed, or relaxed in intellectual companionship with students who chose to profit by his counsel. Nowadays, if a professor has leisure, it is either because his colleagues have discovered by sad experience that he is incompetent outside his own limited sphere, or because he is protectively endowed by nature like the animal of which Hilaire Belloc writes:

> "I shoot the hippopotamus with bullets made of platinum,
> Because if I use leaden ones his hide is sure to flatten 'em."

There are so many things to be done, things that really matter to somebody, things that are demanded every day and every hour with an urgency well-nigh irresistible! While as for any particular piece of research, one realizes that no hungry world is clamoring for it, and if by an infrequent chance it really is relatively important, it will be done independently by two or three other people in the course of the next few years. The fact that there is a personal satisfaction in doing it, does not meet the issue. The investigator who used to boast that his researches amused only himself, and were of no possible interest to anybody else, may still investigate, but he does not boast. When the gentleman of leisure was the ideal of society, a gentleman whose activities were harmless was a good citizen. But this democracy which has happened so unexpectedly demands that you do something useful, or at least make out a case for the usefulness of what you do. More than that, if you are not satisfied with evasion, it demands a degree of usefulness at least comparable with the alternatives that press for attention. To those of us who have to earn a living, and through force of circumstances would not be gentlemen of leisure even if that profession still enjoyed its old prestige, there is significance in Gottlieb's words in "Arrowsmith": "Why should the world pay me for doing what I want, and what they do not want?" The relentless question for the individual is not why he should do mathematics, but how he can persistently refuse to do a hundred other things that leave no time or energy for mathematics.

Well, in the first place, I am not abashed to say that I believe that mathematical training has a great deal to do with thinking straight. The force of this contention has been so much exaggerated on the one hand, and so much depreciated on the other, that stress on its residual merit is not out of place. Let it be granted that prejudice and error have been known to survive exposure to courses in geometry. It is no less true that malignant organisms in the body frequently survive exposure to the most approved medical treatment. Let it be granted that the decisions of every-day life are seldom reached by the formal application of logic, and that those who know how to think logically need constantly to be on their guard not to let their reasoning faculties corrupt their better judgment. Still it is of inestimable benefit to any nation to have a few men within its borders who know the difference between a sound argument and a specious one, and to have abroad in its population a wide-spread consciousness that there are norms of thought transcending factional expediency. It is my firm conviction that our average in this respect is higher for the presence among us of those of our colleagues and pupils, however few they may be, who consciously look for the wider significance of the standards with which their mathematical experience has familiarized them.

Mathematics not only demands straight thinking; it grants the student the satisfaction of knowing when he is thinking straight. There is no other field in which he can so soon stand on his own feet and face the world with confidence, assured that on his own ground his authority is final. Not many of your students, to be sure, ever reach this stage. Most of them keep on running to you with each successive bit of work, to ask you if it is right. But now and then, with your encouragement and guidance, one of them comes of age, having learned the meaning of self-criticism without self-distrust, and takes his place among those who know whereof they speak. He is potentially a man fit to exercise authority, with a full realization of the responsibility that goes with it.

A distinguishing characteristic of mathematics is its universality, its independence of time, place, and circumstance. The philosophers have made a great deal of this. They have ascribed to mathematical truth an absolute quality, transcending human experience and human existence. Perhaps they are right; nobody can prove the contrary. The alleged absoluteness of mathematics has always made it a refuge for souls who wanted to shut themselves away from the world. Probably the science always will owe much to those who see in it an escape from human limitations. But the universal persistence of mathematics has an appeal also for those of us who can not grasp so profound a metaphysical formulation. Even if we cannot attain anything beyond human experience, there is a compelling power in processes of thought which we know we have in common with intelligent men of all races and all times. Throughout the bewildering diversity of human nature, we have in the recognition of the facts of mathematics something as dependable and predictable as any phenomenon in the physical universe. By the exercise of our mathematical faculties we are brought into community of experience with Thales and Newton and their every-day associates, and their successors whose great-grandfathers have not yet been born. To this extent, the universal brotherhood of man is not a hypothesis or an aspiration, but an intimately accessible fact.

Still another phrase that needs to be rescued from abuse and restored to its proper place of dignity is "the enlargement of human knowledge." The words cover, if not a multitude of sins, at any rate a multitude of futilities. I may take this occasion to report, as a result of my own research (executed by a few seconds' manipulation of a calculating machine), that the product of the numbers 63471928 and 84023963 is 5333162929810664. I am confident that this fact is not to be found in the literature, and that it was never known until I discovered it. If there is exalted merit in such an addition to knowledge, I am ready to propose an unlimited number of topics for investigation.

But the point that I really want to make is that this *reductio ad absurdum* is as far from being final as the shibboleth against which it is directed. Collectively and in the long run we are finding out things that the world wants to know, however ephemeral most of our individual contributions may be. To mention specific instances, the methods in exterior ballistics developed by Moulton and Bliss during the war depend on the Picard process of successive approximations and the introduction of an adjoint system of differential equations, familiar products of modern mathematical research. The preliminary work on which the theory of relativity is based was as much mathematical as physical. The quantitative relations which make the theory of statistics forbiddingly abstruse to the non-mathematician are commonplaces in the study of orthogonal functions and approximation by means of them. While the inventors of the statistical measures in question do not appear to have been extensively familiar with this particular part of the mathematical background, and while their achievement is all the more remarkable on that account, the general theory is available for the enlightenment of the student.

More fundamentally, the calculus is an essential part (one hesitates to say an integral part) of our civilization, and it is the research of the last hundred years that has given us a real understanding of the meaning of limits, and assured us that the processes of the calculus will bear critical examination. I believe that in fifty years more we shall have achieved a similar clarification of the concepts of probability. As in the case of the notion of limit, which can be defined in thirty seconds, but can be really appreciated, even by gifted graduate students, only after months of study, and took decades in reaching general acceptance, it will not be a question merely of a few formulations and theorems, but of years of accumulated experience, the gradual suppression of what is irrelevant and the emergence of what is essential. Incidentally, in going about our business toward the attainment of this end, we shall not only be offering the products of our own study, but shall be bringing together the biologist, the psychologist, the economist, and the physician, and making available to each the common elements in the experience of all.

No less important perhaps than what we contribute to the technique of the natural sciences is what we learn directly about the workings of our own minds. When the attempt has been made to analyze the processes of rational thought, the case of mathematical reasoning has always been cited as typical. Our activity is a continuing laboratory experiment, which is no nearer finality than the rest of the science of psychology, and I believe there is much still to be learned from it. We gradually recognize, for example, that the achievement of a mathematical proof in our consciousness is an experience rather than an act, and that clarification is as likely to accompany relaxation following effort as to result immediately from the effort itself; but we can only guess how far these observations may ultimately be of theoretical and practical importance. Again, in trying to trace the inheritance of mental traits, we are likely to come down to specific cases by looking for something that can be identified as mathematical ability. One does not expect that the relations will be direct and simple. On finding that the highest score in an examination in my calculus class was made by a sister of a girl who had been similarly distinguished in the corresponding class a year or two earlier, and being informed that a third sister was taking freshman mathematics, I inquired as to her progress there, and was told that she wasn't doing so well in trigonometry, but got a B + in English for a theme about trigonometry as the course that she disliked most. But however individual manifestations vary, it may turn out that there are fundamental differences in the chemical constitution of the nervous system which make sustained intellectual concentration practicable for one person, and insuperably distasteful to another. And whatever the answer may be, we are as likely as anybody to have something to offer toward the solution.

On the plane of social rather than individual psychology, there is a fascinating subject of inquiry in the relation between mathematical advances and the general consciousness of the age that produces them. Is there any underlying reason why Newton should have been a product of the age following Shakespeare, and Gauss a contemporary of Beethoven? Conversely, the main currents of mathematical progress must continue to have a profound influence on the habits of thought of the race. Who can estimate how much the emancipation of the intellect owes to the putting of the circle-squarer in his proper place, and the evidence of Gauss, Bolyai, Lobachewsky, and Riemann that the authority of Euclid did not finally seal the book of knowledge? The modern mathematician, like his predecessors, has done his part for the liberation of the human spirit.

In conclusion, let me emphasize that the advancement of mathematics is a great human enterprise in which we are all participants, and not mere spectators. You are making a contribution if you carry through a piece of research, or if you accept a committee appointment which leaves a colleague free to carry on his research, or if you study what he has done and give him the benefit of your appreciation and criticism, or if you merely work over the existing material and add to the fund of experience from which comes the collective assurance that we are on the right track. The science shares the vitality of all men in whose consciousness it exists.

35(1928), 406–411

Dr. J. R. RITT, of Columbia University, has been promoted to an assistant professorship of mathematics.

28(1921), 332

Eliakim Hastings Moore
[With permission from the University of Chicago Mathematics Department]

How Can Interest in Calculus Be Increased?[1]

Roscoe Woods

As teachers I presume that all of us have certain aims in our work. There is no doubt but that these aims vary with the individual. However, one of the important facts to keep before ourselves is that mathematics, due to its inherent nature, needs a teacher more than any other subject. If we do not have definite aims of a high type as to our functions as teachers, we are not going to create much enthusiasm for our subject. Unless we are convinced that the study of mathematics can do something for the student beyond the acquisition of certain mechanical skills and facts required for the next course and unless we believe that the student acquires certain fundamental thought processes from mathematics, we will not engender a great deal of interest in our students.

The study of mathematics ought to help the student to express himself in clear, forcible, and concise language. It should unfold his rational faculties and teach him how to use them. In the various subjects that we set before him and in the problems that he solves, he ought to be able to penetrate from the surface to the central idea. We as teachers should be convinced and should teach that one of the primary aims of all learning, whether of a mathematical nature of not, is to see relations that exist between things and to correlate facts with principles that seem to be unrelated. This is another way of saying that mathematics should be so taught that the student is brought to realize that he has faculties that can discern the good and the bad intellectually and that this discernment is the foundation of the development of character for which all education should exist.

In the teaching of the calculus these general aims should be more nearly attained than in any other single subject. Perhaps, to be more specific, we ought to ask ourselves why we teach calculus, to liberal art students, in preference to some other subject which would contribute toward their culture and mental development. The answer to this question presents a multiplicity of notions. In a broad way we are accustomed to think of the practical side of calculus, when engineers are mentioned; of its applications, when other science students are named; and of the cultural side, when liberal arts students are considered. Let us enumerate some of the aims for teaching calculus. Do we teach it with the idea that the student will recognize forms and acquire facts and technique? Do we teach it with the idea that it unifies all that the student has learned before and that it is the key stronghold from which all other fields of mathematics are to be explored? Do we teach it solely as a preparation of the engineer and of the teacher of elementary mathematics? Do we teach it as a means of interpreting physical phenomena? Do we teach it

[1]Read before the Iowa Section of the Mathematical Association of America, May 4, 1928.

simply as a tool whereby other achievements are accomplished? Or do we teach it as a formal exercise in logic with the idea that it may stimulate certain thought processes?

Some years ago Professor Osgood of Harvard,[2] in an address before the American Mathematical Society, maintained that the calculus should be taught as a means of interpreting physical phenomena. He claimed that there was only one calculus for all students, no matter what the ultimate aim or profession of the student. He presents the course formally and uses the differential as soon as it is possible to introduce it. Professor Rietz has likewise written an article, which appeared in this Monthly,[3] in which he points out that the aim in teaching calculus should be to implant certain notions that the student should remember ten years afterward. Professor Bliss of Chicago[4] has written a monograph on the subject, *The function concept and the calculus*. All of these papers are well worth reading and will help you to decide why the calculus should be taught.

At first sight it seems that it ought to be a fairly simple matter for us to inculcate in the student the three central ideas of the calculus; namely, the derivative, the anti-derivate or the indefinite integral, and the definite integral interpreted as the limit of a sum. But the appreciation of these ideas presupposes a considerable amount of preparation on the part of the student. For the study of the calculus, work, intelligence and a certain spirit of adventure are needed. To pursue this study properly and with the greatest pleasure, the student must also possess the spirit of exploration. A few students possess this spirit but the general attitude is a passive one.

Some of the factors that give rise to the lethargic attitude are: (1) the trend of high school instruction, (2) preparation of the teachers of elementary mathematics, (3) training of college freshman, (4) the fact that no organized attempt is made to acquaint students with the content and purposes of the calculus, (5) the type of text book at our disposal, and (6) the general attitude and habits of the American student. Let us examine these six reasons briefly.

(1) The best index to the trend in high school instruction is to note the quality of the product produced. The way in which the high school graduate handles his complex simplification problems, the manner in which he analyzes problems in story form, and his feeble attempt to give definitions and logical explanations accurately usually indicate a lack of training rather than a lack of intelligence. Also there are certain habits that are indicative of training. For instance, the failure to read carefully the text is very common and usually leads to poor grades in college. This probably arises from the fact that the problems assigned in high school were of a single type and needed only the application of the illustrative exercise. Let us give another illustration of habit formation. Very few high school students have any idea how to study for recitation or how to prepare for a written lesson or how to use their time during the examination period. In other words, the fad of omitting the hard topics or of giving them little emphasis seriously handicaps the student's future.

(2) In general our elementary teachers are not highly trained. For instance only 47 percent in this state have had training in calculus. We should blush with shame when we compare the training of our elementary teachers with that of teachers in Europe.[5] There is no surer way of dampening a student's ardor for mathematics than to give him a teacher who is unprepared and who has no inherent interest in mathematics. We know that it is not uncommon for a high school principal to ask a teacher who majored in English, for instance, to take a mathematics class in the

high school. There seems to be an impression abroad that anyone can teach mathematics and just as long as this impression prevails, just so long will this vicious practice of assigning majors in other fields to teach mathematics classes continue.

(3) There are very few freshmen who are adequately prepared for and who can successfully master college algebra, trigonometry, and analytical geometry in one year, four periods per week. It requires time to acquire mathematical ideas and much training for mastery. To be frank we shall have to admit that college algebra is a conglomeration of topics. We admit that they are needed and badly needed; but they are not built around a central idea. I refer to the rational functions. In trigonometry we give the student a smattering of the circular functions, but we do not let him hear of their inverses. Analytical geometry offers the first opportunity to the student to use all his mathematical knowledge. It is the proving ground. The average student has just about enough time to learn the terminology connected with the course, but not enough of the technique required for enjoyment. He learns that there are such things as the straight line and the standard forms of the conics. He is allowed a glimpse of polar coordinates, higher plane curves, and transformation of coordinates. His solid analytical geometry is sealed with a great seal. He finishes his course with no idea of what the fundamental problem of analytical geometry is. He has graphed a few curves but he has not been given time to study the graphs after they have been made.

(4) We never point out to the student that all his work thus far has to do with functions of different kinds. We never classify them for him. He has been learning some of their properties through algebraical and graphical methods, but he has not been brought face to face with the tangent problem in his analytics, nor have his algebra and trigonometry convinced him that he needs to have a method for studying the variation of these functions. At no stage in the freshman's preparation do we acquaint him with the aims and purposes and the grandeur of the calculus. If he elects mathematics in the second year, he faces the same kind of useless (?) computation that he has done all his life with no apparent reason or incentive. Mathematical enthusiasm does not thrive in such soil.

(5) The absence of good exposition in our calculus texts is apparent. In order to meet the demand of the technical schools and of economy our texts are a sort of a hybrid between a good text and a mere handbook. If the teacher is not careful his students finish calculus not knowing the classes of the functions that have been studied. Unless the teacher is diligent, the student has not even found out the fundamental problems of the course, and his attitude toward advanced calculus is similar to that which he had for his calculus when he finished analytics.

(6) Finally, the student's attitude toward learning for learning's sake needs a revision. What student thinks of studying when not in school? Money can be made with little study, and a position can be secured through channels other than that of scholastic attainment. The question uppermost in every student's mind is, "Of what practical value is this?" The information he seeks is that which he thinks will enable him to make ready money. But so long as the colleges and universities grant diplomas by the wholesale, why should we blame the student if he desires a diploma rather than the possession of that for which the diploma stands.

How can interest in calculus be increased? Clearly by trying to remedy the conditions set forth above. The following suggestions are offered. (1) Take every opportunity available to convince our fellow teachers outside of mathematics that we have a live, growing science. (2) Make more use in our elementary classes of the history of mathematics and the biographies of mathematicians both dead and living. (3) Be sure that our mathematics majors and advanced students know the various divisions of our science and its importance in the world. (4) Study carefully the problem of the dependence of all the sciences upon mathematics and make these results available. (5) Rearrange our courses to meet changed conditions, giving more credit where credit is due. (6) Prepare a series of charts, on various topics, that can be used in the classroom, especially one showing divisions of our science. (7) Assemble all our students by classes for general lectures along broad and unifying lines. (8) Consider some means on the part of the Iowa Section of the Association of recognizing the ability of any outstanding student who has finished the calculus. (9) Finally, let the Mathematical Association of America, the mathematics sections of the State teachers' associations, and the National Council of Teachers of Mathematics bend every effort to keep mathematical instruction in the hands of those prepared to give it.

36(1929), 28–32

FIRST CHAUVENET PRIZE.

The committee on the award of the first Chauvenet Prize for excellence in mathematical exposition, Professors W. C. GRAUSTEIN, ANNA PELL WHEELER, and A. B. VAN VLECK, chairman, recommended that the award be made to Professor G. A. BLISS of the University of Chicago for his paper on "Algebraic functions and their divisors," published in the *Annals of Mathematics*, volume 26, Numbers 1 and 2, September and December 1924. The Trustees voted to approve this choice and the thank the members of the committee for their arduous but very valuable efforts. The award was announced at the business meeting and the prize of one hundred dollars, furnished by a member of the Association, was presented to Professor Bliss following the meetings.

33(1926), 177

Professor F. N. COLE, of Columbia University, for twenty-five years secretary of the American Mathematical Society and editor of the *Bulletin*, died on May 26, 1926, at the age of sixty-four. Death was caused by heart failure brought on by an infected tooth. He was to have retired from teaching on September 20, his next birthday.

33(1926), 238

Professor H. B. FINE, of Princeton University, was fatally injured by an automobile on the evening of Friday, December 21 and died about one A.M. on December 22, 1928. He was seventy years of age.

36(1929), 118

Recent Publications

Warren Weaver

A Debate on the Theory of Relativity. By R. D. Carmichael, W. D. MacMillan, M. E. Hufford, and H. T. Davis. The Open Court Publishing Co., 1927. 154 pages.

The volume under review contains a report of the debate on the theory of relativity held at Indiana University in May, 1926 under the auspices of the Indiana chapter of Sigma Xi. The most considerable direct attempt made by any of the speakers to define what a theory is or what properties a theory should have is contained in Professor Carmichael's opening speech. He points out (p. 36) that a theory "must be in suitable agreement with the facts of nature, it must have those aesthetic qualities which render it pleasing to the human spirit, and it must furnish what is to us the most agreeable theory from the point of view of convenience."

It seems advisable to state here, in somewhat explicit form, a view of the nature of a physical theory.[1] The statement is made because only in this way does it seem possible to estimate what this debate accomplished and what it did not. The remarks will refer to an ideal[2] rather than to any actual physical theory; and will be so brief as to contain, of necessity, certain crudities and inaccuracies.

An ideal physical theory starts with a set of undefined elements and certain primary unproved propositions concerning these elements. By the use of the canons of logic, there is then deduced a further set of propositions concerning the elements. These deduced propositions may be called the secondary propositions or the theorems of the theory. The undefined elements are then identified with certain physical entities, part of which at least are experimentally measurable. The secondary propositions or theorems thereby become proposed laws of nature, since they state quantitative relationships of uniformity in the external world which may be tested by experiment. The "theory" is then this entire logical structure,—undefined elements, undefined or primary propositions, the deduced theorems, and the proposed laws of nature which result from the theorem due to the physical identification of the elements.

One may then note three qualities which a theory may or may not possess. It may or may not possess the quality of *self-consistency*, according as the primary propositions are or are not consistent and according as there has or has not been an error in the purely logical structure of the theory.

Secondly, the theory may or may not possess the property of *correctness* or *truth*, according as the proposed natural laws do or do not turn out to be actual natural laws. It is doubtful whether the validity of a law can ever be rigorously established by experiment. Experimental evidence has, however, shown that some proposed laws must be discarded, and it has permitted others to be (temporarily, at least) retained. There will, moreover, always be border-line cases concerning which scientific men will disagree; and this disagreement is an inevitable and desirable result of the subjective judgments involved.

Thirdly, the theory may or may not possess the property of *value*. A little thought will convince anyone that a theory may be self-consistent and true, in the senses just discussed, but still be utterly valueless. The value of the physical theory seems to lie in its ability to "explain" the natural laws which are associated with it, and which are its fruition. This "explanation" is chiefly of two sorts, explanation by simplification and explanation by analogy; and the theory is, in the case of these two types of explanation, likely to be called a "mathematical theory" or a "mechanical theory" respectively.

The value which attaches to a theory which explains laws by simplification seems to arise from the subjective sense of satisfaction with which one recognizes that the several laws to which the theory leads appear as united and hence simplified through the central logical structure of the theory. In the case of explanation by analogy, the value of the explanation results partially from the same, and partially from other no less subjective considerations. There are important instances of theories in which there is possible a double identification of the undefined elements. One of these identifications leads to the theory itself, and the other to the analogy. Thus in the kinetic theory of gases certain elements are identified with the molecules of the gas and these elements are also thought of as (that is, identified with) minute perfectly elastic spheres. This second identification furnishes a mechanical model, and the double theory, considered as a whole, is a mechanical theory which explains by analogy. The satisfaction that any one person feels as a result of such an explanation is clearly a subjective matter. A great school of physicists has enjoyed mechanical theories and their explanations by analogy. The tendencies of modern theoretical physics have convinced a goodly number of persons that we should be very skeptical of most explanations by analogy. Thus, one who really believes in the electrical constitution of matter recognizes that the laws of ordinary mechanics are statistical consequences of the laws for individual electrodynamic action; so that a mechanical explanation of electrodynamics is without meaning or importance.

These remarks concerning the nature of a physical theory lead to three conclusions, each important for our present purpose. First, unless one is raising very fundamental questions concerning the validity of the canons of logic, there is a reasonably definite way of determining whether or not a theory is self-consistent.[3] Secondly, the question as to the truth of a theory must be attacked through experiment. After the experimental evidence is obtained, each person must decide for himself as to whether this evidence convinces him that the laws of the theory have or have not been established, this judgment being of an obviously subjective nature. Thirdly, the question as to whether a theory is valuable is, to a still greater extent, a subjective matter. It depends upon one's personal decision as to whether or not the theory explains its laws. Thus a theory can be true for one and untrue for another, and it may certainly be valuable to one and of little value to another. If one judges a theory to be valuable, he is the more likely to also judge it true, and vice versa. It has even been seriously suggested that some theories are doubtless valuable even though we judge them to be definitely untrue.

These remarks seem to the reviewer to make clear what such a debate could and could not accomplish, and they make it easy to report what actually did occur. In support of the theory, Professor Carmichael, opening for the affirmative, explained in an exceedingly clear and interesting way the nature of the restricted and the general theory of relativity. Only penetrating familiarity and long experience with the doctrines of relativity could produce so illuminating a review of its most important features. He called attention to the naturalness of the ideas, to the diverse laws which are obtained without the introduction of *ad hoc* hypotheses, and to the primitive simplicity of a space-time event as compared with the artificial separation into space and time. In the light of the remarks above, we should thus say that the first speaker expounded the theory and expressed and justified his subjective judgment that it has great value. He also affirmed his belief that the theory is true, but left arguments on this point to the other member of the affirmative team.

Professor MacMillan, opening for the negative, confessed that he is relatively indifferent as to whether or not the theory is true because, in his estimation, it is without value. It lacks value to him because its explanations do not explain. His intuitions are outraged by the abandonment of Euclidean space and Newtonian time, and the result is to him confusion rather than simplification. As one of the great thinkers on cosmological problems, he has on many happy occasions helped to enlarge our vision of those vast stretches of time during which our solar system has been evolving, and such considerations have helped him to gain a sort of cosmic patience. He is willing that science wait a few score or a few hundred years, if necessary, for a theory which will meet all our new experimental facts, but which will, at the same time, hold to the "postulates of normal intuition." He is unwilling to argue about the extremely delicate experimental tests because, true or not true, the theory makes him scientifically unhappy. It is a valuable thing that sound and sincere ideas of this sort be expressed. To seek to modify them through argument is idle and improper. The only way to change these opinions is to make Professor MacMillan over into some other person, and that is a project which American mathematicians will not support.

Following these two opening addresses, Professor M. E. Hufford, for the negative, and Professor H. T. Davis, for the affirmative, considered the experimental evidence for the truth of the theory of relativity. The Kaufman-Bucherer experiment, the Trouton-Noble experiment, the Michelson-Morley-Miller experiment, the Michelson-Gale experiment, the star deflection test, the spectrum shift test, the advance in perihelion test, and the evidence from spectral line structure are all discussed, pro and con. These delicate and subtle experiments cannot, in the reviewer's opinion, be profitably discussed before a popular audience and in an hour's time. It is not scientifically sound to represent that a discrepancy of one per cent or twenty per cent, or even fifty per cent between theory and experiment is decisive evidence for or against a theory; nor is it safe to make the somewhat naive assumption that the minute one begins to talk of experimental evidence he is on "solid ground." Only those two great winnowers for the truth, time and the theory of probability, will be able to ultimately convince scientists that the first, second, ... order effects of the theory of relativity are or are not confirmed by experiment. Professors Hufford and Davis had a difficult if not a hopeless task. Their discussion is interesting but it is, of necessity, not very enlightening or convincing.

The debate closed with short rebuttals, by Professors MacMillan and Carmichael, in which are restated and amplified their estimates of the value of the theory and its probable philosophical influence. The book, as a whole, is interesting and stimulating. It contains a useful summary of a good bit of experimental evidence. It contains illuminating and suggestive confessions of scientific faith. But if one wants to know whether the theory of relativity is, for him, a true and a valuable theory, he must steep himself in its doctrines and then search his own mind.

36(1929), 38–42

The original manuscript of Professor Albert Einstein's latest published research, "Zur Einheitlichen Feld-Theorie," is now in this country in the possession of Wesleyan University, Middletown, Connecticut, where it will be permanently kept in the Olin Library and exhibited to those interested to see it. The manuscript consists of eight pages of closely knit lines, all in Einstein's handwriting, together with mathematical calculations and interlineations. Certain portions have been crossed out and do not, therefore, appear in the published paper.

The first seven pages of the manuscript contain the scientific treatise, the results of six years of Einstein's deepest thought, and the mathematical statement of a scientific theory which it has been said that not more than twelve men understood at the present time. It is said that Professor Einstein considers this theory of more scientific significance than his famous theory of relativity and that he believes it will be years before the world of science will be able to grasp fully all the details and implications of his theory and check up on his calculations. The manuscript is autographed at the end of page seven. The eighth page contains expressions of thanks to Professor Einstein's co-workers.

36(1929), 296

KNOTS

Professor Alexander discussed the problem of finding sufficient invariants to determine completely the knot type of an arbitrary simple closed curve in space of three dimensions. He outlined the derivation of one of these invariants which takes the form of a polynomial $\Delta(x)$ with integral coefficients, where both the degree of the polynomial and the coefficients are functions of the curve with which it is associated. He pointed out that the problem was not completely solved and suggested several problems related to it.

J. R. KLINE, *Secretary*

36(1929), 304

Recent Publications

James Byrnie Shaw

The Pastures of Wonder. By Cassius Jackson Keyser. Columbia University Press, New York, 1929. xii + 208 pages. $2.75.

In this book the author has carried to the extreme the tendencies shown in his previous writings concerning mathematics, and has attached here also his views of science. He has departed widely from his view, once expressed, that science is "a sublimated form of play, the austere and lofty analogue of the kitten playing with the entangled skein or of the eaglet sporting with the mountain winds."

He discusses two "realms": that of mathematics, and that of science. That of mathematics, in brief, he asserts is nothing more than deductive logic. That of science he proposes to limit to the establishment of categorical propositions. Mathematics is an edifice of hypotheticals, science an edifice of categoricals. With the first idea many will take issue. The proposal many will reject. That some mathematicians and some others have wished to limit mathematics to its purely logical aspect is of course true, but mathematicians generally refuse to agree that what is thus excluded is not mathematics. That the deductive reasonings from hypotheses in science do not constitute a legitimate part of science few will agree with. There are of course scientists who would limit science to accurate and ordered accounts of phenomena. But this limitation appeals to few. The author undertakes to meet certain challenges to his definitions, and of course disposes of them to his own satisfaction. The book is a natural outcome of the notions of the "logistic" development by Russell and others. The antidotes for this disease to be found in the writings of Poincaré and others need to be more widely spread. One of the most serious challenges to logistic is the simple question: What constructive mathematics has it ever produced? To confine mathematics to logistic is worse than to confine it to "the study of integers and what can be got out of them, and nothing else."

To quote an example which the author himself gives, the proposition that "the square on the hypotenuse of a right-angled triangle is equal to the sum of the squares on the other two sides" is a categorical proposition, and is not therefore mathematical. This is, according to the author, the conclusion from a set of unstated "Euclidean" postulates, and a conclusion by itself is not a mathematical proposition. The mathematical proposition is the statement that this conclusion follows from the original postulates. On this basis all conclusions in mathematics would seem to be in science, not mathematics. The only thing left to mathematics is the barren statement that the conclusion has been "proved." The premises may be false, they may be mere empty symbols, and the conclusion may be nowhere applicable, but if the logic is correct (*and this is a big assumption*) then we have a mathematical statement. It is the old and absurd statement of Russell, that in mathematics we do not know what we are talking about nor if our conclusions are true. But one needs only to notice that we have struck here an endless sequence from which we cannot escape. For we should say above; *if the process of logic used in demonstrating Pythagoras' theorem is valid, then the conclusion is a consequence of the postulates.* Now this is itself a hypothetical proposition, and *its* conclusion is a *categorical proposition.* Thus the original statement that the theorem of Pythagoras follows from the postulates of Euclid is categorical and so not mathematical. This can be extended ad libitum. In fact, every hypothetical proposition is also a categorical proposition, and the distinction the author desires to make breaks down.

The statement of Benjamin Peirce, made years before Russell's work, that mathematics is the science that draws necessary conclusions, is the basis for these logistic arguments. But the other statement of Peirce is almost invariably omitted. It is that the processes of logic cannot be applied without being transmuted into various forms, and this transmutation is the mathematical process in the inquiry. This is very important, for a large part of mathematical investigation does not consist in drawing conclusions, but in building structures of thinking. Most of mathematics consists in the development of ideal structures. The brilliant son of Benjamin Peirce, Charles Saunders Peirce, gave as the definition of mathematics, in the Century Dictionary, "the study of ideal constructions (often applicable to real problems), and the discovery thereby of relations between the parts of these constructions, before unknown." He adds, "The observations being upon objects of imagination merely, the discoveries of mathematics are susceptible of being rendered quite certain." Such observations are categorical assertions.

The author of the book under review makes much of the *forms* of propositions and elaborates this with various examples of syllogism. But he evidently fails to see that forms themselves are objects, and assertions about them are categorical assertions. It was Kempe who wished to define mathematics as the science of pure form, but he was quite clear in his vision of forms as ideal objects, in practically the same sense as C. S. Peirce. As the author says on page 99, mathematical verities are eternally true, but not because they are mere logic. Mere logic itself is an ideal creation of the human mind, and may be utterly changed. Witness the work of Brouwer on the law of excluded middle. They are true because the human mind has the ability to see the universal characters in its own creations.

As for the proposed definition of science, we may leave the scientist to accept or reject it. The discussion of the proposal with imaginary scientists, of course frames answers for the objections already set. The proposal seems to have as one aim to exclude mathematics from science. This has already been done if we accept the common modern view that science is based on the observation of the natural world. Also he tries to exclude the mathematical ideals of methodology and thought from science. This is being rapidly accomplished by modern physics in some directions, but is being equally rapidly shown to be impossible in others. The reviewer is referring to Bridgman's *Logic of Modern Physics,* and to all the present exposition of atoms as matrices. The author of course admits the necessity for the scientist to use mathematics, and probably would not deny the fact that many advances in mathematics have originated in scientific problems.

The title, "Pastures of Wonder," is ascribed in the preface to the existence of two great realms, the categorical and the possible (*equivalent to hypothetical?*) and the wonder these two create. But the two worlds are also called propositional, and one may well ask if the universe consists merely of propositions, for human experience would certainly deny this statement. And propositions later are defined to be only those statements which can be tagged as true or false. Are there no others of use? And the old, old question, which Russell gave up, still remains: What does true or false mean, and why choose one rather than the other?

The universe is indeed filled with pastures of wonder. One of these is the pasture of beauty. One can defend the play of the mathematician as the loitering of an enchanted spirit in this pasture. And this pasture is not the rarefied ether, empty of all content. Mathematics would have died of inanition long ago were it that. For all creative mathematicians are chiefly concerned in making categorical assertions about the ideal world in which they live. This is what the "educated layman" to whom the author addresses his essay should understand as the meaning of the term mathematics. If the layman will pick up almost any book on mathematics he will find that the writer is chiefly concerned with the properties, the characters, the structure, the relations of the mathematical objects he is considering. And he should not be misled into thinking that the only feature of the book which is *mathematical* is the empty assertion that the conclusions follow from the premisses. He will find indeed that the greater part of the book is derived from direct intuition and not from logic at all. If he succeeds in seeing the structures discussed as the writer sees them he will be filled with wonder at their beauty.

37(1930), 81–84

The corner stone of the new Eckhart Hall on the main quadrangle of the University of Chicago was laid on July 12. This Hall will be occupied by the departments of physics, mathematics, and astronomy.

36(1929), 407

The Fine Memorial Mathematics Hall, which will be erected at Princeton University at a cost of $400,000 in memory of the late Henry B. Fine, for many years a professor of mathematics and dean of science, will be started in the near future.

36(1929), 453

Princeton University announces the appointment of Dr. J. Von Neumann, of the University of Berlin, and Dr. E. Wigner, of the Kaiser Wilhelm Institut für Physikalische Chemie, as visiting lecturers in Mathematical Physics for the second semester of the present academic year.

37(1930), 103

Gilbert Ames Bliss
[With permission from the University of Chicago Mathematics Department]

An Unusual Use of the Nodal Cubic in the Plane

Bessie I. Miller

Until recently manufacturers of lamp reflectors usually employed a reflector approximating the mathematician's paraboloid of revolution. Such a surface throws light in parallel rays provided there is a point-source of light at the focus. Practically the manufacturer found it satisfactory to use a V-shaped filament lying in a horizontal plane on the axis of the reflector. The axis of the V coincides with the axis of the reflector; the vertex of the V is directed away from the vertex of the reflecting surface. This provides for some latitude in "fore and aft" focussing. Now the manufacturers wish to use two filaments one above the other, so arranged that one can be used for distance lighting during fast travel, the other for conditions occurring in city driving or in passing vehicles moving in the opposite direction. The parabolic reflector of the past fails in this case, since any displacement above or below the axis causes the light to break into divergent beams.

Wm. H. Wood, M. D., of Cleveland, Ohio, who is a specialist in the treatment of disorders caused by the improper functioning of the glands and who is a well-known inventor of commercial and scientific instruments concerned with the reflection of light, has recently added to his long list of patents. He has constructed a new generating curve for an automobile lamp reflector to replace the parabola of the older reflectors. His method was wholly experimental, and the surface eventually obtained was the result of an extraordinarily delicate process used in the polishing of the surface point by point, until the desired corrections on the parabolic reflector were obtained. The engineers to whom was submitted a plane section of the surface obtained by measurement were unable to analyze the curve, but were able to determine a construction for it. It was however not difficult mathematically to find the equation of the curve which, unexpectedly to the inventor, I found to be a nodal cubic, a thing of which he had never heard.

The relation between the old parabola and the new nodal cubic can easily be seen if their equations,

(1) $$y^2 = 4px$$

and

(2) $$y^3 - cy^2 - 4pxy + 4cx^2 = 0,$$

are graphed. The constant c in (2) is the distance above the axis on the latus rectum of (1) at which the second filament is to be placed. The node of (2) is at the vertex of (1). The loop lies almost wholly but not quite in the first quadrant. The upper branch of the loop and a portion of the curve in the fourth quadrant is used in the reflector. The result is that upward divergence of rays from the axis is entirely removed and the beam itself is directed slightly downwards from the horizontal. Hence an on-coming driver is not annoyed.

The manufacturer cannot conveniently use a cubic curve for a generator of a surface, so after determining the cubic it was necessary for me to determine a parabola which approximated the cubic within the region in which it is used. This was done. The formulas both for the particular cubic and parabola have now been patented and the reflectors based on the cubic are on the market.

37(1930), 240–241

A Peculiar Function

By J. P. Ballantine, University of Washington

I sometimes show my beginners in calculus the following function:

A pie reposes on a plate of radius R. A piece of central angle θ is cut and put on a separate plate of radius r. How large must r be? Obviously r is a function of θ. It turns out to have the following formula:

$$r = 0 \qquad \theta = 0$$
$$r = \tfrac{1}{2}R \sec \tfrac{1}{2}\theta \qquad 0 < \theta \leq 90°$$
$$r = R \cos \tfrac{1}{2}\theta \qquad 90° \leq \theta \leq 180°$$
$$r = R \qquad 180° \leq \theta \leq 360°.$$

The function has one discontinuity at $\theta = 0$, and its second derivative has various discontinuities.

Note by the Editor

The preceding example by Professor Ballantine seems an uncommonly good one to use when introducing the notion of discontinuous functions to an elementary class as illustrating how naturally such functions arise. It might be of considerable interest and value to make a collection of some more examples of this sort. Can anyone suggest another one equally simple and equally free from an appearance of artificiality?

R. E. G.

37(1930), 250

The Rockefeller Institute has given funds to endow a new institute of mathematics at the University of Göttingen. It will be directed by Professor R. Courant.

37(1930), 267

Professor Florian Cajori died suddenly of pneumonia on August 14, 1930, at his home in Berkeley, California. He was a charter member of the Mathematical Association of America and was one of an original group of four (later enlarged to twelve) representatives of mid-western universities and colleges who made possible the re-establishment of the American Mathematical Monthly on a sound financial basis. A detailed account of his historical researches will be published in the *Monthly* in due course.

37(1930), 392

3. A MATURING *MONTHLY*, 1931–1940

Eric Temple Bell, President of the MAA 1931

A Maturing Monthly

Journals are living beings. They are born, experience cycles of change and rebellion, and (sometimes) mature. By the early 1930's, the Mathematical Association was an influential and respected force in American mathematics. The Monthly, which brought the Association to life, was becoming older and (sometimes) more mature.

In 1931, the Empire State building was opened to the public, and a jury convicted Al Capone of income tax evasion. The nation became increasingly frustrated by an economic depression that seemed to settle in for a long stay. Within two years Franklin Delano Roosevelt would become the 32nd president, the first woman would be appointed to the U.S. cabinet, the U.S. would foreswear armed intervention in the Western Hemisphere, and prohibition would end. In 1935, Congress passed the Social Security Act; in 1938, it enacted the first minimum wage.

While FDR campaigned for the White House, mathematicians held their International Congress in Zürich, and for the first time in 20 years it was truly international in character. Emmy Noether gave a plenary address—so did Ludwig Bieberbach. Early the next year, Hitler came to power in Germany: Emmy Noether would leave for the U.S. and Bieberbach would stay to become infamous. The great migration of mathematicians would soon begin, and American mathematics would change forever.

A new generation of mathematicians was moving into prominence. First Marshal Stone and then a newly arrived von Neumann took giant strides in Functional Analysis. Norbert Wiener was at the height of his powers. As the decade progressed, these would be joined by hundreds of the finest mathematicians in the world, all fleeing the turmoil in Europe: Artin, Bochner, Brauer, Chevalley, Courant, Eilenberg, Friedrichs, Kac, Levy, Noether, Pólya, Szegö, Ulam, Weil, Weyl, Wigner, and Zorn. The list is longer. Many found a temporary home at the new Institute for Advanced Study in Princeton.

At the same time, a combination of the continuing depression and the influx of competition made employment for young mathematicians increasingly difficult. The Association formed a Commission (one of many over the next 60 years) to study the problem. They concluded that about one-fourth of those seeking jobs had not found suitable employment, but because fewer than one thousand of the 3,500 college mathematics teachers held the Ph.D. the commission predicted that the demand would soon exceed the supply. It was one of many such predictions, which almost always held true—for awhile.

In 1938, there were 65 new Ph.D.'s in mathematics. Their prospects in academic life were grim. Starting salaries were about $1800, which was lower than salaries at the beginning of the decade. The normal teaching load at Purdue was 18 hours, but that was higher than most (where it was 12 to 15 hours). Promotions were slow, and most faculty were at the bottom. Departments were almost universally undemocratic, ruled by a head with an indefinite term.

The combination of events—few jobs and many immigrants—led to much debate in the mathematical community about hiring foreign mathematicians rather than young Americans. There were bigots and antisemites, but there also were people who sincerely agonized about difficult choices. It was a difficult time for mathematicians, as America assumed leadership of mathematics in the world.

The Monthly became more scholarly, in part because of writers such as E. T. Bell and J. W. Young (who died unexpectedly in 1932). It published major articles on current mathematics, including the revolutionary work of Gödel. It began to publish short, illuminating articles of a certain kind that later would be called Notes. The Notes were written by respected big shots and unknown young newcomers. In addition to articles on teaching and pedagogy, the Monthly published material on the whole profession, and even published research. The Monthly was becoming respectable, and as proof it published the first of many articles on Big Game Hunting, poking fun at itself and at mathematics. Every editor wrote about the new Monthly in his first issue, and restated the desire to keep *part* of the Monthly directed towards the general reader. The Monthly had grown up and fought to stay fresh.

Emmy Noether died (during an appendectomy) in April of 1935, while George Gershwin was completing Porgy and Bess. In 1936, the first Fields Medals were awarded (to Lars Ahlfors and Jesse Douglas) at the International Congress in Oslo; shortly after, Margaret Mitchell published *Gone Wtih the Wind*. Herbert Ellsworth Slaught, who had served both the Monthly and the Association so well for so many years, died in 1937; two months later, Amelia Earhart was lost in the Pacific. The University of Toronto won the first William Lowell Putnam Mathematical Competition in April of 1938, competing against 66 other institutions; a short time later, Orson Welles broadcast a radio dramatization of *War of the Worlds* that produced a national panic. And in 1939, while Birkhoff and Mac Lane were finishing their textbook *A Survey of Modern Algebra*, Europe was sinking into war.

Recent Publications

E. T. Bell

Contributions to the History of Determinants, 1900–1920. By Sir Thomas Muir, D.Sc., LL.D., F.R.S., C.M.G. Blackie and Son Ltd., London and Glasgow, 1930. pp. xxiii + 408.

In this sequel to his classic four volume work on the history of determinants from 1693 to 1900, Sir Thomas Muir maintains the high level of thoroughness and interest which make the earlier volumes a delight. The plan of the present book is uniform with that of its predecessors, to which it is an indispensable adjunct. The publishers have spared no pains to make the easy, open pages attractive and intelligible at a glance, and they, with the author, are to be congratulated on a fine book.

Whether the reader is interested for its own sake in the theory of determinants, or whether he uses determinants as the merest incidents in his own work, he will find much in this volume to induce him at least to browse for a spell and enjoy the author's occasional pungent remarks on the miscellaneous fare offered. As Dr. Johnson observed of the incomparable haggis, "there is much fine, confused feeding" in a mathematical history, and in a historical account of determinants the tremendous feast resembles a Chinese banquet rather than a single, well cooked dish. To those who have acquired a more sophisticated palate, trained to the dry flavor of the greater modern algebraic theories, the theory of determinants as revealed by its history may seem rather lacking in coherence and more like an assortment of tidbits than a coordinated theory. The abstract part of it all, that which alone is entitled in a modern sense to the high designation "theory," can be stated in a couple of pages; the rest is mere repetition and variation, endlessly, on a theme which is neither intricate nor subtle.

In the twenty years covered in this volume the same duplication of results by one writer after another, as was noted frequently by the author in earlier volumes, is as conspicuous as ever. Seasoned mathematicians writing on determinants appear to be as negligent as the merest beginners in acquainting themselves with what their predecessors have done. There may be some merit in establishing known theorems all on one's own, but there is more in consigning the results to the wastebasket after discovering that they have already been printed three times or more. Some of the duplications reported in this volume seem almost incredible, and could hardly have been perpetrated honestly by anyone who had taken the trouble to consult Muir's earlier works. For this reason, if no other, the complete set of Muir's histories together with his eleven supplementary papers should be in every mathematical library. More important, anyone who itches to write anything on determinants in the classical notation might profitably spend half an hour running down the cause of his distress in Muir's histories; the odds are that he will be cured.

Another caution of a different kind is rubbed in repeatedly by a reading of Muir's historical studies: it is not the recurrent which is of greater interest, but the generating function or difference equation that gave rise to the recurrent. A determinant for the nth Bernoulli number, for example, may be, and probably is, as futile as it is pretty. Only a second and more misguided Wronski imagines he has solved the riddle of the universe when he evolves an utterly unmanageable determinant of the nth order to express the supreme law he thinks he has discovered. It makes no difference whether the determinant be symbolized in Greek or Latin characters, or in the more mysterious looking Hebrew which some specialists on determinants favor; if the determinant is of the nth order and not almost pathologically degenerate, it is no advance over the humble generating identity of which it is a pretentious disguise.

The sequence of chapters (with eleven omissions noted and treated elsewhere by the author) follows the plan of the earlier volumes so closely that there is no need to retail it here. The extremely interesting interpolated Chapter I (α), devoted to Hadamard's approximation-theorem from 1900 to 1917, raises a point of particular historical interest, as Muir insists that Lord Kelvin's name has a claim to be attached to the theorem. Among other judicial remarks, intended no doubt to adjust priority, those introducing Chapter VI (alternants), comparing the determinantal aspects of East Prussia and the State of New York, cannot fail to amuse American readers. One would think that determinants should be the last thing on earth to inspire chauvinism or envy, hatred and malice, but apparently it is not so. Let us hope that the enthusiastic investigators mentioned, neither of whom is longer writing for publication through the usual media, are not cutting one another on the Elysian fields.

Muir's longer writings are always lightened by a human touch, which may take the form of anything from a dry joke to a warm appreciation. Nor is the reader left in any doubt as to what is or is not "important." This, to the reviewer, seems to be a strategic error in writing a scientific history. What the professional user of such a history wants is the facts, and nothing but the facts, presented to him in the most concise form consistent with clarity, and without the historian's personal opinions on those facts, even in footnotes; it does not greatly matter what the "general reader" or the dilettante would prefer. If there is any reason for supposing historians of scientific subjects to be less fallible than historians in any other field, it is not evident, and the documents in the case of the older-fashioned literary type of history seem to show that the personal opinions and interpretations of historians serve no useful purpose, except to reveal to curious psychlogists the prejudices of the historians.

The entire history inevitably suggests comparison with the only other one of recent times devoted exclusively to a branch of pure mathematics. This comparison may be irrelevant, but as another reviewer ("A. C. A." in *Nature*, November 29, 1930) invites it in the words "if there exists anywhere a more detailed and comprehensive history of any branch of theoretical knowledge, one would be interested to hear of it," the material for a comparison may be briefly indicated. In round numbers, Dickson takes 1600 pages to present, in minute detail, the history of the theory of numbers from Pythagoras (or before), 500 B.C., to 1923; Muir takes 2400 to tell less minutely the incomparably simpler and shorter story of determinants from Leibnitz, 1693 to 1920, or, allowing for the difference in type, about the same space. In range and in detail there is no common measure for the two histories. Nor is there in style; Dickson's, for a serious student, has the complete efficiency of a battery of machine guns; Muir's is like the leisurely sprinkling of a shady lawn with a garden hose. Those who object to having a hail of facts shot at them without mercy may consider how far they would be likely to penetrate hostile territory, bristling with difficulties, if armed only with a garden hose or a watering pot.

As the author frequently points out, there has been but little diminution, if any, in the rate of production of writings on determinants in the twenty years covered by this volume. It may be suggested, however, that the counting of titles is not a reliable index when, as in this history, any incidental use of determinants in a paper whose main object often is far other than a contribution to determinants, is emphasized by isolation as the one fact of apparent interest. And, in passing, it appears to the reviewer that such a dislocation of secondary matter frequently gives an erroneous impression of a writer's motives in attempting to contribute to mathematical literature at all. In some instances it seems that no more relevant estimate of a paper by picking out the determinants it contains is possible than if the compiler were to emphasize, let us say, the incidental use of linear equations or algebraic division. Nor is the matter greatly helped when the historian supplements the record of secondary facts with his personal opinion on their importance or the lack of it.

One gets the impression on reading this book that the author is somewhat chagrined by the dearth of striking contributions to determinants in recent years. A possible explanation for this lack may be the simple fact—if it is one—that the theory of determinants, as a theory, has petered out. One most striking contribution to the theory does not seem to be mentioned by the author. The name of Ricci does not occur in the list of authors reported in this book or in the preceding volumes of the history. Possibly a perusal of Ricci's paper of 1899, in which the "Systems E" appear, seemingly for the first time, may suggest to future historians the year in which the *theory* of determinants expired and was buried under the rich loam of an infinitely wider and more fertile field. Those who claim that determinants are a comparatively trivial incident in the vaster and simpler theory of tensor algebra seem to have the right on their side.

A modern reading of one paper abstracted in the present volume might have suggested to the historian that the dearth of fundamental advances he deplores is due to the nature of the subject rather than to a lack of imagination on the part of the investigators. For, it seems to the reviewer, the entire significance of E. H. Moore's "Fundamental Remark Concerning Determinantal Notation, etc.," of 1900, has been completely missed by the historian. Between them, Ricci and Moore buried the theory of determinants in 1899–1900. Subsequent progress seems to show that this funeral is the fundamental development which the historian misses in his researches.

It may be too optimistic to hope that determinants will fade out of the mathematical picture in a generation; their notation alone is a thing of beauty to those who can appreciate that sort of beauty. But it would seem to be well worth someone's trouble to write a short, accessible tract on determinants from a modern point of view, revealing their true simplicity and their strictly incidental character, as a historical pendent to Spottiswoode's paper of 1851 in which the now apparently obsolete theory was first didactically presented. The proof of the multiplication theorem by tensor algebra, for example, is a matter of two short lines, both obvious.

Whatever the future of determinants is to be, there can be no doubt that Sir Thomas Muir's history will have no serious rival in its own field. In bringing to a close this fascinating story of a subject which claimed a considerable share of the attention of the great algebraists of the nineteenth century, Sir Thomas Muir has made all who are interested in the scientific history of science his permanent debtors.

38(1931) 161–164

ONE HUNDRED PER CENT MEMBERSHIP

The membership in the Association is now about 2200, including 137 institutional members. These figures indicate one hundred percent increase over the total charter membership in 1916. Presumably this also indicates that among the mathematics faculties in many institutions the Association membership has doubled during these fourteen years. However, in some institutions, probably in very many, *all* members of the mathematics staff and the institution itself have belonged to the Association from the outset. Such a one hundred percent membership has been maintained, for instance, by the University of Chicago. It is urgently desired by the membership committee to ascertain all of the institutions of which this is now true, and to this end the committee on membership requests the cooperation of each mathematical staff. Will the secretary or some representative of each such department transmit to the Secretary of the Association at Oberlin, Ohio, the following information:

(1) Has your department at present a one hundred percent individual membership in the Association?

(2) Has your institution an institutional membership in the Association?

(3) Will you cordially invite any non-members in your group to join the Association?

(4) Will you present to your institution the desirability of becoming an institutional member of the Association?

Membership in the Association is a mark of professional standing and a contribution to the promotion of mathematical interests in America. Members are entitled to receive all publications of the Association at cost, including the Carus Monographs and the Rhind Mathematical Papyrus. The Association now has eighteen sections distributed over the country so that any member may attend a meeting within reasonable distance; and all meetings, sectional and national, are fully reported in the American Mathematical Monthly.

37(1930), 563

The Human Aspect in the Early History of the American Mathematical Monthly[1]

B. F. Finkel

Since, by request of the program committee, the human aspect of the early history of the American Mathematical Monthly is to be emphasized, I hope I may be pardoned for speaking more often in the first person than a due sense of modesty would permit.

Why the American Mathematical Monthly? It may be of some interest to the younger members of the Mathematical Association of America, and perhaps to the older members as well, to know something of the early mathematical background of its founder. A number of the older members of this Association have been personally known to me throughout the existence of the Monthly, but I judge that most of the members here today have reached their majority since the founding of the Monthly. Many of these know very little of its early history and still less of its founder.

. . .

A superficial survey of the early volumes of the Monthly will, perhaps, readily disclose the fact that the founder was not a mathematical genius; for such, unlike Antaeus, loses his strength when touching earth, but shines with effulgent glory when soaring among the clouds of mathematical research. I did know, however, that the fundamental branches of mathematics were poorly taught in our high schools.

. . .

It soon became apparent to me that the teachers of mathematics in our high schools and academies and normal schools felt no need for such a journal as the Monthly. Whether these teachers were unaware of their lack of equipment for the work they were attempting to do, or whether they scorned the Monthly because of the presumption on the part of its editor that they were in need of such a stimulus, I never was able to learn. Nevertheless I think I am safe in saying that during the nineteen years when the Monthly was in my possession not more than a dozen high school teachers were on our subscription list at any one time. Thus it came to pass in due process of time, that the field which the American Mathematical Monthly was designed to cultivate for the benefit of high school mathematics teachers particularly, became occupied by a more virile race of mathematicians, namely the teachers of college and university mathematics, particularly the former. The

---

[1] A paper read before the Mathematical Association of America at its meeting at Cleveland, Ohio, January 1, 1931, by invitation of the program committee.

Monthly soon adapted itself to the needs of the field of collegiate mathematics and in that field it has made its most noteworthy contributions. It was my purpose from the first that the Monthly should become a sort of repository of mathematical material of lasting value, and with that in view that it should not contain reports or other material which diminished in value with time. Thus in the early volumes no reports of any kind appear.

The first number of the Monthly appeared the latter part of January, 1894.

· · ·

It is very gratifying to me to relate that one of the contributors to the first number of the Monthly was a young man in his nineteenth year doing graduate work under Dr. Halsted in the University of Texas, and giving promise at that age of becoming one of the foremost mathematicians of the world. The title of his article is, "Lowest integers representing the sides of a right-angled triangle." The author is Leonard E. Dickson. Young Dickson also contributed an article on "The simplest model for illustrating the conic sections" in the August number of Volume I, and one on "The inscription of regular polygons" in the October, November, and December numbers. We shall have more to say of him later. Professor Robert J. Aley contributed a "Bibliography on the history of mathematics" and a list of mathematical periodicals. In the March number of Volume I, David Eugene Smith of the State Normal School, Ypsilanti, Michigan, later to become the well-known historian of mathematics and one of the leading teachers of mathematics in Teachers College, Columbia University, and author of many books on mathematics, contributed a critical note on "J. K. Ellwood's Remarks on Division," thus opening up the first controversy to be published in the American Mathematical Monthly. In this number also Professor Halsted began his series of articles on "Non-Euclidean Geometry, Historical and Expository," a series continued through many succeeding numbers.

· · ·

Having been assigned a graduate scholarship in the University of Chicago in the summer of 1895, I attended the second summer session, and it was then that I became personally acquainted with Leonard Eugene Dickson. Dickson had been appointed to a University Fellowship and was doing graduate work towards the degree of Doctor of Philosophy. During this and the following summer Mr. Dickson and I had many friendly conversations about the Monthly and its future. I speculated with him at that time as to the possibility of the University's taking it over, thus to insure its permanency. After taking his degree at the University of Chicago, studying in Europe, teaching in the University of California and in the University of Texas, Dr. Dickson was called in 1900 to the University of Chicago as assistant professor of mathematics. On my way home to Springfield in September, 1902, I went through Chicago to call on him and to invite him to join me in the editorship of the Monthly. Not seeing his way clear at the time to give me a definite answer, he withheld his reply until he could consider the matter more fully. After some meditation he wrote me saying that he would accept the co-editorship with me. The day of his decision was a red-letter day in the history of the Monthly. His official connection with the Monthly began with the October, 1902, number.

· · ·

During Professor Dickson's connection with the Monthly, the University of Chicago contributed a subsidy of $50.00 per year, thus helping to meet the expense of publication. All cost of publication over receipts from subscriptions and this

University subsidy was borne by me personally. No help of any kind was employed and no expenses incurred except those of printing, binding and mailing the Monthly. Mrs. Finkel helped me read practically all of the proof and often addressed all the wrappers herself. For the three summer numbers of Volume II she read the proofs alone. It was some task to proof-read type matter set up by the inexperienced type-setters at Kidder. The first galley sheet was so full of marks that no room was left on the margin to make further corrections. This often necessitated my going to the printing office as many as three times for each galley proof in order to insure comparative freedom from typographical errors.

In severing his official connection with the Monthly, Professor Dickson suggested that his mantle be placed upon the shoulders of the aggressive, indomitable, and persevering Professor H. E. Slaught. This move was strongly supported by Professor E. H. Moore. Both Professor Slaught and I agreed to the suggestion, and thus was inaugurated a second red-letter day in the history of the Monthly. Soon after his connection with the Monthly began, Professor Slaught secured through the influence of Professor Townsend an annual subsidy of $50.00 from the University of Illinois and Professor G. A. Miller as editorial representative.

One of the objects set forth in the founding of the Monthly was that it should reach its readers regularly each month. During the five or six years preceding 1909 I was often obliged to state that an issue was delayed for one reason or another. The chief reason was that our printers took on more work at times than their office force could handle, and as a consequence the Monthly was laid aside until the rush was over. This happened frequently even though I had a contract that drew a forfeit of five dollars for each day the Monthly was delayed beyond the scheduled mailing date. As the type-setting for the Monthly was done by Mr. Dixon himself, and could not be done by anyone else in his office, any indisposition on his part for any reason whatsoever delayed the type-setting and consequently the mailing of the issue. When the Monthly was delayed thus on account of illness or other unavoidable causes, I was very lenient and did not demand the forfeit. On one occasion on account of the illness of Mr. Dixon I remitted the forfeit. Later in the same year when the Monthly was delayed a week or two I demanded the forfeit. Its payment, however, was refused on the ground that I had violated the contract by having remitted the previous forfeit. I then asked Mr. Dixon if that was the basis on which he was going to transact business and he said it was. I said, "All right, I think we understand each other perfectly." I assured him that there would be no more remitting of forfeits under future contracts and there never was, even though in several cases the forfeit covered the cost of printing the delayed number. Mr. Dixon was a fine gentleman, and he and I transacted all our business on the most friendly basis, never having had any unfriendly words during the whole of the more than seventeen years of business relations. However, the exaction of the forfeits did not improve materially the regularity of the publication of the Monthly, and it became apparent to me that the time was approaching when Mr. Dixon could no longer afford to publish the Monthly. If, then, the Monthly was to continue its existence, some arrangement would have to be made to forestall its discontinuance.

· · ·

In the prosecution of my work as editor, I have been both blamed and praised. In editing a journal, as in other affairs of life; one cannot please everybody. It was my aim to deal generously and justly with our contributors, keeping in mind that truth and accuracy should not be sacrificed under any circumstances. My correspondence with circle-squarers, angle-trisectors, cube-duplicators, and Fermat's Last Problem solvers was voluminous, varied and interesting, and if compiled would add an interesting chapter to De Morgan's *Budget of Paradoxes*. I regret that I did not keep all my correspondence in this connection. I shall give two samples:

A few years ago I received in pamphlet form a demonstration of the trisection of an angle. The author informed me that if I could show him that his demonstration was wrong he would send me a turkey for my Christmas dinner. I pointed out the weak spot in this demonstration, but I never received the turkey, since the author was judge, advocate, and jury in the matter.

Another angle-trisector in 1927 wrote me as follows: "I have solved the problem of trisecting a rectilinear angle, of trisecting the arc of a circle,... This despite the theories and quack formulae advanced by half-baked mathematicians in an attempt to prove the possible impossible." He stated that he had referred his construction to professors at Northwestern University and the University of Michigan, one of whom admitted that he found no flaw in the construction and another passed the buck by suggesting that the construction be sent to me. This trisector goes on to say that "the skepticism that exists as to the possibility of trisecting a rectilinear angle is largely due to the number of people who have submitted what they thought to be solutions but which proved not to be. To a person who has any real knowledge of mathematics this seems strange, but human experience notes that in most mathematics classes there is not even one mathematician, and that in a class in which there is *one*, there are generally one, two or three others. But those who are not mathematicians 'get by' and some of them become teachers of mathematics." He then asks the question, "But is it not assuming a great deal to claim that a rectilinear angle, or the arc of a cycle, cannot be trisected?" As this angle-trisector had all his drawings copyrighted and was considering propositions from publishing houses, he said he would be willing to loan them to me on my agreeing not to publish them. I let the whole matter drop at this stage.

If the Monthly did not accomplish all that it should have done during the first nineteen years of its existence, Goldschmid's criticism of a noted painting is apropos: "The painting could have been improved if the painter had taken more pains." In the case of the Monthly, be it remembered that the work on it was my avocation. During most of the time while I was conducting the publication of the Monthly I was teaching from nineteen to twenty-seven hours per week; for the first seven years in Drury College I was secretary of the faculty and registrar of the college; I served on various committees; was director of summer sessions for three years; and was nominally librarian of the college for ten or twelve years. No consideration was ever given me by the college because of the work the Monthly entailed and I never asked for any such consideration.

38(1931), 305–319

The fifteenth annual meeting of the Association was held at Cleveland, Ohio, on Wednesday and Thursday, December 31, 1930, and January 1, 1931, in affiliation with the American Association for the Advancement of Science and the American Mathematical Society. Two hundred sixty-one were in attendance at the meetings, including one hundred ninety-seven members of the Association.

38(1931), 121

MEAN VALUE OF THE ORDINATE OF THE LOCUS OF THE RATIONAL INTEGRAL ALGEBRAIC FUNCTION OF DEGREE n EXPRESSED AS A WEIGHTED MEAN OF $n + l$ ORDINATES AND THE RESULTING RULES OF QUADRATURE

By BENJAMIN F. GROAT

38(1931), 212

Functions of the Mathematical Association of America

J. W. Young

The American Mathematical Society is devoted to research—it will voluntarily have nothing to do with anything else. This singleness of purpose is doubtless a source of strength to the Society. It would, however, be a source of weakness to mathematics as a whole, and, indeed, to the development of research itself were not other important phases of mathematical enterprise taken care of elsewhere. It makes the definition of the functions of the Association very easy: Everything that is worth doing for mathematics, other than research, is a function of the Association. The latter is the proverbial "George" of the mathematical family.

. . .

In any case, to attract men and women of ability to our subject is one of our important functions; we do so, presumably, by attempting to arouse and maintain a vital interest in mathematics among those properly qualified. The establishment of undergraduate mathematical clubs is a means to this end. Some of us, perhaps many of us, can testify that the problem department of the Monthly a generation ago was a powerful stimulus in this direction. During my own undergraduate days I was a subscriber to the Monthly and used to have great fun in attempting to solve the problems proposed. The programs of the clubs just referred to show that there are many topics of interest to undergraduates. Occasionally, indeed, papers presented at such meetings would merit publication if a suitable organ were available. Think of the added stimulus in the preparation of such a paper, if the author could look forward to the possibility of its appearing in print.

And so I suggest the publication under the auspices of the Association of a new magazine devoted to the mathematical interests of undergraduates and to others, as will appear presently. A prominent feature of this new magazine would be a department of problems within the range of ability of undergraduates, including some to the solution of which a freshman could aspire. They should be interesting, unusual, not of the conventional text book variety. Such problems exist, though they may not always be easy to find or to invent. The magazine would contain articles of interest to undergraduates, some perhaps, as has been suggested, written by undergraduates. It would take too long to elaborate in detail the plans for such a magazine—nor am I qualified to do so. A further source of articles for such a magazine will appear presently. At this point I wish merely to venture the assertion that this project would probably not require any additional funds. Such a magazine would I believe become self-supporting and might indeed be a source of additional income to the Association. It need not, in fact I think it should not, be a large magazine. Its subscription price should be low, not more than $2.00 annually for say ten numbers published during the college year. But its subscribers should be many. We should expect subscribers among the students and faculties of every college of reasonable standing in the country; the mathematical clubs would be our agents. It should get subscribers from among the more enlightened, progressive, ambitious secondary school teachers; it should do for the mathematical interests of such teachers what the Mathematics Teacher does for their pedagogical interests. It would, I venture to say, get subscribers from the general public. The kind of magazine I have in mind would appeal to all those, relatively few in number but in the aggregate numerous, who have mathematical interests. It should be, according to my vision, an agency for the popularizing of mathematics. But of that more presently—I am getting a bit ahead of my story.

Does the project strike a responsive chord in the minds and hearts of some of you? Among our more than two thousand members does the man or woman exist who has the ideals, the enthusiasm, the courage, the tact, the organizing ability to carry this project to success as editor-in-chief? Do we have in our membership the men and women able and eager to cooperate? It is a large an difficult project; it will have to command the enthusiastic and active support of a considerable number to make it go.

. . .

We can't all be research men. Some of us do not even want to be. I have sought to show that there exist wide fields of enquiry and activity other than research that are important, interesting and worthy of intensive cultivation, and that are being sadly neglected. I have incidentally attempted to combat the attitude, if and wherever it exists, that would make of research a fetish, that proclaims that the only worthy function of a mathematician is research and that other activities are to be looked on with contempt. There is fortunately very little of this sort of self-righteous snobbery in our two organizations. The great majority of our research men themselves realize that the roots and trunk and branches of the tree of mathematics are quite as important as the blossom or the fruit; and that the former must exhibit healthy life if the latter are to be produced at all. Any other attitude is so utterly stupid as hardly to merit attention. To change the simile the star back on a football team who advances the ball and receives the plaudits of the multitude knows, unless his head has been completely turned by an exaggerated sense of his own importance, that he would be quite helpless were it not for a strong line and able interference.

It is probably impossible to determine the relative value of a guard and a back on a football team. It is probably even more difficult to determine the relative value, to the mathematical organism as a whole, of the research man and the man who labors to improve the conditions which make research possible and which give it significance. Both are essential. But I am quite clear in my own mind, that if entry into the mathematical heaven depends on what a man has done for mathematics during his life on earth, the record of such a man as our good friend Slaught will far outweigh that of most mere research men. There is important work for all of us. The sin of the mathematician is not that he doesn't do research, the sin is idleness, when there is work to be done. If there be sinners in my audience I would urge them to sin no more. If your interest is in research, do that; if you are of a philosophical temperament, cultivate the gardens of criticism, evaluation , and interpretation; if your interest is historical, do your plowing in the field of history; if you have the insight to see simplicity in apparent complexity, cultivate the field of advanced mathematics from the elementary point of view; if you have the gift of popular exposition, develop your abilities in that direction; if you have executive and organizing ability, place that ability at the disposal of your organization. Whatever your abilities there is work for you to do—for the greater glory of mathematics. And this, I think, is the nearest I have ever come to preaching a sermon.

39(1932), 6–15

Moderne Algebra, Teil I. B. L. v. d. Waerden: Die Grundlehren der mathematischen Wissenschaften in Einzeldarstellungen, Vol. 33. Julius Springer, Berlin, 1930. 243 + 8 pages.

The domain of abstract algebra has at present a flourishing period in Germany, and a number of recent textbooks on the subject gives evidence of this development. The present book by v. d. Waerden gives a very clear and satisfactory treatment of the main ideas of the theory of groups, fields and some of the fundamentals of the theory of rings and ideals. It shows many points of connection with the recent book by O. Haupt. The second volume promises to be of still higher interest; it will, according to the plans of the author, deal with general ideal theory and elimination, linear algebras and hypercomplex systems, and the so-called theory of representation. A more complete review of this book will be printed in the *Bulletin*.

OYSTEIN ORE

38(1931), 226

Dr. N. H. McCoy has been appointed to an assistant professorship at Smith College.

38(1931), 546

Logarithmetica Britannica. Being a Standard Table of Logarithms to Twenty Decimal Places. By Alexander John Thompson. Part V, Numbers 50000 to 60000 Issued by the Biometric Laboratory, University of London, to Commemorate the Tercentenary of Henry Briggs' Publication of the *Arithmetica Logarithmica*, 1624. Subscription Issue. Cambridge, The University Press, 1931.

This is the fifth part (the fourth not yet published) of this tremendous undertaking. It consists of twenty-place logarithms of numbers of five digits, accompanied by values of second and fourth differences. The project speaks for itself; it is sufficient to say that the result is all that is to be expected of any product of the Cambridge Press.

R. A. J.

38(1931), 407

The Junior Mathematics Club of the University of Chicago.

The officers for 1930–1931 were Mr. Arnold E. Ross, President; Mr. Saunders MacLane, Secreatary-Tresurer.

38(1931), 421

Studies in the Theory of Numbers. By Leonard E. Dickson. The University of Chicago
   Science Series, 1930. x + 230 pages. $4.00.

This important volume has two claims to distinction: it contains an amazing number
of new results in the theory of quadratic forms; and it represents a systematic
treatment of the arithmetic theory of quadratic forms, starting from first principles.
Either one of these accomplishments by itself would entitle the author and his
students and collaborators (Arnold Ross, Gordon Pall, A. Oppenheim) to the lasting
gratitude of all interested in the theory of numbers; the combination makes the book
of quite outstanding value.

   It would seem to call for some explanation why a systematic treatment *ab ovo* of an
apparently well developed field such as the arithmetic theory of quadratic forms
should be hailed as a noteworthy achievement. It will probably be to many readers, as
it was to the reviewer, a shock to learn that in spite of the eminence of the
mathematicians who have contributed to the theory (Gauss, Seeber, Smith, Zolatareff,
Markoff, Frobenius, Minkowski, Eisenstein; to mention only some of those no longer
living) and in spite of the very large number of textbooks on theory of numbers, we
have no satisfactory exhaustive treatment of this field. In particular, the volumes of
Bachmann "Die Arthmetik der Quadratishchen Formen" are shown to be in
important respects unreliable.

   One can but admire the courage of an author who will undertake to rebuild the
whole structure rather than to patch up the unsound portions. One can only guess at
the amount of labor covered by the modest words of the preface; "It was no small task
to write a satisfactory exposition." On the other hand, we know, in this country as well
as in Europe, how much the theory of numbers owes to the insistence of Dickson on
precision in the statement of theorems and to his uncanny ability to detect, and to
mend, unsound arguments; it seems therefore only fair that to him and his students
should belong the credit of writing the first reliable treatment of the arithmetic
theory.

40(1933), 40

INSTITUTE FOR ADVANCED STUDY

   In describing the new Institute for Advanced Study at Princeton, Processor Veblen
said that a few years ago Mr. Bamberger decided to devote his wealth to some useful
purpose and through the influence of Mr. Abraham Flexner decided to devote it to a
project for the furtherance of pure scholarship. The plan contemplates a small group
of mathematicians who will be free to do scientific work involving no bestowal of
degrees, large liberty being allowed to the professors in conducting their activities in
the form of seminars or formal lectures of none, as they may wish. It is expected that
the students will be beyond the stage of the usual graduate student and that
mathematicians will come to the Institute for limited periods of time for the purpose
of doing some particular piece of work, for writing a book, etc.

40(1933), 128

A Photo-Electric Number Sieve

D. H. Lehmer

Until recently, the mathematician has been considered the only scientist fortunate enough not to need any laboratory equipment to carry out his researches. The last decade, however, has shown a tendency among mathematicians to adapt the existing commercial calculating machines to their computations,[1] and in rare cases to invent devices of their own to perform special operations.[2]

There is an important class of problems in the theory of numbers which cannot be solved readily by any commercial calculating machine, so that the number-theorist has had to resort to a kind of graphical method (similar to the celebrated Sieve of Eratosthenes) in handling problems of this sort. By the generous cooperation of the Carnegie Institution of Washington it has been possible to construct a new kind of calculating machine, applicable to this class of problems, in which modern physics has made its contribution to the oldest and least practical branch of mathematics.

In order to make clear the details of this machine, it is desirable to say a few words about the kind of problems it can solve and the underlying principles by means of which it solves them. We shall not attempt to describe completely this class of problems, since this would require the introduction of various concepts and notations unfamiliar to the general reader who is not well acquainted with the theory of numbers. It will suffice to illustrate with the following problem which is a typical representative.

Let it be required to find an integer x for which $ax^2 + bx + c$ is a square number y^2. Here a, b, and c are given integers and determine the problem.[3] Simple as this problem seems at first sight, it contains as special cases problems of extreme difficulty. Thus $a = 1$, and $b = 0$ gives $c = y^2 - x^2 = (y - x)(y + x)$. This problem, then, is equivalent to factoring the number c and is one of the central problems of the theory of numbers. From the fact that a, b, and c are small we must not conclude that x is small. In fact, when $a = 1549$, $b = 0$, and $c = 1$, we find the smallest positive value of x to be

$$x = 12223 \quad 09542 \quad 82674 \quad 74959 \quad 34242 \quad 68334 \quad 63805$$

$$08818 \quad 07626 \quad 31786 \quad 81966 \quad 09867 \quad 28279 \quad 63220.$$

To give an elementary explanation of the method by means of which the machine solves such problems, we may introduce the idea of a finite arithmetic

---

[1] Dr. L. J. Comrie of Great Britain's Nautical Almanac Office is an exponent of this procedure. See Monthly Notices of the Royal Astronomical Society, 92, (1932), 523–541.

[2] The mechanical intergraphs recently developed by Dr. V. Bush of the Massachusetts Institute of Technology may be cited in this connection. See Journal of the Franklin Institute, vol. 212, (1931), 447–488.

[3] The problem is not solved by showing that x exists, nor by giving inequalities which x must satisfy. The problem is to exhibit x.

which deals only with the numbers $0, 1, 2, 3, \ldots, p - 1$. We may perform addition, subtraction and multiplication in the ordinary manner, but in every case the result is divided by p and the remainder alone preserved. In this way we remain inside our system of numbers. The number of multiples of p which we discard in reducing the answer to a number in our system is as immaterial to us as the number of complete revolutions of the wheel is to the roulette player.

Let us imagine a problem in ordinary arithmetic requiring the solution in integers of a certain equation. This equation may be translated bodily into our finite arithmetic by merely replacing the coefficients of the equation by their remainders on division by p. To solve the problem in the finite arithmetic is simple enough because the values of the unknown are restricted to lie among the p numbers $0, 1, 2, \ldots, p - 1$. If p is small we can find all the solutions by actual trial. But why is it useful to consider a real problem in one of these artificial arithmetics? What interpretation can we give to the solutions of the problem in such a system? To answer these questions let us consider one of the answers to the original problem. If we divide this answer by p, the remainder will be one of the solutions of the problem in the finite arithmetic. The desired answer then is some one of these artificial answers plus a certain multiple of p. This is not much information, but it is easy to obtain. Moreover we have not committed ourselves to the choice of p. In fact, we may choose as many different p's as we like, and if they have no common factor, the information offered by each arithmetic will be independent. The combination of these various bits of information leads to the solution of the original problem. It is this combination which the machine is designed to effect.

In order to obtain a mechanical picture of the situation let us imagine a disk gear with p teeth having a small hole opposite each tooth. These holes are at a constant distance from the periphery of the gear and are numbered from 0 to $p - 1$. Let us plug up all the holes except those which correspond to answers to our problem considered in the finite arithmetic. If a light is set behind the gear, then, as the gear rotates, it will transmit the light when and only when the number of teeth turned past has a chance of satisfying the problem as far as this particular gear is concerned. To get the combined effect of using several p's, we may set up several gears parallel to each other. If the gears are mounted so that they have a common line of tangency and are driven at the same linear speed, then the light which shines through a gear will be transmitted or blotted out by the next gear. In this way the gears are allowed to pass judgment upon the eligibility of each number as it turns past. The decision of each judge is not influenced by the rulings of the other judges and his rejection of a candidate is final. When a number is a solution of our problem, however, there will be no dissention among the judges and the beam of light will succeed in running the gauntlet. It is true that other numbers may be thus unanimously elected without being answers to our problem. If enough gears are used, however, these undesirables can be eliminated altogether. In the kind of problems that the machine solves about one half of the holes are stopped up in each gear, and there are 30 gears. This means that we can expect an extraneous alignment of holes about once in $2^{30} = 1073741824$ numbers. By this time the reader has grasped the essential features of the mechanical system. It is clear that the gears may be driven at great speed and still there will be ample time for the light to pass through every gear when the answer arrives. In fact this time interval is one ten-thousandth of a second.

So much for the mechanical system. All that is needed now, is an alert and untiring eye behind the last gear to observe and report the appearance of the tiny

Courtesy of the Carnegie Institution

View of the mechanical system showing one series of gears.

flash of light. This eye is a photo-electric cell. The energy produced by the small amount of radiation falling on the retina of the cell is very small indeed. It must be sufficiently magnified to operate a circuit breaker for the electric motor driving the machine. This is done by means of a six stage amplifier and a three stage relay system. The amplifier magnifies the energy from the photo-electric cell about 700 million times and transmits it to a delicate vacuum tube relay, the first of a series of three which finally stop the machine.

To give a concrete example of how the machine works, let us consider the following problem: to find a value of x for which

(1) $\qquad 91894770302976x^2 + 287722528867021824x + 256527596541064768$

is a perfect square. These large coefficients are not arbitrary numbers. In fact they arise quite naturally in an investigation into the possible factors of the Mersenne number $2^{79} - 1$. The numbers $2^n - 1$ where n is a prime have been the subject of investigation since the time of Euclid. Twelve of these numbers have been proved prime and twelve composite ones have been completely factored. Since 1924 it has been known that if a value of x exists for which (1) is a square then $0 < x < 39110012$. This problem was considered in each of the finite arithmetics corresponding to a prime or a power of a prime $p < 127$, and the appropriate holes in the corresponding gears were stopped up. This presents the problem to the machine, which, canvassing numbers at the rate of 300000 a minute, can cover the above range for x in about two hours without attention. As a matter of fact the power was automatically shut off in 12 seconds, and the machine coasted to a stop. Reversing the machine slowly and substituting the human eye for the photo-electric cell, the light was seen to shine through at $x = 56523$ according to the reading on the revolution counter. Substituting this value of x in (1) we obtain at once the number 309853160646773276521024, which is the square of 556644555032. Hence our problem is solved. Incidentally this leads to the factorization

$$2^{79} - 1 = 2687 \cdot 202029703 \cdot 1113491139767.$$

We take this opportunity to answer a few questions that are almost always asked by those inspecting the machine. The first of these has to do with the period of the machine. Since each gear goes through the same position over and over again, the same is true of the machine as a whole so that the light will pass through the machine periodically. This is quite true in theory. The period of the machine is clearly the least common multiple of the periods of the gears. This means that the machine will return to its original position after

2497 19431 65929 91015 26347 12970 31949 33696 87002 72210

teeth have turned past. Even though the machine runs at the speed of 300000 teeth per minute it would wear out long before it got really started on its period.

Another question that is always asked is: Do you know approximately what the answer is before you start? In the problems we consider, the unknown is lost among millions of consecutive numbers. If we could tell in advance just which million contained the answer it would not have been necessary to construct the machine.

As a final question: What is the exact relation between the machine and mathematics itself? We do not hesitate to apply mathematics to physics, but in applying physics to mathematics there arises the question of the reliability and accuracy of the results. Fortunately, in this case, this question is not a vital one. The accuracy of the mechanical system is absolute. The machine does not depend upon measurements or estimated values. The gears arrive inexorably at their appointed positions with absolute certainty. The photo-electric system, however, with its enormous amplification and its elaborate safeguards against outside disturbances, would not have been possible a few years ago, and even to-day it is easy for the uninitiated to underestimate its reliability. This system is indispensable, however, since any substitute for the weightless beam of light could not be depended upon at the high speeds that some problems demand.

After all, such questions are for the technician. As for the mathematician, he is glad to obtain immediately verifiable answers to his problem no matter how unreliable their source may be. For what sources are more unreliable and yet more indispensable than imagination, intuition or inspiration?

The successful completion of the machine has been made possible first of all by the Carnegie Institution of Washington, which once more came to the aid of the number-theorist by a grant of the necessary funds. We also wish to acknowledge the assistance of Mr. T. J. Palmateer of Stanford University for valuable advice in constructing the mechanical parts of the machine. We are also much indebted to Dr. R. C. Burt of Pasadena who not only constructed the photo-electric system but, on account of his interest in the project, has generously given space in his laboratory where the machine is now operating.

A Simple Continuous Function with a Finite Derivative at No Point

T. H. Hildebrandt

Van der Waerden[1] has given a simple example of a continuous function which at no point has a finite derivative. A slight modification of his example produces a still simpler instance.

Define $f_0(x)$ of period 1, so that

$$f_0(x) = x \quad \text{for } 0 \leqq x \leqq \tfrac{1}{2}$$
$$= 1 - x \quad \text{for } \tfrac{1}{2} \leqq x \leqq 1,$$

$f_0(x)$ forms with the x-axis a set of right-angled isosceles triangles with vertices at $(\pm m, 0)$ and $(\pm m + \tfrac{1}{2}, \tfrac{1}{2})$ $m = 0, 1, 2 \ldots$. Let $f_n(x) = f_0(2^n x)/2^n$. Then obviously $\phi(x) = \sum_0^\infty f(x)$ is a uniformly convergent series of continuous functions and therefore continuous.

To show that $\phi(x)$ has at no point a finite derivative, we make use of the well known lemma: If $\phi'(x)$ exists then

$$\lim_{(h, k) \to (0,0)} \frac{\phi(x + h) - \phi(x - k)}{h + k} = \phi'(x), \qquad h, k \geqq 0.$$

As a consequence if for every x_0 we can determine within every interval enclosing x_0 two pairs of points for which the corresponding secants on $\phi(x)$ differ by unity, $\phi'(x_0)$ cannot exist as a finite number.

We note that $f_n(x) = 0$ for $x = p/2^n$, i.e. the value of $\phi(p/2^n)$ is determined by the functions $f_1(x), \ldots, f_k(x)$, $k < n$. Further since the slope of $f_0(x)$ is constant on $(m, m + \tfrac{1}{2})$, and $(m + \tfrac{1}{2}, m + 1)$ and consequently on any interval of the form $(p/2^m, (p + 1)/2^m)$, for $m > 0$, it follows that the same is true of $f_n(x)$ on any interval of the form $(p/2^{n+m}, (p + 1)/2^{n+m})$ for $m > 0$. Then the slope of the line joining $[(2p + 1)/2^{n+1}, \phi\{(2p + 1)/2^{n+1}\}]$ and $[p/2^n, \phi\{p/2^n\}]$ differs from that of the line joining $[p/2^n, \phi\{p/2^n\}]$ and $[(p + 1)/2^n, \phi\{(p + 1)/2^n\}]$ by the slope of $f_n(x)$ between $x = p/2^n$ and $x = (2p + 1)/2^{n+1}$, viz. 1. A similar statement holds if $2p + 1$ and $p + 1$ are replaced by $2p - 1$ and $p - 1$ respectively. These facts are also immediately evident from the graphical representation of a few of the approximating functions of $\phi(x)$. Since for any x and n, there exists an integer p such that $p/2^n \leqq x < (p + 1)/2^n$, the difference quotient $\{\phi(x + h) - \phi(x - k)\}/(h + k), h, k \geqq 0$, cannot approach a finite limit as $(h, k) \to (0,0)$, for any x, i.e. if $\phi'(x)$ exists, it is infinite.

40(1933), 547–548

[1]Mathematische Zeitschrift, vol. 22 (1930), pp. 474–5.

FROM THE OBITUARY FOR E. H. MOORE

By H. E. Slaught

A re-reading of this presidential address will convince the most casual observer that Professor Moore had given very earnest and deep thought to the question of improvement (yes, of reform) in the teaching of mathematics in the schools and that he had weighed carefully similar movements in Europe, especially in England and Germany. There is now good evidence that he was a seer and a prophet many years in advance of his time. While he recognized the chief responsibility of the American Mathematical Society to be the promotion of research, he nevertheless felt that even research interests in mathematics were bound up with those of mathematical education. He hoped that the Society might give more attention to the pedagogy of mathematics and he felt sure that such a movement would further the highest interests of mathematics in this country. But this movement was delayed for fourteen years and then was brought about by another procedure which will now be fitted into this picture.

It was shown above that Moore's interest in the founding of the MONTHLY was based on his desire for a general extension of mathematical education in this country. This interest again appeared when, in 1908, he urged this writer to undertake the co-editorship of the MONTHLY which Professor Dickson was then vacating to take up other editorial duties. Again in 1912 Moore's cooperation and council were of the greatest importance when the periodical was rescued from financial disaster and taken over by representatives of twelve universities and colleges in the middle west. His joy was like that of a child with a new toy when the first number, in January 1913, of the enlarged publication in its new dress was presented to him. And, finally, his interest and satisfaction were approaching an upper bound when, in January 1916, the MONTHLY became the official journal of the newly organized Mathematical Association of America. Ample evidence cropped up now and then to showthat he was an habitual reader of the MONTHLY—problems and all. The first complete set of bound volumes of the periodical to be deposited in the Association library was the gift of Professor Moore and served as a token of his abiding interest in the movement for which the MONTHLY was to stand.

40(1933), 194

The Place of Rigor in Mathematics[1]

E. T. Bell

I. A Mathematical Truth. As the majority of the members of this Association are vitally interested in the presentation of college mathematics, I decided that a discussion of some topic directly related to mathematical education would be more acceptable than a technical paper on research. Accordingly I chose the question of rigor in mathematics, for that, it seems to me, is the one question in mathematical education which today is of the first importance to both the sanguine educators and the would-be educated.

There is not much to say. For that very reason, by the fundamental law of all public speaking, I shall probably take an interminable time to say it. As a matter of fact everything that it is both necessary and sufficient to say on the place of rigor in mathematics can be said by adding three words to the title. The three words are "is in mathematics." Thus we have the tautology, "The place of rigor in mathematics is in mathematics." This is all I have to say, although I shall keep on saying it for years, or at least for an hour. No mathematical purist can dispute that "the place of rigor in mathematics is in mathematics," for this assertion is tautological, and therefore, according to Wittgenstein, it must be of the same stuff that pure mathematical truths are made of.

Having started with a mathematical truth, it is rather a come-down to proceed to things which confirmed mathematicians—some of whom appear to glory in a bigotry as superstitious and as credulous as that of any mediaeval theologian—will at once brand as outrageous falsehoods, or possibly, in a less irritated mood, ascribe to the fermentations induced in an exhausted brain by the summer heat. Nevertheless I believe that a dispassionate examination of the relevant facts will convince any open-minded person that the main point I shall attempt to make is true, not analytically, of course, but synthetically.

This point is suggested by our tautology: in mathematics as usually and officially presented today there is not the slightest pretension to any rigor which has not been completely stereotyped, necessary so far as it goes, out-of-date, and *comparatively* trivial for at least a decade.

II. Pippa. The present plight of mathematical learning—instruction and research—in regard to the whole question of rigor is strangely reminiscent of Robert Browning's beautiful but somewhat dumb little heroine Pippa in the dramatic poem *Pippa Passes*. Outside of certain innocently sardonic passages in the New Testament it would be difficult to find a more perfect expression of unconscious

[1]Retiring presidential address presented at the meeting at Williamstown, Mass., Sept. 3, 1934.

irony than the famous climax of little Pippa's beautifully inept song—

> *"God's in his heaven—*
> *All's right with the world!"*

Exclamation point and all, this is exactly as Browning wrote it. The incongruity of such a pleasing rotarianism in the bowels of a dramatic poem which is largely concerned with the most ungodly passions and crimes imaginable seems not to have bothered either Pippa or the poet. Perhaps he was pulling his public's leg. Anyway, I like to think he was, as I have always understood that Browning's insight into human nature is rated extraordinarily high by those who go in for that sort of thing. If Browning did not understand what his orthodox public wanted, who in the name of Pippa's heaven did? In Pippa's defense it must be recalled however that she probably had not the slightest idea of all the curious monkeyshines going on all around her. Either she was as green as an April apple or as pure as pure mathematics. Let us go on with the pure mathematics. We shall catch up with Pippa as we go.

Since competent mathematicians began some years ago reviewing mathematical textbooks aimed at the all but infinite college market, there has been one most striking improvement in the better texts by authors who know what they are about. The improved presentations at least strive to attain the beginnings of common intellectual honesty. Instead of attempting—usually with very considerable success —to convince the student that the exponential limit is what it is by an argument that proves nothing but the author's mathematical ignorance and incompetence, reputable authors now state explicitly that a proof is beyond the capacity of the student at his present level. Thoroughgoing modern skeptics might add a footnote to the author's candor, and point out that a proof, according to the latest demands of logical rigor, is also beyond the capacity of the author at *his* present level. However, ignoring the hypothetical footnote, let us merely note that many current texts in elementary analysis, including the integral calculus as usually presented in college, are no longer definitely misleading where a simple confession of inability to do the job properly is all that can be expected under the circumstances.

. . .

A quarter of a century ago mathematical teaching, as represented in advanced textbooks, senior college courses and university lectures, was abreast of what were then the standards of rigor in mathematical proof. About the same time the modern revolution began. Apparently it has not yet registered on mathematical education, which continues to trip along like little Pippa taking the dew-pearled hillside at its too obvious value, entirely unaware that she is greener than the grass she walks on.

> "The year's at the spring
> And day's at the morn";

and it makes not the slightest difference to Pippa how many revolutions may have happened in the night—

> "All's right with the world!"

No sane teacher, I presume, would advocate even hinting to immature young minds that such a thing as a revolution in mathematics is now in full tide. To do so would only confuse where confusion can do no good. Later on a little honest confusion may be of immense benefit to the cocksure or bigoted mind. For the beginners, however, it is no doubt best that they continue to be suckled like young pagans in creeds outworn—if indeed the good old creeds are as outworn as the enthusiastic revolutionaires assure us they are. But up to what age is this state of virginal innocence of the facts of life to continue? Is every research Ph.D. turned out by our great degree mills to begin his real career in well-sanitated ignorance of what has been going on for twenty years, and is still going on, in the very vitals of all mathematics?

Why not take a hint from physics, which thinks no more of a revolution in fundamental concepts than it does of a rather annoying fly in the laboratory? By facing our revolutions fairly, or chasing them and swatting them with joy as physics does, we might find something of more interest in mathematics than we yet have. Such, at any rate, has been the experience of physics. Is it not just possible that some of our exasperating antinomies are beyond resolution so long as we persist in that particular mathematics—the only one we have at present—which is based on Aristotelian logic? Will the difficulties ever be cleared up by traditional reasoning, or are they perhaps waiting for a younger generation, that will not be too respectful of authority, to circumvent the contradictions by building a more inclusive mathematics on a many-valued logic? Whatever is to be the way out and on, we may conjecture that so long as the traditional approach is followed only inappreciable progress will be made. Too many first rate men have tried that way and got nowhere.

In the meantime it seems a pity that practically an entire mathematical generation is being trained to miss what is perhaps the only mathematics of our age that will have any abiding significance for future generations of mathematicians. The brilliant analysis of our own day, as hard, as sharp and as clear as a well cut diamond, may be a beautiful thing to look at, but others before us did relatively as well, if not better. The new light and shade on *all* mathematical reasoning that our generation has seen is something that our predecessors did *not* see. Why ignore it all merely to be in fashion?

41(1934), 599–607

E 36. *Proposed by B. H. Brown, Dartmouth College.*

Show that the thirteenth of the month is more likely to be Friday than any one of the other days of the week.

40(1933), 295

REPORT OF ANNUAL MEETING

It is the consensus of opinion among college teachers of Mathematics (See J. Seidlin, Mathematics Teacher, Dec. 1932) and science that the secondary schools produce graduates with the following general characteristics:

(1) Worn out or weary of mathematics,

(2) No inspiration for individual investigation,

(3) No appreciation of accuracy,

(4) Not able to place a decimal point in its proper place,

(5) Direct and inverse proportions are meaningless.

40(1933), 382

H. E. Slaught, First Editor-in-Chief of the Monthly 1916–17

A Proof of the Fundamental Theorem of Algebra

R. P. Boas, Jr.

This note gives a proof, believed to be new, of the fundamental theorem of algebra; it is obtained by the use of the classical theorem of Picard: If there are two distinct values which a given entire function never assumes, the function is a constant. The proof is extremely simple and may be of interest as an application of Picard's theorem.

The fundamental theorem of algebra may be formulated as follows: An arbitrary polynomial,

$$f(z) = z^n + a_1 z^{n-1} + \cdots + a_{n-1} z + a_n,$$

where n is an integer > 0, and the a_i are constants, has at least one zero (in the complex plane). We shall use in addition to Picard's theorem only the facts that $f(z)$ is an entire function—hence, in particular, continuous—and that $f(z)$ has a pole at infinity.

The proof is indirect. Suppose that $f(z)$ is never zero. I say then that $f(z)$ also fails to take on one of the values $1/k$ ($k = 1, 2, \ldots$). In fact, suppose that there are points z_k such that $f(z_k) = 1/k$ ($k = 1, 2, \ldots$). Since $f(z)$ has a pole at infinity, $|f(z)| > 1$ uniformly outside some circle C. The points z_k all lie within C, and hence have at least one limit point Z within C. Since $f(z)$ is continuous,

$$f(Z) = \lim_{z_k \to Z} f(z_k) = 0.$$

This contradiction allows us to conclude that for some integer k, $f(z)$ fails to take on the value $1/k$. By Picard's theorem, $f(z)$, never assuming the distinct values 0 and $1/k$, must be constant, contrary to the hypothesis that the degree of $f(z)$ was at least 1. This contradiction shows that $f(z)$ must have at least one zero, and the proof is complete.

42(1935), 501–502

41(1934), 592

40(1933), 440

NEWS AND NOTICES

At the next meeting of the International Conference of Mathematicians, to be held in Oslo, in 1936, two gold medals will be awarded to outstanding mathematicians. The award will be made by an international committee appointed for that purpose. The foundation of these medals is due to the efforts of the late Dr. J. C. Fields, F. R. S., Research Professor of Mathematics at the University of Toronto. Dr. Fields was president of the International Congress held at Toronto in 1924, and was the editor of its Proceedings. These Proceedings consist of two large volumes published by the University of Toronto Press. With the proceeds remaining from the sale of these Proceedings, Dr. Fields suggested that a foundation for these medals be established, as a Canadian contribution to the cause of internatioanl scientific cooperation. In 1932 at Zurich, the Congress voted international approval of the foundation of these medals. The task of designing a suitable medal which, according to the wish of Dr. Fields, was to be international in character, was entrusted to the Canadian sculptor, R. Tait McKenzie, who has now completed his work. The medal is two and one-half inches in diameter and on the obverse side shows the head of Archimedes facing toward the right.

41(1934), 199–200

Dr. George David Birkhoff, professor of mathematics at Harvard University has been awarded a prize of ten thousand lire ($825) donated by Pope Pius XI in an international competition for the best book on "Systems for the solution of differential equations." The award was made during the exercises inaugurating Pontifical Hall of Science at Vatican City, on December 17, 1933.

41(1934), 200

Mr. Garrett Birkhoff, of the Society of Fellows, has been appointed to an instructorship at Harvard University.

43(1936), 386

The Ph.D. Degree and Mathematical Research*

R. G. D. Richardson

Recommendations regarding the training and utilization of advanced students of mathematics must be based on specific information concerning the present situation and on a knowledge of how the past has contributed to its up-building. No group of persons can be entirely certain that sound deductions are possible from data as incomplete as those now available. However, queries often raised regarding quality of personnel and regarding supply and demand can be answered with considerable assurance. In so complicated a problem any light that can be shed is undoubtedly welcome, and it is proposed in this report to set forth a variety of facts and to venture partial and tentative answers to certain questions.

We propose questions such as the following: How many doctor's degrees have been conferred on American mathematicians here in the United States and Canada? How many in Europe? What proportion of the present teachers of mathematics in colleges and universities have such degrees? Is there a sufficient number of competent students of mathematics now being enlisted and subjected to proper training by our graduate schools? What percentage of the Ph.D.'s have published considerable research? Is the record of publication improving with the newer crop of Ph.D.'s? What universities have the distinction of the largest average amount of publication by those to whom they have granted degrees? How does the record of the National Research Fellows stand?

· · ·

A careful investigation indicates that the number of Ph.D.'s in mathematics conferred by institutions in the United States and Canada during the period 1862–1934 is 1,286, 168 of which were conferred on women. Graph I exhibits the number of degrees conferred during five-year periods, this number following in general an exponential curve, though interrupted during the world war. The totals by years for the period 1930–34 are 84, 79, 69, 71, 92, respectively.

The figures naturally vary somewhat with the inclusion or exclusion of degrees taken in applied fields such as mathematical physics, mathematical astronomy, mechanics, theory of statistics, etc.; but they are accurate enough for our purposes.

Besides this number of American degrees, the information available indicates that there have been, during the period 1862–1930, 114 degrees conferred by foreign universities on mathematicians who have been active in America.

· · ·

---

*The University of Göttingen accounts for 34.

The upper portion of each block represents American mathematicians with European degrees. The lower portion represents American degrees.

Period	American	Foreign
1862–69	3	2
1870–74	3	0
1875–79	7	4
1880–84	9	1
1885–89	23	10
1890–94	28	12
1895–99	56	11
1900–04	75	21
1905–09	80	13
1910–14	126	18
1915–19	125	3
1920–24	129	9
1925–29	227	9
1930–34	394	7

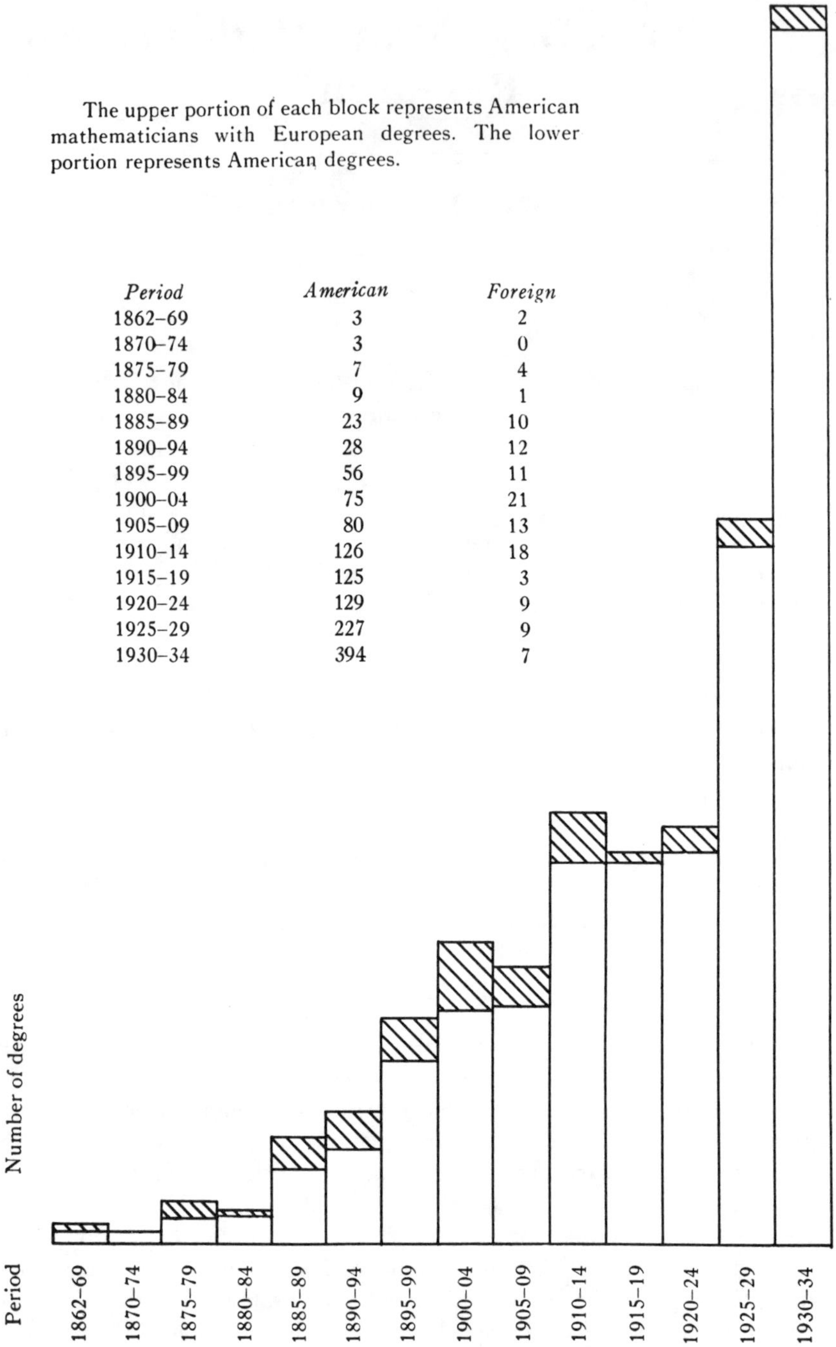

GRAPH I. DISTRIBUTION OF DEGREES CONFERRED IN MATHEMATICS 1862–1934.

Yale University, the first institution in America to confer the Ph.D. degree in course (1861), awarded the degree in mathematics in 1862 to J. H. Worrall. William Watson, later professor at Harvard University, received from Jena in 1862 the first foreign degree of which we have record. The earliest degrees conferred on women were granted to Winifred H. Edgerton by Columbia University in 1886 and to Charlotte A. Scott by the University of London in 1885.

As will be noted from Table I, more than one-sixth of the 1,286 degrees conferred in America have been awarded by the University of Chicago alone. Six institutions—Chicago, together with Cornell, Harvard, Illinois, Johns Hopkins, and Yale—are responsible for more than half. Of the remaining 53 institutions, there have been 7 which have each conferred 25 or more degrees (California, Clark, Columbia, Michigan, Pennsylvania, Princeton, and Wisconsin) and 14 that have each conferred from 10 to 24 degrees (Brown, Bryn Mawr, Catholic, Cincinnati, Indiana, State University of Iowa, Massachusetts Institute, Minnesota, Missouri, Ohio State, Pittsburgh, Rice, Texas, and Virginia). The remaining 94 degrees were given by 32 institutions. During the past ten years only one-half the universities on this list (29 out of 59) have conferred five or more degrees and can thus be considered an important present factor. In the five years 1930–34, Chicago maintains its lead in the number of degrees conferred with 52, while Michigan is second with 35, followed by Cornell and Harvard each with 28.

. . .

Number of College Teachers of Mathematics. Data collected in the autumn of 1935, based on information furnished by the institutions themselves, indicate, as tabulated in Table II, that the number of persons teaching mathematics in colleges, universities, junior colleges, and degree-granting normal colleges in the United States (with its outlying possessions) and Canada is approximately 4,500. This includes some persons who are teaching descriptive geometry, mechanics, and methods in mathematics, as well as some who are teaching part time or who are largely in administrative work or who are emeriti; but on a conservative estimate, 4,000 persons are actually engaged full time in the teaching of mathematics of college freshmen grade or higher. Similar figures were collected in 1932, and there seems to have been a considerable increase in the interim, due chiefly to the growth of junior colleges.

Table II is a statistical study by states of the number of teachers of mathematics in junior colleges, teachers colleges, and other colleges and universities, of the number of men and women teachers, of the number of teachers holding doctor's degrees, and of the number of those who are members of the American Mathematical Society or of the Mathematical Association of America or of both. The best information obtainable indicates that probably somewhat less than 1,300 of the present teachers of mathematics have the Ph.D. degree! Many of the 1,400 listed as having obtained degrees in America or abroad are deceased or have entered fields of work such as government service, banking, or industry. It should be remarked also that mathematics has furnished more than its share of administrative officers to colleges and universities. There have been some doctors who have drifted out of mathematics into other fields of science, and probably more who have correspondingly drifted in.

Of the 4,444 teachers of mathematics listed, 1,292, or 29%, hold the degree of Ph.D., while slightly less (1,263) are members of the American Mathematical Society and slightly more (1,333) are members of the Mathematical Association. In the Summary at the end of Table II the states have been grouped by sections of the

TABLE I

Number of Ph.D. Degrees in Mathematics Conferred by American Universities

Institution	1862–69	70–79	80–89	90–94	95–99	00–04	05–09	10–14	15–19	20–24	25–29	30–34	Total
Boston U.								3					3
Brown											1	13	14
Bryn Mawr			1	1	2			1	1	2	3	2	13
Calif. Inst. of Tech.											5	3	8
California					2		1	5	7	8	11	11	45
Catholic									3		3	9	15
Chicago					4	15	20	25	31	41	49	52	237
Cincinnati								1		1	1	7	10
Clark			4	7	7	2	4	2					26
U. of Colorado							1					1	2
Columbia U.			4	1	3	6	5	11	10	5	6	11	62
Cornell U.		1	3	1	1	7	8	8	8	8	16	28	89
Cumberland				1									1
Dartmouth		1											1
Duke												3	3
Fordham											2		2
Geo. Washington				1								2	3
Harvard and Radcliffe		2	3		5	5	7	11	15	11	16	28	103
Haverford					1								1
Illinois					1			6	8	10	25	23	73
Indiana								2	3		1	5	11
Iowa State											1		1
State U. of Iowa										3	5	9	17
Johns Hopkins		2	12	7	9	8	10	14	7	8	15	11	103
Kansas					1			2			2	1	6
Kentucky												1	1
Lafayette			1	1									2
Marquette												2	2
Mass. Inst. of Tech.										1	4	12	17
Michigan								4	5	3	8	35	55
Minnesota										2	4	4	10
Missouri								1	1	2	4	2	10
Moravian									1				1
Nebraska				2		1					1		4
New York									1		2	2	5
U. of N. Carolina							1					1	2
Ohio State U.												16	16
Ohio Wesleyan			1										1
Otterbein			1										1
U. of Pennsylvania				2	7	3	6	5	7	2	9	11	52
Pittsburgh											1	12	13
Princeton			2	1		1	2	8	3	9	4	18	48
Purdue				1									1
Rensselaer										1		2	3
Rice									1	1	6	5	13
St. Louis U.												2	2
Stanford						1		1			2	3	7
Syracuse U.			1	1		1		2	1	3			9
U. of Texas									1	1	5	4	11
U. of Toronto									1		1	6	8
Tulane				1									1
Vanderbilt		1					1						2
U. of Virginia			1		3	1	3		1			2	11
Washington U.				1								2	3
U. of Washington											2	3	5
West Virginia U.												2	2
Wisconsin					2	1	2	1	3	5	5	20	39
Wooster			1										1
Yale	3	4	3	2	11	11	12	11	4	2	8	8	79
Totals	3	10	32	28	56	75	80	126	125	129	227	395	1286

TABLE III
Analysis of Numbers of Papers Published by American Mathematicians

	Persons taking degrees 1862–1933		Persons taking degrees 1895–1924	
	Number	%	Number	%
No papers	549	46	232	39
1 paper	227	19	109	18
2 papers	100	8	58	10
3–5 papers	131	11	66	11
6–10 papers	70	6	41	7
11–20 papers	69	6	50	9
21–30 papers	20	2	17	3
More than 30 papers	22	2	17	3
Total	1188	100	590	100

country, and we note that 35% of the teachers in the northeast section from Illinois to Maine hold the doctor's degree. Only about 22% of those in the south central states from Kentucky to Texas hold the doctor's degree. In the remainder of the country about 26% hold that degree.

· · ·

Proportion of Ph.D.'s Publishing Research Papers. It goes without saying that the number of papers and the number of pages printed is not an adequate criterion for measuring the influence of a person on mathematical thought. Often the ideas of a scholar appear in papers published by his pupils or colleagues. But a study of the amount of publication is the easiest (perhaps the only) means that is available from a statistical standpoint. There is a great deal of information contained in Graph II and Tables III–V concerning this fundamental matter of the

TABLE V
Amount of Publication by Groups of Five Years after Receiving Degree

Period	First 5 years	Second 5 years	Third 5 years	Fourth 5 years	Fifth 5 years	Sixth 5 years	Seventh 5 years	Total	
American degree	pages	pages	pages	pages	pages	pages	pages	papers	pages
–1894	14.91	8.92	7.38	8.75	4.51	7.03	6.25	6.32	57.75
1895–1899	19.43	16.24	17.50	9.72	15.74	7.39	11.70	7.02	97.72
1900–1904	33.40	26.67	30.80	17.58	9.31	13.33		7.15	131.07
1905–1909	34.07	30.30	12.78	13.80	8.94			5.76	89.89
1910–1914	30.18	19.70	18.31	17.29				6.63	85.47
1915–1919	18.61	10.75	13.66					3.57	43.01
1920–1924	27.40	19.18						3.65	46.58
1925–1929	34.02							2.97	34.02
Foreign degree									
–1894	13.95	44.32	43.74	23.63	30.68	18.26	25.37	14.45	199.95
1895–1899	50.00	47.17	60.00	25.25	19.00	15.00	8.37	16.75	224.79
1900–1904	45.54	23.15	21.85	15.77	15.92	13.69		11.70	135.92
1905–1909	24.10	4.50	4.70	8.50	3.90			3.80	45.00
1910–1914	39.20	33.40	45.10	90.60				17.20	208.30

amount of publication, and, in spite of the reservations just made, the data have real significance.

· · ·

It would be exceedingly interesting to know what proportion of the men and women with mathematical ability of high order are now being drawn into the graduate schools; in other words, how efficiently the nation is using this human material. Are there more persons of mathematical talent in the nation than can well be utilized? But such a study, which would have to begin with high school students, is entirely beyond the power of any single organization such as the Mathematical Association of America or the American Mathematical Society.

· · ·

43(1936) 199–211

UNEMPLOYMENT

Last winter a questionnaire was sent to 50 leading universities in America asking for information concerning persons who already held the doctorate or probably would secure it during 1934 and who were seeking positions for 1934–35. Nearly all of the universities replied, and 120 persons were named who were seeking positions. There were 60 other men and women who received the doctorate in mathematics during 1934. Many of these 180 people held positions, some of which might be considered permanent but others were certainly temporary.

· · ·

The situation might be roughly summarized by stating that there were about 40 or 50 Doctors of Philosophy in mathematics who had not, on October first, found employment reasonably satisfactory to them.

· · ·

A fairly recent survey of 1098 colleges of the United States and Canada, including junior colleges and degree-granting normal schools, showed that they employed 3488 mathematics teachers, of whom 937 held the doctorate and 2551 did not.

42(1935), 143–144

3739. *Proposed by Paul Erdös, The University, Manchester, England.*

Given $n + 1$ integers, $a_1, a_2, \ldots, a_{n+1}$, each less than or equal to $2n$, prove that at least one of them is divisible by some other of the set.

42(1935), 396

An Inquiry. Professor R. A. Johnson asks the question: "Has a computing machine been invented for evaluating determinants by direct process?" Some of our readers can perhaps furnish information on this point. EDITOR.

42(1935), 435

On the Definition of $e^{i\theta}$

G. B. Price

The usual motivation of the definition of $e^{i\theta}$ requires a knowledge of infinite series; the following procedure which does not depend upon series may therefore have a certain interest and usefulness.

Let us begin by making the following assumptions:

(a) $e^{i\theta}$ is a complex number, i.e.,

$$(1) \qquad e^{i\theta} = c(\theta) + is(\theta),$$

where $c(\theta)$ and $s(\theta)$ are functions which are to be determined;

(b) $e^{i\theta}$ obeys the usual laws of exponents;

(c) $e^{i\theta}$ has $e^{-i\theta}$ for its conjugate complex number, i.e.,

$$(2) \qquad e^{-i\theta} = c(\theta) - is(\theta);$$

(d) $e^{i\theta}$ obeys the usual law for differentiation, i.e.,

$$(3) \qquad \frac{de^{i\theta}}{d\theta} = ie^{i\theta};$$

(e) $e^{i\theta}$ is a continuous function of θ for at least one value of θ, e.g.,

$$(4) \qquad \lim_{\theta \to 0} e^{i\theta} = e^0 = 1.$$

From these assumptions it will be shown that the usual definition of $e^{i\theta}$ follows.

From (1), (2), and (b) we have first that

$$(5) \qquad c^2(\theta) + s^2(\theta) = 1.$$

Next by (b) we have $e^{i\theta} \cdot e^{i\phi} = e^{i(\theta+\phi)}$, from which it follows that

$$[c(\theta)c(\phi) - s(\theta)s(\phi)] + i[s(\theta)c(\phi) + c(\theta)s(\phi)] = c(\theta + \phi) + is(\theta + \phi).$$

Then

$$(6) \qquad c(\theta + \phi) = c(\theta)c(\phi) - s(\theta)s(\phi),$$

$$s(\theta + \phi) = s(\theta)c(\phi) + c(\theta)s(\phi).$$

An obvious solution of the functional equations (6) subject to the restriction (5) is $c(\theta) = \cos k\theta$, $s(\theta) = \sin k\theta$. From (d) it follows that $k = 1$. Finally, by means of (e) we can show that this solution is unique.

43(1936), 632–633

REPORT FROM KENTUCKY SECTION

In the educational crisis that has arisen during the depression, mathematics, as an integral part of the secondary curriculum, has found criticism rather harsh and severe. While this criticism has not been entirely unjustified, it has been largely misdirected. The justification for such vehemence is not in the short-comings of mathematics as a subject, which merits intelligent study, but in the deleterious instruction imparted by those teachers of mathematics who are poorly prepared and non-enthusiastic. It is poor teaching whether in the elementary school, secondary school, college, or university, that leaves the impression that mathematics is *merely* a tool subject composed of a conglomerate mass of signs, symbols, and laws of operation. The mathematical preparation of the teacher of secondary mathematics should provide that historical, technical, and liberal information that would enable him so effectively to organize and present the subject matter that those studying under his direction should be confronted with many opportunities to do constructive thinking, to realize the intrinsic worth of mathematics as an interpreter of their environment and as a contributor to their more efficient functioning as members of a civilized social order, and to come into contact with logical processes that should better enable them to discriminatingly integrate those principles and truths which tend to make living more significant and form which might evolve a wholesome philosophy of life.

43(1936), 1–2

TWENTIETH ANNUAL MEETING.

At the conclusion of the Tuesday afternoon program several double-deck busses carried the visiting mathematicians and guests on a sightseeing tour about St. Louis, and then to the Women's Building on the campus of Washington University where the ladies of the faculty and the members of the Washington University chapter of Pi Mu Epsilon served tea in these very pleasant surroundings.

43(1936), 126

Recent Publications*

Ralph Beatley

Challenging Problems in American Schools of Education. By David Eugene Smith, The 1934–35 Sachs Lectures. Bureau of Publications, Teachers College, Columbia University, 1935, 48 pages, 55 cents.

In these two lectures, delivered at Teachers College, Columbia, in February 1935 under the provisions of the Julius and Rosa Sachs Endowment Fund, Dr. Smith arraigns the teachers colleges in no uncertain terms for foisting upon the schools of this country a vast mass of beings labeled "teacher" whose training in pedagogy has been uncoordinated and often ridiculous, whose knowledge of the subjects they expect to teach is quite inadequate, and whose general cultural background is wofully meagre. By way of contrast he depicts the scholarly interests and rich cultural background of that renowned teacher, Julius Sachs, and presents a plan to the teachers colleges whereby, if they will, they can shape their future product after the pattern afforded by Dr. Sachs himself.

Dr. Smith recalls recent efforts in certain quarters to make the training of educational administrators and of college teachers of education more substantial. He would make at least equal provision for prospective teachers in secondary schools. Though he holds up to ridicule a certain teachers college which offers twenty-two courses in the teaching of mathematics alone, he does not decry all training in Education. Though he praises the scholarly and professional spirit of the Ecole Normale Supérieure at Paris, with its great emphasis on command of subject and its slight attention to Education, he would not have our teachers colleges neglect either subject matter or Education, but emphasize both, offering substantial and dignified instruction in each, and adding a third feature, the broad cultural formation of the prospective teacher. Dr. Smith suggests that this last feature can be provided, in part at least, by master teachers giving general courses of broad cultural import, one of which, for example, might be entitled "The Nature of Man and the Universe" and might include such topics as "sociology, economics, conversation, the best music and other fine arts, deportment, and the cultivation of good taste." In this program he would secure proper balance between subject matter and Education by entrusting the administration of the program neither to the faculty of Arts and Sciences nor to the faculty of Education, but to both jointly.

Dr. Smith's insistence on the importance of a sure grasp of subject matter, of substantial and compact instruction in Education, and of broad cultural foundation serves to reinforce the conclusions reached by the recent committee of this Association appointed under the chairmanship of Professor Moulton to consider

---

*Edited by R. A. Johnson, Brooklyn College of the City of New York.

the proper relation between knowledge of Mathematics and knowledge of Education in the formation of the teacher of secondary mathematics. In more concrete form still, the general spirit of Dr. Smith's program and most of the details had already taken shape in the program initiated by President Conant of Harvard University at the very time these lectures were being delivered, whereby a new degree, the Master of Arts in Teaching, has been established, to be awarded jointly by the Faculties of Arts and Sciences and of Education on the basis of proved competence tested by a general examination in subject matter, by a general examination in Education, and by satisfaction of certain requirements of apprenticeship. Though Dr. Smith must have been unaware of these developments as he was speaking, he clearly implies that he would be pleased if a similar, or even somewhat more severe, program should be adopted by Teachers College, Columbia, where both he and Dr. Sachs taught with distinction for so many years. We agree that this would be of tremendous effect in strengthening the tone of mathematical instruction throughout the country; we fervently hope that it may come to pass.

43(1936), 633–635

According to a statement recently issued by Colorado College, this college has become, during recent years, increasingly a center at which visiting research workers have been carrying on their investigations during the summer months. To encourage this movement, the college is prepared to offer during the summer free room accommodations to qualified investigators. Further information may be obtained from the members of the various departments or from the president of the college.

43(1936), 318

Dr. G. A. Miller, professor emeritus of mathematics at the University of Illinois, reports that a new mathematical journal has appeared in Germany. Its title is Deutsche Mathematik. The first issue is dated January 1936. According to Professor Miller, this journal is an attempt to foster nationalism in mathematics instead of internationalism which has heretofore prevailed.

43(1936), 385

Dr. Max Zorn of Yale University has been appointed to an associate professorship at the University of California at Los Angeles.

. . .

The following appointments to instructorships in mathematics are announced:
Cornell University—Dr. Saunders MacLane, Dr. J. B. Rosser.
Harvard University—R. P. Boas, Jr.
Massachusetts Institute of Technology—Dr. W. T. Martin.

. . .

Dr. Julia T. Colpitts, associate professor of mathematics at Iowa State College, died on August 8 at Southampton, England, on her way home from the Congress at Oslo. She was a charter member of the Association, and was active in the Iowa Section of which she had been Chairman.

43(1936), 512

The Indian Mathematician Ramanujan*

G. H. Hardy

I have set myself a task in these lectures which is genuinely difficult and which, if I were determined to begin by making every excuse for failure, I might represent as almost impossible. I have to form myself, as I have never really formed before, and to try to help you to form, some sort of reasoned estimate of the most romantic figure in the recent history of mathematics; a man whose career seems full of paradoxes and contradictions, who defies almost all the canons by which we are accustomed to judge one another, and about whom all of us will probably agree in one judgment only, that he was in some sense a very great mathematician.

The difficulties in judging Ramanujan are obvious and formidable enough. Ramanujan was an Indian, and I suppose that it is always a little difficult for an Englishman and an Indian to understand one another properly. He was, at the best, a half-educated Indian; he never had the advantages, such as they are, of an orthodox Indian training; he never was able to pass the "First Arts Examination" of an Indian university, and never could rise even to be a "Failed B.A." He worked, for most of his life, in practically complete ignorance of modern European mathematics, and died when he was a little over 30 and when his mathematical education had in some ways hardly begun. He published abundantly—his published papers make a volume of nearly 400 pages—but he also left a mass of unpublished work which had never been analysed properly until the last few years. This work includes a great deal that is new, but much more that is rediscovery, and often imperfect rediscovery; and it is sometimes still impossible to distinguish between what he must have rediscovered and what he may somehow have learnt. I cannot imagine anybody saying with any confidence, even now, just how great a mathematician he was and still less how great a mathematician he might have been.

These are genuine difficulties, but I think that we shall find some of them less formidable than they look, and the difficulty which is the greatest for me has nothing to do with the obvious paradoxes of Ramanujan's career. The real difficulty for me is that Ramanujan was, in a way, my discovery. I did not invent him—like other great men, he invented himself—but I was the first really competent person who had the chance to see some of his work, and I can still remember with satisfaction that I could recognise at once what a treasure I had found. And I suppose that I still know more of Ramanujan than any one else, and am still the first authority on this particular subject. There are other people in England, Professor Watson in particular, and Professor Mordell, who know parts of his work very much better than I do, but neither Watson nor Mordell knew

*A lecture delivered at the Harvard Tercentenary Conference of Arts and Sciences, August 31, 1936.

Ramanujan himself as I did. I saw him and talked with him almost every day for several years, and above all I actually collaborated with him. I owe more to him than to any one else in the world with one exception, and my association with him is the one romantic incident in my life. The difficulty for me then is not that I do not know enough about him, but that I know and feel too much and that I simply cannot be impartial.

. . .

The real tragedy about Ramanujan was not his early death. It is of course a disaster that any great man should die young, but a mathematician is often comparatively old at 30, and his death may be less of a catastrophe than it seems. Abel died at 26 and, although he would no doubt have added a great deal more to mathematics, he could hardly have become a greater man. The tragedy of Ramanujan was not that he died young, but that, during his five unfortunate years, his genius was misdirected, side-tracked, and to a certain extent distorted.

I have been looking again through what I wrote about Ramanujan 16 years ago, and, although I know his work a good deal better now than I did then, and can think about him more dispassionately, I do not find a great deal which I should particularly want to alter. But there is just one sentence which now seems to me indefensible. I wrote

> "Opinions may differ about the importance of Ramanujan's work, the kind of standard by which it should be judged, and the influence which it is likely to have on the mathematics of the future. It has not the simplicity and the inevitableness of the very greatest work; it would be greater if it were less strange. One gift it shows which no one can deny, profound and invincible originality. He would probably have been a greater mathematician if he could have been caught and tamed a little in his youth; he would have discovered more that was new, and that, no doubt, of greater importance. On the other hand he would have been less of a Ramanujan, and more of a European professor, and the loss might have been greater than the gain..."

and I stand by that except for the last sentence, which is quite ridiculous sentimentalism. There was no gain at all when the College at Kumbakonam rejected the one great man they had ever possessed, and the loss was irreparable; it is the worst instance that I know of the damage that can be done by an inefficient and inelastic educational system. So little was wanted, £60 a year for five years, occasional contact with almost anyone who had real knowledge and a little imagination, for the world to have gained another of its greatest mathematicians.

. . .

44(1937), 137–155

The following mathematicians received honorary degrees from Harvard University on September 18 in connection with the Tercentenary Celebration: Professors E. J. Cartan, of the University of Paris; L. E. Dickson, of the University of Chicago; G. H. Hardy, of the University of Cambridge, England; and Tullio Levi-Civita, of the University of Rome.

43(1936), 587

Required Mathematics in a Liberal Arts College

W. L. Schaaf

1. Introduction. There is a growing movement throughout the country to eliminate mathematics as a required subject for students in liberal arts colleges. This
unfortunate wave of sentiment is by no means altogether new. Almost ten years
ago the Mathematical Association of America recognized a similar feeling of
dissatisfaction when it appointed a committee to study the problem of collateral
readings in mathematics for first and second year college students. In its report*
this committee said: "It is generally recognized that there has been a strong
tendency to decry the traditional course of instruction in mathematics for freshmen and sophomores in American Colleges. The objections urged are familiar:
... that the drill designed for the specialists is wasted on those who do not
continue mathematical study; that no glimpse of the philosophy or history of the
subject is given to the students who can really assimilate only these features; that
what is needed is a survey of what mathematics aims to accomplish, and not
manipulative speed or problem-solving ingenuity; that the same disciplinary training can be acquired in any other study, while the added gain of lively interest and
ready application makes other subjects more suitable than mathematics; that the
classical languages are being laid aside save for the professional scholar, and
mathematics, which partakes of their character, should share their fate; and so
forth." To be sure, these objections can and must be answered; so, too, with even
more searching criticisms. It must be conceded, however, that the traditional
course usually offered and frequently required, consisting as it does of some
algebra, some trigonometry, and some analytic geometry, can scarcely be regarded
as adequate for those students for whom such a course is to be their last systematic
contact with the subject. It would seem that the extreme formalism into which the
teaching of the subject has fallen furnishes ample argument for the critics who
would abolish it as a required subject.

Before proceeding to an analysis of the complaints lodged against mathematics,
and the formulation of an effective reply to the alleged indictment, it should be
added that not all authorities have yielded to these misdirected attacks or aban-
doned the cause. Thus, in the preface to his recent and charming book, Dresden
[1]† avers his belief that "this movement, in as far as it is guided by sound
educational principles, results from the lack of understanding, among educated
persons in general and among educational authorities in particular, of the essential

---

*This MONTHLY, vol. 35, 1928, pp. 221–228.

†Numbers in brackets refer to references at the end of the paper.

character of mathematics. There is little doubt that for this lack of understanding the teachers of the subject are in a large measure responsible. The fact that such a movement can gain adherence after mathematics has been for many years a required subject of study in schools and colleges points to a serious flaw in the manner in which the subject has been presented. There has been too much emphasis on its formal and narrowly technical aspects, even where technique is not the end to be achieved, and neglect of the wider bearings, of the broad human implications of the subject, and of its more interesting and stimulating problems . . . . Because mathematics has a contribution of fundamental significance to make to the education of our people, we cannot allow false conceptions of the subject to weaken its influence. Mankind may be in a better position to deal with the baffling problems which confront it in the modern world if an understanding of mathematics were the rule rather than the exceptions."

2. Facing facts. Most of the criticisms of mathematics are offered in all sincerity, but many of them are superficial and can readily be answered. Such as are more profound must be recognized as partly or wholly valid. Upon closer analysis it seems to me that most of the objections fall into one or another of the following categories: (1) dissatisfaction with the content of the courses offered, or the manner in which that content is organized; (2) dissatisfaction with the technique of teaching, or the concomitant attitudes on the part of the teachers and students, or both; and (3) the outright denial of any potential educational values resulting from the study of mathematics.

Frankly, the present writer is quite ready to agree with many of the criticisms which fall under the first two headings. In the first place, an honest and courageous appraisal of the freshman mathematics required at present in colleges the country over would compel one to confess that much of it is hopelessly worthless and futile from any point of view. It is deplorable, for example, to contemplate the prevalence even today of traditional, hidebound "trigonometry" and "college algebra," organized in watertight compartments, each a collection of completely unrelated "topics," with excessive emphasis on mechanical manipulative processes and petty details, and with almost no indication whatever of their possible relation to vital problems of physical science, technology, economics, industry, business, and finance. Similarly, we cannot escape the conviction that many of the more recent "survey" courses, "integrated" courses and "appreciation" courses are far from adequate. Many survey courses suffer from attenuation, superficiality, or general mediocrity. The so-called integrated courses frequently attempt to cover entirely too much material; or the basis of correlation is artificial and forced; or else the student fails to see the unification because of the hodge-podge set before him. And the appreciation courses that appear from time to time are often much too general or too perfunctory to be genuinely convincing; students are expected to "appreciate" things with which they are not sufficiently familiar, and so fail to achieve the desired insights and attitudes which make appreciation possible.

In the second place, personal experience and observation again compel the writer to admit that much of the teaching to which freshmen are subjected in their mathematics is sadly in need of reform. Some of it is unbelievable petty and pedantic; much of it is mechanical and uninteresting, as if it were a necessary evil; and not a little of it is unimaginative and uninspiring. There is no use in condoning the situation. When mathematics is taught by individuals without vision, without a love for the subject, and without conviction, then it deserves the fate for which it is apparently headed. Under such circumstances, are students to be blamed if they voice distaste, disgust, or abhorrence for mathematics? I hardly think so. Let us then, in the light of honest self-criticism, make a determined and forthright effort to improve the teaching of freshman mathematics through better selection and organization of content as well as better craftsmanship in teaching.

. . .

44(1937), 445–448

Dr. H. S. M. Coxeter of Washington, D.C., has been appointed to an assistant professorship at the University of Toronto.

44(1937), 122

Professor E. R. Hedrick has been appointed Vice-president of the University of California and Provost of the University of California at Los Angeles, his new duties commencing on March 19th.

44(1937), 274

Professor Richard Courant has been appointed head of the department of mathematics in the Graduate School of New York University.

44(1937), 335

The following appointments to instructorships in mathematics are announced:
University of North Carolina—Dr. Nathan Jacobson.
Princeton University—Mr. R. H. Fox, Dr. N. E. Steenrod.
University of Wisconsin—Dr. C. B. Allendoerfer.

44(1937), 336

The University of Notre Dame announces a symposium on the algebra of geometry and related subjects to be held on February 11 and 12, 1938. Among those who will participate are A. A. Albert, Emil Artin, Garrett Birkhoff, E. V. Huntington, Georges Lemaître, Karl Menger, John von Neumann, Oystein Ore, and M. H. Stone.

45(1938), 63

PUTNAM COMPETITION

For a period of at least three years beginning with the spring of 1938 an intercollegiate mathematical competition is to be held annually designed to stimulate a healthful rivalry in the undergraduate work of mathematical departments in colleges and universities of the United States and Canada. This is made possible by the trustees of the William Lowell Putnam Intercollegiate Memorial Fund, left by Mrs. Putnam in memory of her husband, a member of the Harvard class of 1882. It is hoped that such a competition will further the spirit of intercollegiate scholastic rivalry, in the possibility and value of which Mr. Putnam thoroughly believed. The competition is open only to undergraduates who have not received a degree.

. . .

6. For further encouragement of the Competition, there will be awarded at Harvard University* an annual $1000 William Lowell Putnam Prize Scholarship to one of the first five contestants, this scholarship to be available for one of the two following academic years, at the option of the recipient.

---

*Or at Radcliffe College, in the case of a woman.

45(1938), 64–66

NOTICE ABOUT FORWARDING THE MONTHLY

A member or subscriber wishing to have his copy of the June-July MONTHLY forwarded to a summer address should leave four cents postage with his local post office or carrier, with instructions for remailing.

45(1938), 331

A Contribution to the Mathematical Theory of Big Game Hunting

H. Pétard

This little known mathematical discipline has not, of recent years, received in the literature the attention which, in our opinion, it deserves. In the present paper we present some algorithms which, it is hoped, may be of interest to other workers in the field. Neglecting the more obviously trivial methods, we shall confine our attention to those which involve significant applications of ideas familiar to mathematicians and physicists.

The present time is particularly fitting for the preparation of an account of the subject, since recent advances both in pure mathematics and in theoretical physics have made available powerful tools whose very existence was unsuspected by earlier investigators. At the same time, some of the more elegant classical methods acquire new significance in the light of modern discoveries. Like many other branches of knowledge to which mathematical techniques have been applied in recent years, the Mathematical Theory of Big Game Hunting has a singularly happy unifying effect on the most diverse branches of the exact sciences.

For the sake of simplicity of statement, we shall confine our attention to Lions (*Felis leo*) whose habitat is the Sahara Desert. The methods which we shall enumerate will easily be seen to be applicable, with obvious formal modifications, to other carnivores and to other portions of the globe. The paper is divided into three parts, which draw their material respectively from mathematics, theoretical physics, and experimental physics.

The author desires to acknowledge his indebtedness to the Trivial Club of St. John's College, Cambridge, England; to the M.I.T. chapter of the Society for Useless Research; to the F.o.P., of Princeton University; and to numerous individual contributors, known and unknown, conscious and unconscious.

1. Mathematical methods

1. THE HILBERT, OR AXIOMATIC, METHOD. We place a locked cage at a given point of the desert. We then introduce the following logical system.

AXIOM I. *The class of lions in the Sahara Desert is non-void.*

AXIOM II. *If there is a lion in the Sahara Desert, there is a lion in the cage.*

RULE OF PROCEDURE. *If p is a theorem, and "p implies q" is a theorem, then q is a theorem.*

THEOREM I. *There is a lion in the cage.*

2. THE METHOD OF INVERSIVE GEOMETRY. We place a *spherical* cage in the desert, enter it, and lock it. We perform an inversion with respect to the cage. The lion is then in the interior of the cage, and we are outside.

3. THE METHOD OF PROJECTIVE GEOMETRY. Without loss of generality, we may regard the Sahara Desert as a plane. Project the plane into a line, and then project the line into an interior point of the cage. The lion is projected into the same point.

4. THE BOLZANO-WEIERSTRASS METHOD. Bisect the desert by a line running N-S. The lion is either in the E portion or in the W portion; let us suppose him to be in the W portion. Bisect this portion by a line running E-W. The lion is either in the N portion or in the S portion; let us suppose him to be in the N portion. We continue this process indefinitely, constructing a sufficiently strong fence about the chosen portion at each step. The diameter of the chosen portions approaches zero, so that the lion is ultimately surrounded by a fence of arbitrarily small perimeter.

5. THE "MENGENTHEORETISCH" METHOD. We observe that the desert is a separable space. It therefore contains an enumerable dense set of points, from which can be extracted a sequence having the lion as limit. We then approach the lion stealthily along this sequence, bearing with us suitable equipment.

6. THE PEANO METHOD. Construct, by standard methods, a continuous curve passing through every point of the desert. It has been remarked* that it is possible to traverse such a curve in an arbitrarily short time. Armed with a spear, we traverse the curve in a time shorter than that in which a lion can move his own length.

7. A TOPOLOGICAL METHOD. We observe that a lion has at least the connectivity of the torus. We transport the desert into four-space. It is then possible* to carry out such a deformation that the lion can be returned to three-space in a knotted condition. He is then helpless.

8. THE CAUCHY, OR FUNCTION THEORETICAL, METHOD. We consider an analytic lion-valued function $f(z)$. Let ζ be the cage. Consider the integral

$$\frac{1}{2\pi i} \int_C \frac{f(z)}{z - \zeta}\, dz,$$

where C is the boundary of the desert; its value is $f(\zeta)$, i.e., a lion in the cage.†

9. THE WIENER TAUBERIAN METHOD. We procure a tame lion, L_0, of class $L(-\infty, \infty)$, whose Fourier transform nowhere vanishes, and release it in the desert. L_0 then converges to our cage. By Wiener's General Tauberian Theorem,‡ any other lion, L (say), will then converge to the same cage. Alternatively, we can approximate arbitrarily closely to L by translating L_0 about the desert.§

2. Methods from theoretical physics

10. THE DIRAC METHOD. We observe that wild lions are, *ipso facto*, not observable in the Sahara Desert. Consequently, if there are any lions in the Sahara, they are tame. The capture of a tame lion may be left as an exercise for the reader.

11. THE SCHRÖDINGER METHOD. At any given moment there is a positive probability that there is a lion in the cage. Sit down and wait.

---

*By Hilbert. See E. W. Hobson, The Theory of Functions of a Real Variable and the Theory of Fourier's Series, 1927, vol. 1, pp. 456–457.

*H. Seifert and W. Threlfall, Lehrbuch der Topologie, 1934, pp. 2–3.

†*N.B.* By Picard's Theorem (W. F. Osgood, Lehrbuch der Funktionentheorie, vol. 1, 1928, p. 748), we can catch every lion with at most one exception.

‡N. Wiener, The Fourier Integral and Certain of its Applications, 1933, pp. 73–74.

§N. Wiener, *l. c.*, p. 89.

12. THE METHOD OF NUCLEAR PHYSICS. Place a tame lion in the cage, and apply a Majorana exchange operator‖ between it and a wild lion.

As a variant, let us suppose, to fix ideas, that we require a male lion. We place a tame lioness in the cage, and apply a Heisenberg exchange operator¶ which exchanges the spins.

13. A RELATIVISTIC METHOD. We distribute about the desert lion bait containing large portions of the Companion of Sirius. When enough bait has been taken, we project a beam of light across the desert. This will bend right round the lion, who will then become so dizzy that he can be approached with impunity.

3. Methods from experimental physics

14. THE THERMODYNAMICAL METHOD. We construct a semi-permeable membance, permeable to everything except lions, and sweep it across the desert.

15. THE ATOM-SPLITTING METHOD. We irradiate the desert with slow neutrons. The lion becomes radioactive, and a process of disintegration sets in. When the decay has proceeded sufficiently far, he will become incapable of showing fight.

16. THE MAGNETO-OPTICAL METHOD. We plant a large lenticular bed of catnip (*Nepeta cataria*), whose axis lies along the direction of the horizontal component of the earth's magnetic field, and place a cage at one of its foci. We distribute over the desert large quantities of magnetized spinach (*Spinacia oleracea*), which, as is well know, has a high ferric content. The spinach is eaten by the herbivorous denizens of the desert, which are in turn eaten by lions. The lions are then oriented parallel to the earth's magnetic field, and the resulting beam of lions is focussed by the catnip upon the cage.

45(1938), 447–448

RESULTS OF FIRST PUTNAM

The five persons ranking highest in the examination, arranged alphabetically, were ROBERT GIBSON, Fort Hays Kansas State College; I. KAPLANSKY, University of Toronto; G. W. MACKEY, Rice Institute; M. J. NORRIS, College of St. Thomas; BERNARD SHERMAN, Brooklyn College. Each of these will receive a prize of fifty dollars. The order of the names in this list has no relation to their rank in the examination.

45(1938), 332

‖See, for example, H. A. Bethe and R. F. Bacher, Reviews of Modern Physics, vol. 8, 1936, pp. 82–229; especially pp. 106–107.

¶*Ibid.*

MATHEMATICS EDUCATION

As you see from the list of topics below, I have followed the advice of Professor G. H. Hardy when he said at the conclusion of a lecture before the American Mathematical Society, "The elementary theory of numbers should be one of the very best subjects for early mathematical instruction. It demands very little previous knowledge; its subject matter is tangible and familiar; the processes of reasoning which it employs are simple, general, and few; and it is unique among the mathematical sciences in its appeal to natural human curiosity. A month's intelligent instruction in the theory of numbers ought to be twice as instructive, twice as useful, and at least ten times as entertaining as the same amount of 'calculus for engineers'."

46(1939), 35

LIFE MEMBERSHIP IN THE MATHEMATICAL ASSOCIATION

In accordance with the action of the Association at its Williamsburg meeting, members may obtain life membership in the Association by the payment, at the first of any calendar year, of an amount indicated in the accompanying table. In estimating one's age the birthday anniversary nearest to the first of January when payment is made should be taken.

Age	Fee	Age	Fee	Age	Fee
20.....	$89.49	40.....	$71.72	60.....	$46.74
21.....	88.71	41.....	70.62	61.....	45.40
22.....	87.93	42.....	69.50	62.....	44.06
23.....	87.16	43.....	68.36	63.....	42.73
24.....	86.40	44.....	67.21	64.....	41.40
25.....	85.64	45.....	66.03	65.....	40.08
26.....	84.86	46.....	64.83	66.....	38.76
27.....	84.06	47.....	63.62	67.....	37.46
28.....	83.24	48.....	62.39	68.....	36.17
29.....	82.40	49.....	61.15	69.....	34.89
30.....	81.54	50.....	59.89	70.....	33.63
31.....	80.65	51.....	58.62	71.....	32.38
32.....	79.75	52.....	57.33	72.....	31.16
33.....	78.82	53.....	56.03	73.....	29.95
34.....	77.87	54.....	54.72	74.....	28.76
35.....	76.90	55.....	53.41	75.....	27.60
36.....	75.91	56.....	52.08	76.....	26.47
37.....	74.89	57.....	50.75	77.....	25.36
38.....	73.86	58.....	49.42	78.....	24.28
39.....	72.80	59.....	48.08	79.....	23.22

W. D. CAIRNS, *Secretary-Treasurer*

46(1939), 134

A Theorem on Graphs, with an Application to a Problem of Traffic Control

H. E. Robbins

It is the object of this note to answer a question* which is suggested by the problem of traffic control in a modern city: "When may the arcs of a graph be so oriented that one may pass from any vertex to any other, traversing arcs in the positive sense only?" Any graph with this property will be called orientable. The answer is provided by the following theorem:

THEOREM. *A graph is orientable if and only if it remains connected after the removal of any arc.*

Let us suppose that week-day traffic in our city is not particularly heavy, so that all streets are two-way, but that we wish to be able to repair any one street at a time and still detour traffic around it so that any point in the city may be reached from any other point. On week-ends no repairing is done, so that all streets are available, but due to the heavy traffic (perhaps it is a college town with a noted football team) we wish to make all streets one-way and still be able to get from any point to any other without violating the law. Then the theorem states that if our street-system is suitable for week-day traffic it is also suitable for week-end traffic and conversely. We proceed to give a few definitions and a proof of the theorem.

By a graph G we mean a (finite) set of objects $\langle p, q, \ldots \rangle$ called *vertices*, together with a (finite) set of objects called *arcs*, which join certain pairs of distinct vertices. We shall represent an arc joining p and q by (pq), with the understanding that there may be more than one such arc. G is *connected* if, given any two of its vertices p, q, there exists a chain of arcs joining p and q of the form

$$(1) \qquad (pr_1)(r_1 r_2) \cdots (r_s q).$$

G is *oriented* if a direction is assigned to each arc, symbolized by choosing one of the two representations (pq) and (qp) of the unoriented arc, to be called the *admissible* representation, G is *orientable* if it may be oriented in such a way that any two of its vertices may be joined by a chain (1) of arcs, each in its admissible representation. The definitions of *subgraph*, and of the special subgraph obtained from G by removing one of its arcs, are obvious.

Now referring to the theorem, we see that the necessity of the condition is clear, for if G is disconnected by the removal of an arc (pq), then no matter which representation of (pq) we call admissible in an orientation of G, passage either form p to q or from q to p by a chain of admissible arcs will be impossible, so that

---

*Proposed by S. Ulam. For a simplification in the proof we are indebted to H. Whitney.

119

G is not orientable. It remains to prove the sufficiency of the condition. Suppose G remains connected after the removal of any arc. (Then *a fortiori* G is connected.) Choose a vertex p of G and consider the class of all subgraphs of G which include p and which are orientable. This class is non-void, since the subgraph consisting of p alone satisfies the conditions. Let G' be a sub-graph in this class with maximal number of vertices. We may assume that G' contains all arcs of G whose vertices are in G', since otherwise any such arcs which did not belong to G' could be added, oriented arbitrarily, and G' would remain orientable. Then G' must be identical with G. For suppose p^* is a vertex of G not in G'. Join p^* to p by a chain

$$(2) \qquad (p_1 p_2)(p_2 p_3) \cdots (p_{n-1} p_n), \qquad [p_1 = p^*; p_n = p],$$

and order the arcs of (2) from left to right. Let $(p_k p_{k+1})$ be the last arc of (2) which is not in G'. Then p_k does not belong to G', while p_{k+1} does. By hypothesis, there exists a chain of arcs

$$(3) \qquad (\bar{p}_1 \bar{p}_2)(\bar{p}_2 \bar{p}_3) \cdots (\bar{p}_{m-1} \bar{p}_m), \qquad [\bar{p}_1 = p_k; \bar{p}_m = p_{k+1}],$$

which does not include $(p_k p_{k+1})$. Let $(\bar{p}_s \bar{p}_{s+1})$ be the last arc of (3) which does not belong to G'. Then the subgraph G^* consisting of G' plus $(p_{k+1} p_k)$ and the chain

$$(4) \qquad (\bar{p}_1 \bar{p}_2)(\bar{p}_2 \bar{p}_3) \cdots (\bar{p}_s \bar{p}_{s+1}),$$

where these arcs are to be oriented as written, is clearly orientable, contains p, and has more vertices than G', which is a contradiction and completes the proof.

46(1939), 281–283

Word has reached this country that the Editor of the *Zentralblatt für Mathematik und ihre Grenzgebiete*, Professor Otto Neugebauer, now of Copenhagen, has resigned. The resignation from this mathematical abstracts journal was occasioned by the action of the publisher, Julius Springer of Berlin, in dropping Professor Levi-Civita of Italy from the board without the knowledge of the Editor, as well as by the demand that the Editor give assurance that no emigrants would be allowed to referee articles by German authors. In consequence of this interference with editorial policies, the American associate editors, Professors Tamarkin and Veblen, have tendered their resignations as have also a number of associate editors and collaborators in other countries.

46(1939), 57

The Transcendence of π

Ivan Niven*

Among the proofs of the transcendence of e, which are in general variations and simplifications of the original proof of Hermite, perhaps the simplest is that of A. Hurwitz.† His solution of the problem contains an ingenious device which we now employ to obtain a relatively simple proof of the transcendence of π.

We assume that π is an algebraic number, and show that this leads to a contradiction. Since the product of two algebraic numbers is an algebraic number, the quantity $i\pi$ is a root of an algebraic equation with integral coefficients

$$(1) \qquad \theta_1(x) = 0,$$

whose roots are $\alpha_1 = i\pi, \alpha_2, \alpha_3, \ldots, \alpha_n$. Using Euler's relation $e^{i\pi} + 1 = 0$, we have

$$(2) \qquad (e^{\alpha_1} + 1)(e^{\alpha_2} + 1) \cdots (e^{\alpha_n} + 1) = 0.$$

We now construct an algebraic equation with integral coefficients whose roots are the exponents in the expansion of (2). First consider the exponents

$$(3) \qquad \alpha_1 + \alpha_2, \alpha_1 + \alpha_3, \alpha_2 + \alpha_3, \ldots, \alpha_{n-1} + \alpha_n.$$

By equation (1), the elementary symmetric functions of $\alpha_1, \alpha_2, \ldots, \alpha_n$ are rational numbers. Hence the elementary symmetric functions of the quantities (3) are rational numbers. It follows that the quantities (3) are roots of

$$(4) \qquad \theta_2(x) = 0,$$

an algebraic equation with integral coefficients. Similarly, the sums of the α's taken three at a time are the $_nC_3$ roots of

$$(5) \qquad \theta_3(x) = 0.$$

Proceeding thus, we obtain

$$(6) \qquad \theta_4(x) = 0, \theta_5(x) = 0, \ldots, \theta_n(x) = 0,$$

algebraic equations with integral coefficients, whose roots are the sums of the α's taken $4, 5, \ldots, n$ at a time respectively. The product equation

$$(7) \qquad \theta_1(x)\theta_2(x) \cdots \theta_n(x) = 0$$

has roots which are precisely the exponents in the expansion of (2).

The deletion of zero roots (if any) from equation (7) gives

$$(8) \qquad \theta(x) = cx^r + c_1 x^{r-1} + \cdots + c_r = 0,$$

*Harrison Research Fellow.

†A. Hurwitz, Beweis der Transendenz der Zahl e, Mathematische Annalen, vol. 43, 1893, pp. 220–221 (also in his Mathematische Werke, vol. 2, pp. 134–135).

whose roots $\beta_1, \beta_2, \ldots, \beta_r$ are the non-vanishing exponents in the expansion of (2), and whose coefficients are integers. Hence (2) may be written in the form

$$(9) \qquad e^{\beta_1} + e^{\beta_2} + \cdots + e^{\beta_r} + k = 0,$$

where k is a positive integer.

We define

$$(10) \qquad f(x) = \frac{c^s x^{p-1}\{\theta(x)\}^p}{(p-1)!},$$

where $s = rp - 1$, and p is a prime to be specified. Also we define

$$(11) \qquad F(x) = f(x) + f^{(1)}(x) + f^{(2)}(x) + \cdots + f^{(s+p+1)}(x),$$

noting, with thanks to Hurwitz, that the derivative of $e^{-x}F(x)$ is $-e^{-x}f(x)$. Hence we may write

$$e^{-x}F(x) - e^0 F(0) = \int_0^x - e^{-\xi} f(\xi)\, d\xi.$$

The substitution $\xi = \tau x$ produces

$$F(x) - e^x F(0) = -x \int_0^1 e^{(1-r)x} f(\tau x)\, d\tau.$$

Let x range over the values $\beta_1, \beta_2, \ldots, \beta_r$ and add the resulting equations. Using (9), we obtain

$$(12) \qquad \sum_{j=1}^r F(\beta_j) + kF(0) = -\sum_{j=1}^r \beta_j \int_0^1 e^{(1-r)\beta_j} f(\tau \beta_j)\, d\tau.$$

This result gives us the contradiction we desire. For we shall choose the prime p to make the left side a non-zero integer, and the right side as small as we please.

By (10), we have

$$\sum_{j=1}^r f^{(t)}(\beta_j) = 0, \quad \text{for } 0 \leqq t < p.$$

Also by (10) the polynomial obtained by multiplying $f(x)$ by $(p-1)!$ has integral coefficients. Since the product of p consecutive positive integers is divisible by $p!$, the pth and higher derivatives of $(p-1)!f(x)$ are polynomials in x with integral coefficients divisible by $p!$. Hence the pth and higher derivatives of $f(x)$ are polynomials with integral coefficients each of which is divisible by p. That each of these coefficients is also divisible by c^s is obvious from the definition (10). Thus we have shown that, for $t \geqq p$, the quantity $f^{(t)}(\beta_j)$ is a polynomial in β_j of degree at most s, each of whose coefficients is divisible by pc^s. By (8), a symmetric function of $\beta_1, \beta_2, \ldots, \beta_r$ with integral coefficients and of degree at most s is an integer provided each coefficient is divisible by c^s (by the fundamental theorem on symmetric functions). Hence

$$\sum_{j=1}^r f^{(t)}(\beta_j) = pk_t, \qquad\qquad (t = p, p+1, \ldots, p+s),$$

where the k_t are integers. It follows that

$$\sum_{j=1}^{r} F(\beta_j) = p \sum_{t=p}^{p+s} k_t.$$

In order to complete the proof that the left side of (12) is a non-zero integer, we now show that $kF(0)$ is an integer prime to p. From (10) it is clear that

$$f^{(t)}(0) = 0, \qquad\qquad (t = 0, 1, \ldots, p - 2),$$

$$f^{(p-1)}(0) = c^s c_r^p,$$

$$f^{(t)}(0) = pK_t, \qquad\qquad (t = p, p + 1, \ldots, p + s),$$

where the K_t are integers. If p is chosen greater than each of k, c, c_r (possible since the number of primes is infinite), the desired result follows from (11).

Finally, the right side of (12) equals

$$-\sum_{j=1}^{r} \frac{1}{c} \int_0^1 \frac{\left\{c^r \beta_j \theta(\tau \beta_j)\right\}^p}{(p-1)!} e^{(1-r)\beta_j} \, d\tau.$$

This is a finite sum, each term of which may be made as small as we wish by choosing p very large, because

$$\lim_{p \to \infty} \frac{\left\{c^r \beta_j \theta(\tau \beta_j)\right\}^p}{(p-1)!} = 0.$$

46(1939), 469–471

POSTCARDS ON APPLIED MATHEMATICS

by J. L. Synge

One of the facts which the historian of the future will not fail to note regarding our present epoch is the way in which mathematicians have turned from applied mathematics. Mathematicians may be divided into three classes in respect of their attitude towards applied mathematics: (a) those who have nothing to do with applied mathematics and do not want to, regarding it as an inferior type of intellectual exercise; (b) those who would like to be better acquainted with applied mathematics, but cannot find time for prolonged study of what is not their major interest; (c) those primarily interested in applied mathematics, studying the pure almost solely for its repercussions on the applied. There are, it is true, a few fortunate individuals who cannot be so rudely classified, but their number is so small and their capacity so great that the present writer would not presume to write for or about them; they are the sort that seem able to master the contents of a book by glancing at the cover.

The eighteenth centry was the age of class (c); the twentieth century is the age of class (a). The nineteenth was the age of transition. (Some pure mathematicians may be surprised to learn that Weierstrass made some fundamental contributions to the theory of dynamical stability.) The transition was doubtless greatly helped by Kelvin and Tait's monumental "Natural Philosophy," succulent Irish stew to the physicist, but a bellyache to the mathematician.

46(1939), 155

THE DEADLY PARALLEL

A. J. KEMPNER, University of Colorado

Readers of the MONTHLY may be interested in two excerpts from the daily press:

I. (From the *Rocky Mountain News*, Denver, Friday, February 10, 1939). At the Annual Regional Study Conference of the Progressive Education Association, Denver, Colorado, Dr. Lois Hayden Meek, Professor of Education at Columbia University, is quoted as saying: "Intelligent supervision and encouragement of such group relationship at that time is of far greater importance in the future education of the youth than cut and dried history or mathematics... It is my opinion we should not have so many specialists on our teaching staffs. ... A better method would be to have one teacher teach more subjects and have longer and more frequent contact with pupils, studying more carefully the individual problems of each. *The function of such a teacher would not hinge so much on knowing the subjects to be taught as on knowing how to arrange and present such subjects so they could be most readily learned.*" (Italics mine.)

II. (From the *Denver Post*, Saturday, February 11, 1939, reporting opinions of students at the same conference).

A high school senior from Northern Colorado: "Everything would be fine if we had good teachers. You've been talking a lot about dull subjects that should be dropped from the curriculum. Those subjects are dull because the teachers are dull. If all our teachers were alert and interesting, there would be no such thing as wasted courses. Even a course in Latin would be fun and would accomplish real education if the teacher presented it the right way."

A college student: "It's all a matter of the personality of the teacher. All of these proposed drastic changes in the method of education would be unnecessary if teachers generally had a real understanding of what young people want. We don't resist learning. As a matter of fact we're anxious to learn. All we ask is that there be some pleasure, some intellectual excitement in the process... Every student knows there are some teachers in whose classes he would like to spend much more time than allotted. When the student fidgets and waits impatiently for the bell to ring, it is the fault of the teacher, not the fault of the subject."

· · ·

May it be that we have overlooked our best allies in our fight for a sound education based on fundamentals,—the students themselves?

46(1939), 280

Mathematics Clubs*
An Alignment Chart for the Quadratic Equation†

L. R. Ford

The accompanying nomographic chart, here much reduced, was made for distribution at the 1939 "Open House" of the Armour Institute of Technology. The actual drawing was made by Stephen Kroll, a student. So many requests have been made for the theory which underlies the chart that it seems desirable to write it out in detail.

1. Scales. Examples of scales are the foot rule with marks to indicate inches and fractions thereof, the marks on a thermometer to denote degrees, the minute marks around the face of a clock, and the marks on a protractor. Essentially a scale is a curve expressed in parametric form with the values of the parameter marked at intervals along the curve.

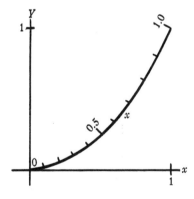

Fig. 1.

We use rectangular coordinates in the plane with X and Y as variables. Consider, for example,

$$X = x, \qquad Y = x^2.$$

This is a parabola, $Y = X^2$. Let us mark on the curve a set of values of the parameter x, as $x = 0, 0.1, 0.2, \ldots, 0.9, 1$. This gives us the scale in Figure 1.

*Edited by E. H. C. Hildebrandt.

†The material in this article was presented before the Men's Mathematics Club of the Chicago and Metropolitan Area, April 21, 1939.

Conversely, if such a scale is given then the X and Y of a point on the curve depend upon (that is, are functions of) the scale number at the point.

2. The alignment chart. Suppose we have three curves in parametric form

$$X = X_1(x), \qquad X = X_2(y), \qquad X = X_3(z),$$
$$Y = Y_1(x), \qquad Y = Y_2(y), \qquad Y = Y_3(z).$$

Three points, one from each scale, lie on a straight line, if and only if

(1)
$$\begin{vmatrix} X_1(x), & Y_1(x), & 1 \\ X_2(y), & Y_2(y), & 1 \\ X_3(z), & Y_3(z), & 1 \end{vmatrix} = 0.$$

We use the chart to solve (1) for one parameter when the other two are given. The line joining a point bearing a given x with a point bearing a given y will have at its intersection with the third scale a z to satisfy (1).

It will be observed that each variable in (1) is restricted to a single row. If we wish to construct an alignment chart to solve an equation in three variables, our first task is to write it in determinant form with each variable appearing in a single row. It is not always possible to do this. There is the further requirement that a column of 1's appear, but this is easily managed.

3. The quadratic equation. We propose to solve

(2)
$$x^2 = ax + b.$$

Here the variables are x, a, and b. We write (2) without difficulty in the form

(3)
$$\begin{vmatrix} x^2, & x, & 1 \\ a, & 1, & 0 \\ b, & 0, & 1 \end{vmatrix} = 0.$$

Let us now multiply through by a non-vanishing determinant whose elements are constants

$$\begin{vmatrix} a_1, & b_1, & c_1 \\ a_2, & b_2, & c_2 \\ a_3, & b_3, & c_3 \end{vmatrix}.$$

Applying the usual rule for the multiplication of two determinants, we get

$$\begin{vmatrix} a_1x^2 + a_2x + a_3, & b_1x^2 + b_2x + b_3, & c_1x^2 + c_2x + c_3 \\ a_1a + a_2, & b_1a + b_2, & c_1a + c_2, \\ a_1b + a_3, & b_1b + b_3, & c_1b + c_3 \end{vmatrix} = 0.$$

Dividing each row by its element in the third column we get the required 1's there. We have then the curves in terms of the parameters x, a, b respectively,

$$C_x: \qquad X = \frac{a_1x^2 + a_2x + a_3}{c_1x^2 + c_2x + c_3}, \qquad Y = \frac{b_1x^2 + b_2x + b_3}{c_1x^2 + c_2x + c_3},$$

$$L_a: \qquad X = \frac{a_1a + a_2}{c_1a + c_2}, \qquad Y = \frac{b_1a + b_2}{c_1a + c_2},$$

$$L_b: \qquad X = \frac{a_1b + a_3}{c_1b + c_3}, \qquad Y = \frac{b_1b + b_3}{c_1b + c_3}.$$

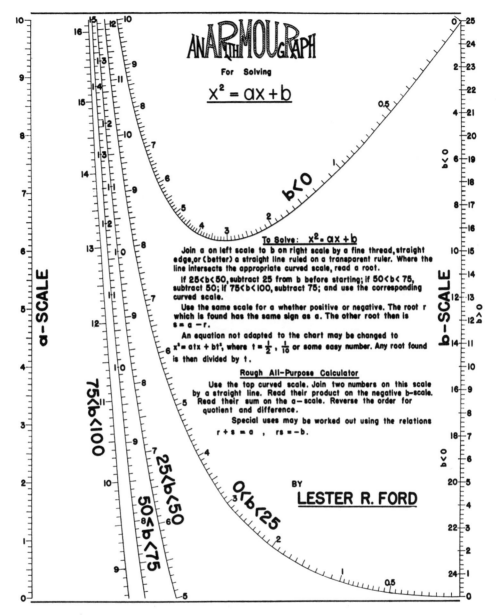

AN ARITHMOUGRAPH

For Solving

$$x^2 = ax + b$$

a–SCALE

b–SCALE

$b < 0$

$b > 0$

b<0

$75 < b < 100$

$25 < b < 50$

$50 < b < 75$

$0 < b < 25$

$b < 0$

To Solve: $x^2 = ax + b$

Join a on left scale to b on right scale by a fine thread, straight edge, or (better) a straight line ruled on a transparent ruler. Where the line intersects the appropriate curved scale, read a root.

If $25 < b < 50$, subtract 25 from b before starting; if $50 < b < 75$, subtract 50; if $75 < b < 100$, subtract 75; and use the corresponding curved scale.

Use the same scale for a whether positive or negative. The root r which is found has the same sign as a. The other root then is $s = a - r$.

An equation not adapted to the chart may be changed to $x^2 = atx + bt^2$, where $t = \frac{1}{2}$, $\frac{1}{10}$ or some easy number. Any root found is then divided by t.

Rough All-Purpose Calculator

Use the top curved scale. Join two numbers on this scale by a straight line. Read their product on the negative b-scale. Read their sum on the a–scale. Reverse the order for quotient and difference.

Special uses may be worked out using the relations $r + s = a$, $rs = -b$.

BY LESTER R. FORD

Nomographic Chart for Solving Quadratics.

Here C_x is a conic and L_a, L_b are straight lines. There is much variety possible owing to the constants at our disposal, but these curves have the general form shown in Figure 2. Here P is the point $X = a_1/c_1$, $Y = b_1/c_1$, which is on all the curves for the values $x = \infty$, $a = \infty$, $b = \infty$. Q is $X = a_3/c_3$, $Y = b_3/c_3$, which lies on C_x and L_b for $x = 0$, $b = 0$. That L_a, C_x are tangent at P is shown readily. Differentiating, we find the slopes of C_s and L_a to be

$$\frac{dY}{dX} = \frac{(b_1c_2 - b_2c_1)x^2 + 2(b_1c_3 - b_3c_1)x + b_2c_3 - b_3c_2}{(a_1c_2 - a_2c_1)x^2 + 2(a_1c_3 - a_3c_1)x + a_2c_3 - a_3c_2},$$

$$\frac{dY}{dX} = \frac{b_1c_2 - b_2c_1}{a_1c_2 - a_2c_1},$$

and the former is equal to the latter when $x = \infty$.

Owing to the constants at our disposal the chart may be put into many forms. We may require that L_a, L_b be the coordinate axes, that C_x be a circle,* and so on.

In our chart we shall require that L_a and L_b be the parallel lines

$$X_a = 0, \qquad X_b = 1.$$

This gives immediately

$$a_1 = a_2 = 0, \qquad c_1 = 0, \qquad a_3 = c_3 = \lambda, \quad \text{say.}$$

Make the a-scale have zero at the origin by setting $b_2 = 0$. Put $b_1 = c_2 = 1$ and $b_3 = \beta$. Our equations now take the form

$$C_x: \qquad X = \frac{\lambda}{x + \lambda}, \qquad Y = \frac{x^2 + \beta}{x + \lambda},$$

$$L_a: \qquad X = 0 \qquad Y = a,$$

$$L_b: \qquad X = 1, \qquad Y = \frac{b + \beta}{\lambda}.$$

In the accompanying chart the units are chosen so that X may run from 0 to 1 and Y from 0 to 10 on the page. We take $\lambda = 5/2$, so that 25 b-units appear.

If $\beta = 0$ we get the b-scale on the right and the hyperbola C_x through the lower right-hand corner of the figure. If $\beta = 25$ we get the hyperbola C_x at the top; $b = 0$ appears at the top of the b-scale and negative values of b run down the scale. For $\beta = -25, -50, -75$, the b's that appear on the figure are those at the right, increased by 25, 50 or 75. The hyperbolas for these three cases are drawn on the chart. Since the equations of L_a are independent of β, we use the same a-scale throughout. The object of breaking the scales in this way is to increase the range of the scales without a corresponding increase in the dimensions of the chart or a decrease in its accuracy.

---

*See Whittaker and Robinson, Calculus of Observation, p. 129, where both these requirements are met.

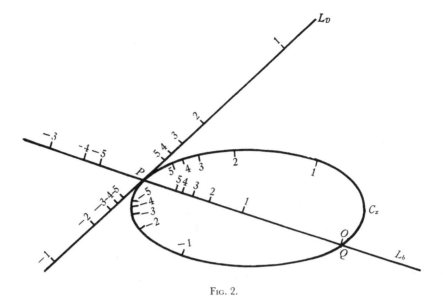

FIG. 2.

4. Remarks. We have in these few paragraphs the essentials of the theory of nomography. The determinant (1) is the basic notion of the theory. The use of the determinant (4) as a multiplier provides the transformation theory. We thus achieve the results of the general collineation in a very elementary and comparatively painless way.

46(1939), 508–511

Mathematical Snapshots. By H. Steinhaus. (Translated from Polish.) New York, G. E. Stechert and Company, 1938. 4 + 135 pages. $2.50.

Mathematical Snapshots is hardly to be classified under any of the usual categories. Indeed, the temptation is to make a new category—An Evening's Amusement—and to let this book be number one. The author uses pictures, figures, and models in addition to text to call attention to things of a mathematical or semi-mathematical character that are frequently overlooked or misunderstood. There is nothing like formal proof. The style is delightful, although there is an occasional lack of clarity, apparently introduced by translation. In fact, this book was more interesting to the reviewer than any book on so-called mathematical recreations that has ever come to his attention. Anyone, mathematician or what not, will enjoy turning its pages, looking at the pictures, and handling the models which are enclosed in a cover pocket, also reading the text. It is recommended to all, especially to those who are fond of puzzles and have a flair for mathematical recreations. The person who has never seriously studied mathematics, but who did well in high school mathematics and has a sneaking notion that he would have done well in more advanced work, will be delighted. Angle trisectors, circle squarers, and all such will find in it profitable channels for their abilities. Introduce them to it.

TOMLINSON FORT

46(1939), 354

John Wesley Young (1879–1932)

4. BATTLES AND WARS, 1941–1950

C. C. MacDuffee, President of the MAA 1945

Battles and Wars

"It is true that there are branches of applied mathematics, such as ballistics and aerodynamics, which have been developed deliberately for war...but none of them has any claim to rank as 'real'. They are indeed repulsively ugly and intolerably dull; even Littlewood could not make ballistics respectable, and if he could not, who can? So a real mathematician has his conscience clear; there is nothing to be set against any value his work may have; mathematics is...a 'harmless and innocent' occupation." Those words were written by G. H. Hardy in 1940. Paradoxically, they were a call to arms; they presaged a battle yet to come in American mathematics.

By early 1941, German troops had occupied Denmark, Norway, the Netherlands, and had defeated France. The Battle of Britain (in the air) was over, preventing a quick victory for the Axis powers. Germany was secretly preparing to invade Russia.

In that same year, the War Preparedness Committee of the Association published a long report in the Monthly that gave advice about War Mathematics. (Don't emphasize advanced topics; stick to classical material rather than specialized cookbook courses.) W. D. Cairns was serving his last year as Secretary-Treasurer of the Association, a position he had held since its founding in 1915. Math Reviews was a struggling new journal at Brown University, run by Tamarkin and Neugebauer. By the end of the year, Pearl Harbor brought America fully into the war as well.

Mathematicians everywhere were affected. Many accepted jobs in war related industries, or working for the government, or teaching special courses for the Army and Navy. The demand caused a shortage of mathematics teachers at every level, and colleges responded by increasing teaching loads and retraining people in related fields.

In 1943, the War Preparedness Committee was replaced by the War Policy Committee, which was joint between the Association and the Society. Panels and committees seemed to be everywhere...the Office of Scientific Research and Development (OSRD), the National Defense Research Committee (NDRC), the Applied Mathematics Group (AMG), the Statistical Research Group (SRG), and the Bombing Research Group (BRG)...everyone used only the initials in the current literature. Early in the war, the American Mathematical Society had urged Vannevar Bush, who led the national scientific mobilization, to make better use of mathematicians. When Bush appointed Warren Weaver to head a crucial panel, Marshall Stone, the president of the AMS, protested. The panel was unrepresentative, he complained. Weaver defended his appointments by pointing out that he needed "team players" rather than "prima donnas" who are "ornaments of a peaceful civilization."

It was the culmination of years of resentment, and the beginning of a gentle tension in American mathematics for the next 50 years. In 1938, G. D. Birkhoff had complained about the disregard of applied mathematics, and in 1943 Richardson at Brown advised (in the Monthly) that "steps should be taken to strengthen one of the weakest links in the American scientific chain." The result was a renewal of interest in Applied Mathematics, and the founding of special programs (the Courant Institute at NYU and the Applied Mathematics Department at Brown). At the same time, applied and pure mathematicians alike referred to Hardy's comments to lend evidence to their case.

While mathematicians were squabbling, the Monthly was thriving under one of its greatest editors, Lester R. Ford. In 1943, the Monthly published the first of a new series of articles, *What is the Ergodic Theorem?* by G. D. Birkhoff. It was a phenomenally successful series, and many more gems would follows during the next half century. The Monthly reported on the migration of mathematicians that had taken place during the preceding decade. It published a regular feature, the Department of War Information by C. V. Newsom. And when the war was over, it reported on the G.I. Bill and the dramatic changes in colleges and universities throughout the country.

In 1945, Eilenberg and Mac Lane were defining categories and functors in a paper that appeared in the Transactions of the American Mathematical Society as the war drew to a close. American mathematics would be changed forever. Vannevar Bush was crafting his famous report *Science, The Endless Frontier*. Plans were being made to create permanent government funding of research, and 5 years later, the National Science Foundation would come into existence. In the meantime, the Office of Naval Research (and Mina Reese) would provide continued government funding of mathematics. Jackie Robinson broke the color barrier in major league baseball in 1947, while George Dantzig was inventing the techniques of linear programming. Alger Hiss came to trial in 1949, just after Atle Selberg produced his elementary proof of the Prime Number Theorem.

At the close of the decade, the Monthly published a beautiful, long paper by Andre Weil called *The Future of Mathematics*. It also published a short note in computer science by Claude Shannon called *A Symmetrical Notation for Numbers*. Both papers foretold the future of mathematics.

On the Modern Development of Celestial Mechanics[*]

C. L. Siegel

Celestial mechanics deals with the problem of n bodies, or in other words, with the theory of the motion of n particles $P_1, \ldots, P_n$ in three-dimensional euclidean space attracting each other according to Newton's law of gravitation. If we denote by m_k the mass of the particle P_k and by r_{kl} the distance $P_k P_l$, the potential of gravitation of the system is

$$-U = - \sum_{1 \leq k < l \leq n} m_k m_l r_{kl}^{-1}.$$

Let x_k, y_k, z_k be the rectangular cartesian coördinates of P_k; then the differential equations of the motion of P_k are

$$m_k \ddot{x}_k = \frac{\partial U}{\partial x_k}, \qquad m_k \ddot{y}_k = \frac{\partial U}{\partial y_k}, \qquad m_k \ddot{z}_k = \frac{\partial U}{\partial z_k}, \qquad (k = 1, \ldots, n).$$

This is a system of $3n$ ordinary differential equations of the second order. If we introduce the components of velocity u_k, v_k, w_k, the equations of motion may be written as a system of $6n$ differential equations of the first order, namely,

$$(1) \qquad \dot{x}_k = u_k, \dot{y}_k = v_k, \dot{z}_k = w_k, \dot{u}_k = \frac{1}{m_k} \frac{\partial U}{\partial x_k}, \dot{v}_k = \frac{1}{m_k} \frac{\partial U}{\partial y_k}, \dot{w}_k = \frac{1}{m_k} \frac{\partial U}{\partial z_k}.$$

We consider the $6n$ real values $x_k, y_k, z_k, u_k, v_k, w_k, (k = 1, \ldots, n)$, as the coördinates of a point Q in a space of $6n$ dimensions and we denote by S the manifold of all points Q for which the $n(n + 1)/2$ distances r_{kl} are different from 0. The theorem of existence for the solutions of differential equations asserts that through any point Q of S passes exactly one curve of motion. It is the main problem of celestial mechanics to study the topological and analytical properties of this manifold of stream lines in the $6n$-dimensional space S. The complete solution of this problem seems to be far beyond the power of the known mathematical methods, but interesting special results have been obtained during the last 60 years, since the original discoveries of Hill in lunar theory. I will try to give an account of some of the more important of these modern results. They are connected with the names of Bruns, Poincaré, and Sundman.

Let us begin with the investigations of Bruns. They are concerned with the integrals of the system of differential equations (1). If

$$(2) \qquad \xi_k = f_k(\xi_1, \ldots, \xi_m, t), \qquad (k = 1, \ldots, m),$$

[*]Delivered at Rutgers University, February 11, 1941, as a symposium lecture on celestial mechanics, given during the celebration of the 175th anniversary of the founding of the university.

is a system of differential equations of the first order, then an integral of this system is any function $\phi(\xi_1, \ldots, \xi_m, t)$ which is constant for all solutions of (2). From the condition $\dot\phi = 0$ we infer the relationship

$$\frac{\partial \phi}{\partial t} + \sum_{k=1}^{m} f_k \frac{\partial \phi}{\partial \xi_k} = 0;$$

hence an integral of (2) is any solution ϕ of this partial differential equation. It is proved in the theory of differential equations, that the system (2) of m differential equations of the first order reduces to a system of only $m - 1$ differential equations of the first order, if we know any integral which is not identically constant. More generally, if we know r independent integrals of (2), this system may be replaced by a system of only $m - r$ differential equations of the first order, and if we find m independent integrals, then (2) is completely solved.

Since the researches of Euler and Lagrange we know 10 independent integrals of the system (2), namely, the 6 integrals of momentum, the 3 integrals of angular momentum, and the integral of energy. The integrals of momentum are

$$\phi_1 = \sum_{k=1}^{n} m_k u_k, \qquad \phi_2 = \sum_{k=1}^{n} m_k v_k, \qquad \phi_3 = \sum_{k=1}^{n} m_k w_k,$$

$$\phi_4 = t\phi_1 - \sum_{k=1}^{n} m_k x_k, \qquad \phi_5 = t\phi_2 - \sum_{k=1}^{n} m_k y_k, \qquad \phi_6 = t\phi_3 - \sum_{k=1}^{n} m_k z_k;$$

they assert that the center of gravity of the system of particles $P_1, \ldots, P_n$ moves on a straight line with constant velocity. The integrals of angular momentum are

$$\phi_7 = \sum_{k=1}^{n} m_k(y_k w_k - z_k v_k), \qquad \phi_8 = \sum_{k=1}^{n} m_k(z_k u_k - x_k w_k),$$

$$\phi_9 = \sum_{k=1}^{n} m_k(x_k v_k - y_k u_k),$$

and the integral of energy is

$$\phi_{10} = T - U,$$

where

$$T = \frac{1}{2} \sum_{k=1}^{n} m_k(u_k^2 + v_k^2 + w_k^2)$$

denotes the kinetic energy of the system of particles. These 10 integrals of (1) have the special property that they are algebraic integrals, *i.e.*, algebraic functions of the variables $t, x_1, \ldots, w_n$. For a long time, mathematicians and astronomers tried in vain to find other simple integrals. Finally Bruns proved that there are no other independent algebraic integrals of the problem of n bodies; in other words, that any algebraic integral of the problem of n bodies is an algebraic function of the known integrals $\phi_1, \ldots, \phi_{10}$. This theorem of Bruns shows us that a further reduction of the problem of n bodies cannot be obtained by algebraic methods. The proof of Bruns's theorem is rather difficult; it uses the same ideas which led Liouville to his theorem that the elliptic functions cannot be expressed as a finite combination of exponential, logarithmic, and algebraic functions.

The researches of Sundman deal only with the special case $n = 3$, the problem of three bodies. Since the right-hand sides in (1) are analytic functions of the $6n$

variables $x_1, \ldots, w_n$, we conclude from Cauchy's existence theorem that the solutions of (1) are analytic functions of the independent variable t. We choose a fixed real value t_0 of t and consider the coördinates x_k, y_k, z_k, $(k = 1, 2, 3)$, on any curve of motion for increasing real values of $t > t_0$. There are two possibilities: Either these coördinates are regular for all values of $t > t_0$ or there exists a finite number $t_1 > t_0$ such that all coördinates are regular for $t_0 \leqq t < t_1$, but at least one coördinate is singular for $t = t_1$. Let us now investigate the behaviour of the motion of the three bodies for $t \to t_1$. Sundman proved that in the moment $t = t_1$ we have either a simple collision or a general collision; this means that either two of the three bodies dash together or all three of them dash together, at a certain point of the space. Moreover, he found that a general collision can only occur if the three integrals of angular momentum have the value zero. If we assume that this is not the case, we have a simple collision for $t = t_1$. Sundman proved that then the coördinates, as functions of t, have a branch-point of the second order for $t = t_1$; this means that they can be represented in the neighborhood of $t = t_1$ by convergent power series of the uniformizing variable $(t - t_1)^{1/3}$ with real coefficients. Hence, we may consider the analytic continuation of these functions beyond the branch-point $t = t_1$. According to the three possible determinations of the cube root, we find three different analytic continuations beyond the branch-point, and exactly one of these branches will be real for real values $t > t_1$. Choosing this real branch for every one of the nine coördinates, we obtain a real analytic continuation of the motion beyond the point of simple collision. Of course this analytic continuation has no physical meaning, but it is important for the mathematical investigation of the differential equations.

Consider now the behaviour of the analytic functions x_k, y_k, z_k for increasing real values of $t > t_1$. There are again two possibilities: Either they are regular for all finite $t > t_1$ or there exists a first singularity $t = t_2$. Since we have assumed that the integrals of angular momentum are not all zero on our orbit, the singularity $t = t_2$ is again a simple collision; that means a branch-point of the second order, and we may construct the real analytic continuation of the motion beyond this branch-point $t = t_2$. It may happen that we find in this manner an infinite number of times $t_1, t_2, t_3, \ldots$ of simple collisions. Now Sundman proved that this infinite increasing sequence $t_1, t_2, t_3, \ldots$ does not tend to a finite limit; in other words, that the times of the single collisions do not cluster at a finite value of the time. Consequently the motion may be continued for all real finite values of the time greater than the initial value $t = t_0$. The same is obviously true for decreasing values of $t < t_0$. Therefore we have a real analytic continuation of the motion for all finite real values of the time t with the following property: If $t = \tau$ is no point of collision, then the coördinates are power series of the variable $t - \tau$ in the neighborhood of $t = \tau$; if $t = \tau$ is a point of collision, then the coördinates are, in this neighborhood, power series of the variable $(t - \tau)^{1/3}$.

These power series will not converge for all values of t, but only in a certain neighborhood of the special value τ. Sundman made the important discovery that the whole motion may be represented by one single power series, if we introduce instead of $t - \tau$ or $(t - \tau)^{1/3}$ a certain new uniformizing variable s defined by

$$(3) \qquad s = \int_{t_0}^{t} (U + 1) \, dt.$$

If t runs from $-\infty$ to $+\infty$, the new variable s does the same. The coördinates x_k, y_k, z_k are now regular functions of s, for all finite real values of s, and the same holds for t as a function of s. Sundman proved that the singularities of x_k, y_k, z_k, t as functions of the complex variable s do not cluster towards the real s-axis; in other words, that there exists a certain strip containing the real s-axis which is completely free from singularities of those functions. The proof of this statement depends upon two lemmas which are of special interest. The first lemma asserts that throughout the whole motion the perimeter of the triangle $P_1 P_2 P_3$ has a positive lower bound not involving the time t, and the second lemma is the following one: Consider for any moment $t = \tau$ that point P_k which is opposite to the smallest side of the triangle, then the velocity $(u_k^2 + v_k^2 + w_k^2)^{1/2}$ of this point has a finite upper bound not depending upon τ. Applying these two lemmas, Sundman proved the existence of a strip $-\delta < I(s) < \delta$ not containing any singularity of x_k, y_k, z_k, t; here δ is a positive number depending only upon the initial conditions, and $I(s)$ is the imaginary part of s. By the substitution

$$(4) \qquad p = \frac{e^{\pi s/2\delta} - 1}{e^{\pi s/2\delta} + 1}, \qquad s = \frac{2\delta}{\pi} \log \frac{1 + p}{1 - p},$$

the strip is conformally mapped onto the unit circle $|p| < 1$. The segment $-1 < p < 1$ corresponds to the real s-axis, hence also to the real t-axis, and x_k, y_k, z_k, t are regular functions of the uniformizing parameter p in the whole unit circle $|p| < 1$. Therefore they may be expressed by power series of the variable p converging for $|p| < 1$. If p runs from -1 to $+1$, the time t runs from $-\infty$ to $+\infty$ and the whole motion is represented by those power series. This is Sundman's final result:

If the three integrals of angular momentum are not all zero, then the coördinates x_k, y_k, z_k, $(k = 1, 2, 3)$, and the time t can be represented by power series of the parameter p defined in (3) and (4). The power series converge for $|p| < 1$, and we obtain the whole curve of motion for $-1 < p < 1$.

I have spoken rather explicitly of the methods and the results of Sundman, because his important papers have been studied by only very few people. The researches of Poincaré are more widely known; they were published in his famous *Méthodes nouvelles de la mécanique céleste*. It is impossible to give in brief a complete account of the different ingenious and fertile ideas of his work, and I will restrict myself to a sketch of his investigations concerning periodical orbits.

We consider again a system of differential equations of the first order,

$$(5) \qquad \xi_k = f_k(\xi_1, \ldots, \xi_m), \qquad\qquad (k = 1, \ldots, m),$$

and assume now that the functions f_k do not involve explicitly the independent variable t and that they have continuous partial derivatives of the first order, in a certain domain D. Moreover, we assume that there exists an integral not depending upon t, i.e., a function $\phi(\xi_1, \ldots, \xi_m)$ which is constant for any solution of (5); let also ϕ have continuous partial derivatives in any point of D. These conditions are fulfilled in the special case of our system (1). The general solution of (5) for the initial conditions $t = 0, \xi_k = \alpha_k, (k = 1, \ldots, m)$, has the form

$$\xi_k = g_k(t, \alpha_1, \ldots, \alpha_m), \qquad g_k(0, \alpha_1, \ldots, \alpha_m) = \alpha_k, \qquad (k = 1, \ldots, m).$$

If we know in D a periodical solution with the period $\tau > 0$ and the initial values $\xi_k = \beta_k, (k = 1, \ldots, m)$, then the relationship

$$(6) \qquad g_k(\tau, \beta_1, \ldots, \beta_m) = \beta_k, \qquad\qquad (k = 1, \ldots, m),$$

holds. By the theorem of uniqueness, the condition (6) is also sufficient for periodicity with the period τ. We consider all orbits through points in the neighborhood of the point $Q_0 = (\beta_1, \ldots, \beta_m)$ and try to find other periodical solutions in D with a slightly different period σ. Let us assume that the given closed orbit through Q_0 is not tangential to the plane $\xi_1 = \beta_1$; this means that $f_1(\beta_1, \ldots, \beta_m) \neq 0$. The solution passing for $t = 0$ through the point $Q = (\beta_1, \alpha_2, \ldots, \alpha_m)$ of that plane cuts it for a second time $t = \sigma > 0$ in a point $(\beta_1, \xi_2, \ldots, \xi_m)$, and σ lies in an arbitrarily small neighborhood of τ, if only the differences $\beta_k - \alpha_k$, $(k = 2, \ldots, m)$, are sufficiently small. This orbit through Q will be closed if $\sigma, \alpha_2, \ldots, \alpha_m$ satisfy the m equations $h_1 = 0, \ldots, h_m = 0$ where $h_1, \ldots, h_m$ denote the following functions of $\sigma, \alpha_2, \ldots, \alpha_m$:

$$h_1 = g_1(\sigma, \beta_1, \alpha_2, \ldots, \alpha_m) - \beta_1, \qquad h_k = g_k(\sigma, \beta_1, \alpha_2, \ldots, \alpha_m) - \alpha_k,$$

$$(k = 2, \ldots, m).$$

If we assume that not all partial derivatives $\partial\phi/\partial\xi_k$, $(k = 1, \ldots, m)$, of the integral ϕ vanish at the point Q_0, we infer from the relationship

$$\phi(h_1 + \beta_1, h_2 + \alpha_2, \ldots, h_m + \alpha_m) = \phi(\beta_1, \alpha_2, \ldots, \alpha_m)$$

that one of the m equations $h_k = 0$ follows from the other $m - 1$, for sufficiently small $\beta_k - \alpha_k$. Consequently we have only to solve $m - 1$ equations $h_k = 0$ for the m unknown quantities $\sigma, \alpha_2, \ldots, \alpha_m$, and we know the particular solution $\sigma = \tau$, $\alpha_k = \beta_k$, $(k = 2, \ldots, m)$. By a well known theorem concerning implicit functions, our system of equations has for any given σ in a sufficiently small neighborhood of τ a uniquely determined solution $\alpha_2, \ldots, \alpha_m$, if the functional determinant of the $m - 1$ left-hand sides h_k as functions of the variables $\alpha_2, \ldots, \alpha_m$ does not vanish for $\sigma = \tau, \alpha_2 = \beta_2, \ldots, \alpha_m = \beta_m$. Under this last assumption we obtain a one-parameter manifold of closed orbits in the neighborhood of the given closed orbit.

If we want to apply this method of Poincaré, we have to know already a periodical solution, and the problem arises how to find such an initial solution. This problem is of a different character, and Poincaré tried to solve it by topological methods. Let us consider the solution of (5) through any point Q of the surface $\xi_1 = \beta_1$ for increasing values of t, and let us assume that it cuts again this surface at a point Q'. In this manner a topological mapping of the surface onto itself is defined, and obviously the periodical solutions correspond to the fixed points $Q = Q'$ of this mapping. The problem of finding closed orbits is therefore transformed into the problem of proving the existence of fixed points under a topological mapping of a surface onto itself. Poincaré suggested that under certain conditions a fixed point will really exist, and Birkhoff later proved this suggestion. In his researches on surface transformations, Birkhoff obtained several other results which have interesting applications to dynamical problems.

I hope to have explained that some important steps have been made since the first ingenious researches of Hill. However, there remain still a great number of unsolved problems in celestial mechanics, *e.g.*, the problems of stability and transitivity, and it seems that the solution of the main problems will require new methods of analysis.

TRIVIA MATHEMATICA

According to reports from the recent Columbus meetings of the Association and the Society, a new publication, *Trivia Mathematica*, is about to emerge from its chrysalis. At one of those profound meetings which last far into the night, learned savants sired and damned this new idea. Reports, though meager, place Brown and Princeton as focal points, with Cornell and M.I.T. as vortices of the hyperboles. Apparently without reMORSE, Princeton has added HURWITZ to those of divine Providence to FLOOD Cambridge and Ithaca with puns. No one has been judged best at this game, we understand, but doubtless by general consent they place WIENERworst.

E.J.M.

647(1940), 43

The Division of Mathematics at Harvard University has awarded the William Lowell Putnam Memorial Scholarship for 1940 to A. M. Gleason of Yale University.

47(1940), 330

For the summer quarter, 1940, Stanford University has appointed Professor J. D. Tamarkin of Brown University and Professor Emil Artin of the University of Indiana. Professor Tamarkin will lecture on "Modern Theory of Integration" and "Selected Topics from Elementary Mathematics", while Professor Artin will give a course on "Algebraic Numbers."

47(1940), 121

E402 [1940, 48]. *Proposed by Irving Kaplansky, Harvard University.*

If n, r, and a are positive integers, the congruence $n^2 \equiv n \pmod{10^a}$ obviously implies $n^r \equiv n \pmod{10^a}$. (When such a number n has only a digits, it is called an automorphic number.) For what values of r does $n^r \equiv n \pmod{10^a}$ imply $n^2 \equiv n \pmod{10^a}$?

47(1940), 572

The Addition Formulas for the Sine and Cosine

E. J. McShane

Off and on, for some years, I have tried to find a proof of the addition formulas for the sine and cosine which would have the following three properties. (1) It should be valid for all angles, and not involve any discussion of the quadrants in which the angles lie. (2) It should not require previous knowledge of the formulas for the functions of $n \cdot 90° \pm A$ in terms of functions of A. (3) It should not be too difficult for first-year students to follow. The proof below, which I have not seen published, satisfies (1) and (2), and perhaps comes as close to satisfying (3) as any other. It requires a knowledge of the distance formula, of the general definitions of the trigonometric functions, and of the equations

$$\cos^2 \theta + \sin^2 \theta = 1,$$
$$\cos 0° = \sin 90° = 1, \qquad \cos 90° = \sin 0° = 0.$$

From the definitions of the sine and cosine we deduce at once that if P is a point whose distance from the origin is r and for which the angle of OP with the positive x-axis is θ, the coördinates of P are $(r \cos \theta, r \sin \theta)$.

Let A and B be any two angles. With a vertex O and a half-line OW as a beginning we construct angles A and B, and on their terminal half-lines we choose points P and Q respectively, each at distance 1 from O. Let d denote the distance from P to Q. We shall now make two computations for d^2, using first OW and then OQ as x-axis.

First using OW as x-axis, we find that the coördinates of P and Q are $(\cos A, \sin A)$ and $(\cos B, \sin B)$ respectively, since they each have distance 1 from the origin O and the half-lines OP, OQ make the respective angles A, B with the positive x-axis. Hence

$$d^2 = (\cos A - \cos B)^2 + (\sin A - \sin B)^2$$
$$= 2 - 2[\cos A \cos B + \sin A \sin B].$$

Next we use OQ as positive x-axis. The half-lines OP, OQ now make the respective angles $A - B, 0$ with the positive x-axis, so P and Q have coördinates $(\cos (A - B), \sin(A - B))$ and $(1, 0)$ respectively. Hence

$$d^2 = (\cos (A - B) - 1)^2 + \sin^2(A - B)$$
$$= 2 - 2\cos(A - B).$$

Equating the two expressions for d^2 yields

(1) $$\cos(A - B) = \cos A \cos B + \sin A \sin B.$$

If we wish to use the formulas for the functions of $n \cdot 90° \pm A$ in terms of functions of A, the formulas for $\cos(A + B)$ and $\sin(A \pm B)$ can quickly be deduced from (1). However, we do not need to use the formulas for the functions of $n \cdot 90° \pm A$; the equations mentioned in the introduction, together with (1), are enough.

In (1) we set $A = 0$, obtaining

$$(2) \qquad\qquad \cos(-B) = \cos B.$$

Again, by setting $A = 90°$ in (1) we find

$$(3) \qquad\qquad \cos(90° - B) = \sin B.$$

In (3) we set $B = 90° - C$; this yields

$$(4) \qquad\qquad \sin(90° - C) = \cos C.$$

By (3), and (4) and (1), we have

$$(5) \qquad \sin(A + B) = \cos[90° - (A + B)]$$
$$= \cos[(90° - A) - B]$$
$$= \cos(90° - A)\cos B + \sin(90° - A)\sin B$$
$$= \sin A \cos B + \cos A \sin B.$$

In (3) we set $B = -A$, obtaining

$$(6) \qquad\qquad \sin(-A) = \cos(90° + A).$$

Now from (6) and (1), we deduce

$$(7) \qquad\qquad \sin(-A) = \cos[A - (-90°)]$$
$$= -\sin A.$$

By (1), (2), and (7) we have

$$(8) \qquad \cos(A + B) = \cos[A - (-B)]$$
$$= \cos A \cos(-B) + \sin A \sin(-B)$$
$$= \cos A \cos B - \sin A \sin B;$$

and likewise by (5), (2), and (7) we find

$$(9) \qquad\qquad \sin(A - B) = \sin A \cos B - \cos A \sin B.$$

Since we have not used the formulas for the functions of $n \cdot 90° \pm A$ in deducing (1), (5), (8), and (9) we can use the latter, with the known values of $\sin(n \cdot 90°)$ and $\cos(n \cdot 90°)$, to deduce the former.

What is the Ergodic Theorem?

G. D. Birkhoff

The integral of Lebesgue (1901), founded upon Borel measure, has been a dominating weapon in the striking advance of Analysis during the present century. Perhaps the Ergodic Theorem (1931) is destined to hold a central position in this development. Indeed, Wiener and Wintner in a recent article* refer to it as "the only result of real generality established for the solutions of dynamical systems."

To understand the theorem and the nature of its applications it is necessary first of all to say something about (Borel-Lebesgue) measure, *i.e.*, "probability" in the sense sketched by Poincaré in the third volume of his *Méthodes Nouvelles de la Mécanique Céleste*. We restrict ourselves to the case of a line segment of unit length with coördinate x, $0 \leqq x \leqq 1$. Suppose that we have a set of non-overlapping intervals, finite in number and of total length $l < 1$ in this segment. The probability in a certain intuitive sense that a point, *taken at random*, lies in one of these intervals, is l; and the probability that it lies in the complementary set is of course $1 - l$.

Now suppose that we are given a point set M containing an infinite number of points, which can be enclosed within an infinite set of non-overlapping intervals of lengths $l_1, l_2, \ldots$ of total length.

$$l_1 + l_2 + l_3 + \cdots = l < 1.$$

Then clearly the probability that a point, taken at random, lies in M, cannot exceed l; and the probability that it lies in the complementary set is at least $1 - l$. If now M is of such a nature that it can be enclosed in an infinite set of intervals of total length not exceeding an arbitrarily small quantity ϵ, it is apparent that the probability of a random point falling in M does not exceed ϵ, *i.e.* the probability is 0. Such a set M is said to be of measure 0.

For instance, the set of rational points $x = m/n$ which is everywhere dense on the line segment, is of measure 0. In fact these points may be arranged in order

$$0, 1; \tfrac{1}{2}; \tfrac{1}{3}, \tfrac{2}{3}; \tfrac{1}{4}, \tfrac{3}{4}; \tfrac{1}{5}, \tfrac{2}{5}, \tfrac{3}{5}, \tfrac{4}{5}; \cdots$$

and the nth one of these points may obviously be enclosed within an interval of

---

*On the ergodic dynamics of almost periodic systems, American Journal of Mathematics, vol. 63, 1941. For an introduction to the literature see Eberhard Hopf's "Ergodentheorie," Ergebnisse der Mathematik und ihrer Grenzgebiete, Berlin, Springer, 1937. Our discussion here deals only with the "Ergodic Theorem," and not at all with the "Mean Ergodic Theorem" of von Neumann, which stimulated me to reconsider some old ideas, and so led me to the discovery and proof of the Ergodic Theorem, embodying a strong, precise result which, so far as I know, had never been hoped for.

length $\epsilon/2^n$. Since we have

$$\frac{\epsilon}{2} + \frac{\epsilon}{4} + \frac{\epsilon}{8} + \cdots = \epsilon,$$

it is evident that this set of rational points is of measure 0.

More generally, if we have a set M such that it can be enclosed within a set of intervals of length $l_1, l_2, \cdots$ with

$$l_1 + l_2 + \cdots \leqq l + \epsilon$$

while the complementary set $\overline{M}$ can be enclosed similarly within intervals $\bar{l}_1, \bar{l}_2, \ldots$ with

$$\bar{l}_1 + \bar{l}_2 + \cdots \leqq (1 - l) + \epsilon$$

for $\epsilon > 0$ arbitrarily small, then $\overline{M}$ is said to be measurable of measure l; and its complementary set M will then clearly be measurable of measure $1 - l$. In this case the probability that a random point falls in M is obviously to be regarded as l.

All ordinary infinite sets specifically defined by analytic methods are found to be measurable in this sense.

The gist of the Ergodic Theorem can now be illustrated by means of our line segment.

Suppose that there is given any one-to-one *measure preserving* transformation T of the line segment $0 \leqq x \leqq 1$ into itself; T may have a finite or infinite number of discontinuities. A first simple example is the following: Imagine the line segment $0 \leqq x < 1$ bent into a circle of circumference 1, without any stretching; the first transformation T is merely a rotation of this circle through a certain angle α. A second example is the following: The line segment is divided into the infinite set of intervals,

$$0 \leqq x < \tfrac{1}{2}; \tfrac{1}{2} \leqq x < \tfrac{3}{4}; \tfrac{3}{4} \leqq x < \tfrac{7}{8}, \ldots$$

and then the second interval is interchanged with the first, the fourth with the third, etc., thus defining the transformation T. In both cases T is evidently of the stated type, and measure is preserved.

The Ergodic Theorem then says: *For any such measure-preserving transformation T, and for each individual point P (except possibly an exceptional set of measure 0), there is a definite probability that its iterates under T, from P on, namely*

$$P, T(P), T^2(P), \ldots \quad \text{and} \quad P, T^{-1}(P), T^{-2}(P), \ldots$$

fall in a given measurable set M.

In other words the proportion of n of these points (beginning with P) which lie in the set M tends toward a definite limit μ_p, as n approaches infinity in either direction.

More generally, a line segment may be replaced by a finite volume M of n-dimensions, $n > 1$, and the points of M may be assigned a variable (integrable) positive weight, $w(P)$. The generalized theorem would then assert that the corresponding weighted means tend toward a limit μ_p. In the simple special case first stated, this weight is 1 for the points of M and 0 for the points not in M. Or, again, for $n > 1$ the discrete transformation T may be replaced by a steady measure-preserving flow T_t in time t, and the analogous theorem holds.

To illustrate this last possibility, suppose that in the square $0 \leqq x < 1, 0 \leqq y < 1$, the points move with a uniform velocity in a fixed direction, making an angle α with that of the x axis, and leaving the square to return at the homologous point

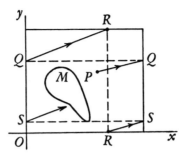

FIG. 1.

(see the adjoining figure). Evidently such a transformation T_t is area-preserving. Let now M be any selected measurable part of the square, and let P be any point of the square—aside always from a possible exceptional set of measure 0. On the basis of the same theorem, there is a definite probability in infinite time, $t \geq 0$ or $t \leq 0$ that $P_t = T_t(P)$ falls within M, and this probability is the same in both directions. More generally a weight $w(P)$ may be introduced in the case of a "flow" as well as in the discrete case.

In more analytic garb, the theorem states in the two cases respectively that for $n \to \pm\infty, T \to \pm\infty$:

$$\frac{w(P) + w(T(P)) + \cdots w(T^{n-1}(P))}{n} \to \mu_P; \qquad \frac{1}{T}\int_0^T w(P)\,dP \to \mu_P.$$

The kind of applications to dynamical systems which the Ergodic Theorem affords are exceedingly varied and interesting. Take the simple example of an idealized convex billiard table on which an idealized billiard ball P moves with velocity 1. In the figure let $\phi = \text{arc } OA$, ϕ_1, $= \text{arc } OA_1$, $l = AP$, $l^* = AA_1$. We have a transformation $(\theta_1, \phi_2) = T(\theta, \phi)$ defined over a rectangle

$$0 < \theta < \pi; \qquad 0 \leq \phi \leq p, \qquad (p = \text{perimeter of table})$$

in the $\theta\phi$-plane, associated with the motion. It is not hard to prove that T is measure-preserving in the sense that the double integral

$$\int\int \frac{\sin\theta}{\sin\theta_1}\,d\theta\,d\phi$$

has the same value when extended over any measurable part of this rectangle as over its image under T; indeed it would be possible to deform the rectangle so that, over the new region, ordinary areas are preserved.

Furthermore it is clear that, if we associate with any "state of motion" of the billiard ball, as of P, the three coordinates θ, ϕ, l then a steady flow T_t is defined in the corresponding region of three-dimensional $\theta\phi l$-space:

$$0 < \theta < \pi; \qquad 0 \leq \phi < p, \qquad 0 \leq l \leq l^*$$

in which the following volume integral is preserved:

$$\int\left(\int\int \frac{\sin\theta}{\sin\theta_1}\,d\theta\,d\phi\right) dl.$$

Thus the theorem applies to this flow.

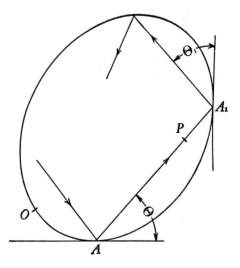

Fig. 2.

Here are three obvious applications to this simple but typical dynamical problem:

(1) the average length of n successive chords of the path tends to a definite limit, the same whether the time t increases or decreases;

(2) the average angle θ at n successive collisions tends to a definite limiting value;

(3) the billiard ball tends in the limit to lie in any assigned area of the table a definite proportion of the time.

There is one especially interesting case, which may in fact be the "general case" as far as we know: It may happen that all of the points of our volume behave in essentially the same way in the mean (aside always from the excepted set of measure 0, of course). If they do not so behave, the underlying space can be subdivided into *invariant* measurable sets; thus for an elliptical table, the motions lying wholly in the ring outside a smaller confocal ellipse form such a closed invariant set; and this is an integrable problem—a limiting case of geodesics on a flattening ellipsoid.

What the Ergodic Theorem means, roughly speaking, is that for a discrete measure-preserving transformation or a measure-preserving flow of a finite volume, probabilities and weighted means tend toward limits when we start from a definite state P (not belonging to a possible exceptional set of measure 0), and, furthermore, the limiting value is the same in both directions.

The Ergodic Theorem applies to manifold deep problems of analysis and of applied mathematics—as well to the solar system as to our simple billiard ball problem! Thus in G. W. Hill's celebrated idealization of the earth-sun-moon problem (the restricted problem of three bodies) we can at once assert (with probability 1) that the moon possesses a true mean angular state of rotation about the earth (measured from the epoch), the same in both directions of the time.

The Migration of Mathematicians

Arnold Dresden

In 1785, Joseph Priestley was elected to membership in the American Philosophical Society. In June 1794, at the age of 61, he came to America and settled in Northumberland, Pennsylvania. At its meeting of June 20, 1794, the American Philosophical Society appointed a committee to prepare a congratulatory address to Dr. Priestley. At the same meeting the committee reported a draft, which was adopted "and the officers of the Society with as many of the other members as can conveniently attend are directed to meet at the Hall tomorrow afternoon at 1 o'clock in order to present the same." In the minutes of the meeting of July 18, 1794 it was "reported that a number of the officers and members...waited on Dr. Priestley 'and presented the address.'" It is as follows:

To Joseph Priestley, LL.D., &c.

The American Philosophical Society, held at Philadelphia, for Promoting Useful Knowledge, offer you their sincere congratulations on your safe arrival in this country. Associated for the purposes of extending and disseminating those improvements in the sciences and the arts, which most conduce to the substantial happiness of man, the Society felicitate themselves and their country, that your talents and virtues have been transferred to this Republic. Considering you as an illustrious member of this institution, your colleagues anticipate your aid, in zealously promoting the objects which unite them; as a virtuous man possessing eminent and useful acquirements, they contemplate with pleasure, the accession of such worth to the American Commonwealth; and looking forward to your future character of a citizen of this your adopted country, they rejoice in greeting, as such, an enlightened Republican.

In this free and happy country, those unalienable rights, which the Author of Nature committed to man as a sacred deposite, have been secured. Here, we have been enabled, under the favour of Divine Providence, to establish a government of laws and not of men; a government, which secures to its citizens equal rights and equal liberty; and which offers an asylum to the good, to the persecuted, and to the oppressed of other climes.

May you long enjoy every blessing, which an elevated and highly cultivated mind, a pure conscience, and a free country are capable of bestowing.

By order of the Society,
DAVID RITTENHOUSE *Pres*.

PHILADA, June 20th 1794

It is the spirit of this congratulatory address which I should like to recapture in this brief report on the mathematicians from abroad who have come to America in consequence of the events of the last ten years. It is meant to be congratulatory both to the newcomers and to us who were here to receive them, having come at birth or later. American mathematicians would like to wait on the good who have sought asylum from persecution and oppression, whose worth has acceded to the American Commonwealth. And this welcome is to be extended not only to those of high eminence in the profession, but to all who intend to make their contribution to American life as professional mathematicians. It is but natural that great honor is accorded to those who have already made contributions of great value and whose

147

association with American mathematics gives more than promise for the future. But it is a truism that rank and file men are needed as well as captains, colonels and generals. American mathematicians are "associated for the purposes of extending and disseminating" the beauty and value of their science and they "felicitate themselves and their country" that so many "talents and virtues" have become united with them.

Thus no attempt is made to evaluate the significance of the individuals whose names appear; in each group the names are placed in alphabetical order. While it is a regrettable fact that in 1942 it is still necessary to open an asylum to the persecuted and oppressed, we have good reason to congratulate ourselves that the shortage of teachers of mathematics, needed for the instruction of members of the armed forces, may be met, at least in part, by using the talents of those who have come, whatever be the level of eminence to which these talents entitle them.

No one should minimize the difficulties caused by the arrival during a short period of time of a large number of Priestleys. It requires all the resources of administrative statesmanship which we can command to make practically effective the welcome which we want to give them. This problem constitutes a challenge to the strength of our organization, and it is a source of satisfaction that to such a large extent we have succeeded in solving it.

The record which follows indicates how far we have succeeded. It will be seen that several of the newcomers have not yet been placed; and that many more have not found situations in which they can contribute their gifts and accomplishments most effectively. Perhaps this report will help to bring about more satisfactory placements. There is no doubt that the increase in our resources is an event of the first magnitude in the history of American mathematics. If properly utilized it should give an impetus to the development of American mathematics, whose effects will carry us forward to great heights.

As with most historical events of importance, there were forebodings of what was to come, for some years before 1933. There are many who can be looked upon as forerunners of the great migration which began in 1933. The names of Bohnenblust, von Kármán, Landé, von Neumann, Radó, Schoenberg, Seidel, Shohat, Struik, Tamarkin, Uspensky, Wigner, Wintner, Zariski, have long been familiar to American mathematicians. They are men who have become completely absorbed in the scientific life of the country and who have enriched it in a significant way.

49(1942), 415–429

THEORY AND PRACTICE

I have given you one reason why this bottleneck has arisen. There is another reason which goes very deep. It is our national suspicion of theory, on the part of the general public. We are perilously lowbrow. This is dangerous in a democracy where the great motivating forces must come from the people. One result has been a lack of coöperation between the theoretically-minded scientist and the practically-minded scientist. The pure scientists have intensified their study of science for science's sake, and the applied scientists have adhered to "common sense" and the laboratory. It is one of the problems of education to show that the more mature and socially-minded way is to respect both theory and practice, and particularly their combination.

Marston Morse and William L. Hart

48(1941), 294

What is a Curve?

G. T. Whyburn

1. Introduction. When the searching light of modern mathematical thinking is focused on the classical notion of a curve, this idea is found to involve elements of vagueness which must be clarified by accurate and exact definition. Fortunately this has been made possible and relatively simple by development in the field of set-theoretic topology. We shall endeavor to set forth below, first the need for explicit definition of a curve, then the definition itself, and finally several illustrations of types of simple curves which can be completely characterized by their topological properties and which more nearly approach the classical notion of a curve.

2. The classical notion. The concept of a curve as the "path (or locus) of a continuously moving point" usually is accompanied by intuitive notions of *thinness* and *two-sidedness*. When the curve is in a plane, these were thought to be consequences of the rather vaguely formulated definition of a curve as just given.

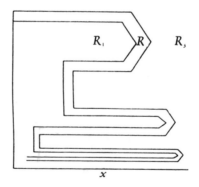

Fig. 1.

That the path of a continuously moving point is not necessarily a thin or curve-like set was shown by Peano and somewhat later by E. H. Moore, who demonstrated the remarkable fact that a square plus its interior can be exhibited as the continuous image of the interval. In other words, if S denotes a square plus its interior, we can define continuous functions $x(t)$ and $y(t)$ on the interval $0 \leq t \leq 1$ so that as t varies from 0 to 1, the point $P[x(t), y(t)]$ moves continuously through all the points of S.

A still more striking result in this direction is the remarkable theorem proved independently by Hahn and Mazurkiewicz about 1913. This theorem asserts that in order for a point set M (in euclidean space of any number of dimensions) to be representable as the continuous image of the interval $0 \leq t \leq 1$, it is necessary and sufficient that M be a locally connected continuum. (A *continuum* in euclidean

149

space is a closed, bounded, and connected set; and a continuum M is *locally connected* provided that for any $\epsilon > 0$ a $\delta > 0$ exists such that any two points x and y of M at a distance apart $< \delta$ can be joined by a subcontinuum of M of diameter $< \epsilon$). Thus since obviously not only a square but also a cube, an n-dimensional interval, an n-dimensional sphere and a multitude of other sets are locally connected continua, any such set M can be represented as the path of a continuously moving point in the sense that we can define continuous functions

$$x_i = x_i(t) \qquad\qquad 0 \leq t \leq 1, i = 1, 2, \ldots, n,$$

such that as t varies from 0 to 1 the point P with coordinates $(x_1, x_2, \ldots, x_n)$ moves continuously through all the points of M.

Even when a set is sufficiently "thin" or "1-dimensional" that we would probably call it a curve it may be in a plane and still not be two-sided. To illustrate we note that in Figure 1 any point on the base of the continuum, such as x, is a boundary point of each of the three regions R_1, R_2, R_3 into which the continuum divides the plane. Hence there are *three sides* of the base of this continuum. (Clearly we could add extra oscillating curves to the figure so as to make an arbitrarily large number or even an infinite number of regions each having all base points x on their boundaries). Nevertheless our continuum is a thin 1-dimensional set made up of an infinite number of line segments. Now it is possible to construct in a plane a continuum which is thin in the sense that it will not contain the interior of any circle and yet is so unusual that it will divide the plane into any finite number or an infinite number of regions and, further, it will be the boundary of each one of these regions. Also a plane continuum can be constructed which not only itself cuts the plane into infinitely many regions but has the remarkable property that every subcontinuum of it (any "piece" of it) also cuts the plane into infinitely many regions.

3. Dimensionality. General definitions of curve, surface, solid. Undoubtedly sufficient evidence has been given of the necessity of being precise in our definitions and statements concerning curves, surfaces, etc., and of the unreliability of our intuition concerning these concepts.

We leave aside the continuous traversibility of the set as a criterion characterizing or distinguishing between curves, surfaces, solids, etc., since we have seen how it fails in this respect, and concentrate on content or dimensionality of the set as a guide.

Hence it seems natural and adequate to define a *curve* as a 1-dimensional continuum, a *surface* as a 2-dimensional continuum and a *solid body* as a 3-dimensional continuum.

These definitions are satisfactory provided we give an adequate definition of dimensionality of a set. To this end let us concentrate our attention on compact sets, *i.e.*, sets K which have the property that any infinite subset has a limit point belonging to K, sets which are closed and bounded if they lie in a euclidean space.

We then define the dimensionality of the empty set to be -1 and agree the dimensionality of any other set is to be ≥ 0. Assuming, then, that we have defined the dimensionality concept for dimensions $\leq n - 1$, by induction we define a set K to be of dimensionality n provided (1) every pair of distinct points p and q of K can be separated in K by some set X of dimensionality $\leq n - 1$, *i.e.*, $K - X$ falls into two separated sets K_p and K_q containing p and q respectively; and (2) some pair of points of K cannot be separated in K by a subset of K of dimensionality $< n - 1$. Thus for $n \geqq 0$, a set K is of dimension n provided n is the least integer such that every pair of distinct points of K can be separated in K by the removal of a subset of dimension not greater than $n - 1$.

According to this definition, then, a compact set K is of dimension 0 provided every two points of K can be separated in K by omitting the empty set, *i.e.*, provided they are already separated in K. Hence a 0-dimensional set is one which is non-empty but is totally disconnected in the sense that its only connected subsets are single points. A compact set K is 1-dimensional provided any two points can be separated in K by omitting from K a 0-dimensional or totally disconnected set but some two points cannot be separated without omitting some points from K. A compact set K is 2-dimensional provided each pair of points of K can be separated in K by omitting a 1-dimensional set but not every pair can be separated by omitting a 0-dimensional set, and so on.

Stated in other terms, if we accept our definition that a curve is a 1-dimensional continuum, a surface is a 2-dimensional continuum, and a solid body is a 3-dimensional continuum, we see that a non-empty compact set K is 0-dimensional if every pair of its points are separated in K. The set is 1-dimensional at most provided we can (with shears if you like) separate any two of its points by cutting the set along a 0-dimensional set, *i.e.*, by cutting out only single points as connected pieces. The set is 2-dimensional at most provided we can separate any two points by cutting the set along a 1-dimensional set, *i.e.*, by cutting out only curves as connected sets. The set is 3-dimensional at most if we can separate any two points by cutting (with a saw perhaps) the set along a 2-dimensional set, *i.e.*, by cutting out only surfaces as connected sets.

4. Some simple types of curves. Having defined exactly the notions of curve, surface, and solid in terms of their topological properties in such a way that they correspond roughly to the geometrical notions of line, plane, and space, we consider now some interesting particular kinds of curves which may be similarly characterized.

Take first a straight line interval ab joining two points a and b and ask the question "What properties of a set make it essentially like an interval?" or "When are the points in a set associated together like those in the interval ab?" For example, if ab is a taut string and we release the tension and let it go back but do not allow it to loop over onto itself, it is no longer straight but it retains its same essential structure. It can still be severed by cutting out any one of its points other than a or b; and it is this property in particular which characterizes the interval completely from the topological point of view. In other words, if we understand by

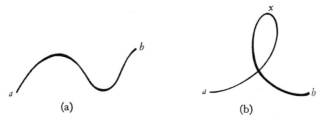

(a)   (b)

Fig. 2.

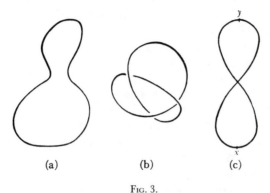

(a)   (b)   (c)

Fig. 3.

a *simple arc* any set of points which is topologically equivalent to an interval in the sense that its points can be put into one-to-one and continuous correspondence with the points of an interval, then *in order that a continuum T be a simple arc it is necessary and sufficient that T contain two points a and b such that the removal of any point of T other than a or b will disconnect T*. Thus in Fig. 2, (a) a simple arc, but (b) is not a simple arc because the removal of neither a, b, nor x will separate the set (*i.e.*, will make it fall apart).

Consider next a circle C and let us ask similar questions. If C is distorted, as was our interval, by letting it slacken and bend but not fold onto itself or be broken violently, it is seen to retain its essential set structure. It retains the property, for example, of being severed by the removal of any two of its points whatever. Here again the property mentioned is characteristic for the type of curves which are topologically equivalent to the circle. In other words, if we define a *simple closed curve* as a set which can be put in one-to-one and continuous correspondence with a circle, then *in order that a continuum C be a simple closed curve it is necessary and sufficient that C be disconnected by the omission of any two of its points*. Thus in Figure 2, (a) is not a simple closed curve since the removal of both a and b leaves the set connected. In Figure 3, (a) and (b) are simple closed curves but (c) is not a simple closed curve because the removal of x and y leaves the set connected.

A curve which is made up of a finite number of simple arcs which overlap with each other only at end points of themselves is called a *graph* or a *linear graph*. A graph, then could be regarded as being constructed by putting together in any one of numerous ways a finite number of simple arcs so that no two of the arcs will overlap anywhere except possibly at an end point of both. All of the curves illustrated in Figs. 2 and 3 are graphs; and of course many more complicated structures could be made which would still be graphs. However, if a graph is in a plane it, like the simpler curves previously discussed, will have the classical property of 2-sidedness which does not belong to all curves.

Finally, we mention two further types of curves which in general are not graphs and yet whose structure is interesting and simple, namely the *dendrite* or *acyclic curve* and the *boundary curve*. A *dendrite* is a locally connected continuum which contains no simple closed curve. It may contain infinitely many simple arcs [See Fig. 4 (a)]. In fact it may be impossible to express it as the sum even of countably many arcs, and yet it has the property that any two of its points are end points of one and only one arc in the curve. A *boundary curve* is a locally connected continuum

FIG. 4.

which can be so imbedded in a plane that it will be the boundary of a connected region of the plane. Although it is true that every dendrite is a boundary curve, in general a boundary curve will contain one and may contain infinitely many simple closed curves [See Fig. 4 (b)]. However, it is interesting to note that no such curve could contain a cross bar on a simple closed curve. In other words, the most that any two simple closed curves can overlap is in a single point (point of "tangency"). Thus any boundary curve breaks up into so called cyclic elements which are either single points or simple closed curves, no two of these have more than one common point, and these fit together to make up the curve and give it a structure relative to these elements which is very similar to that of a dendrite. [Compare Fig. 4 (a) with Fig. 4 (b)].

5. Conclusion. We have touched but a few of the many interesting aspects of the fundamental theory of curves. The subject has an extensive literature, particularly from the topological point of view, which the explorative reader will find fascinating as well as instructive. The field is a live one and it is currently receiving important contributions. Interesting and difficult problems remain unsolved. There is much to attract and repay the student who will expend the effort necessary to acquire a knowledge of these problems and to master the methods which have been devised for attacking them.

49(1942), 493–497

E 469 [1941, 266]. *Proposed by Virgil Claudian, Bucharest, Roumania.*

Show that the exradii and circumradius of a triangle satisfy the identity

$$\sum \frac{a^2(b^2 - c^2)}{r_a(r_b^2 - r_c^2)} = 4R.$$

49(1942), 123

Purdue University announces the following promotions and appointments: Dr. M. W. Keller and Dr. J. W. T. Youngs have been promoted to assistant professorships; Dr. Leonidas Alaoglu, B. H. Arnold, N. J. Fine, Dr. J. H. Giese, Dr. Michael Golomb, Dr. Ivan Niven, Dr. Maxwell Reade, and Dr. G. S. Young have been appointed as instructors.

. . .

Professor C. R. Adams has been made chairman of the department of mathematics in Brown University.

. . .

Dr. P. R. Halmos has been appointed associate at the University of Illinois.

49(1942), 489

WAR MATHEMATICS

(Three hours per week for one year; suggested by F. L. Griffin)

PURE MATHEMATICAL TOPICS

Chapter   I. Preliminary Ideas.
           II. Trigonometric Functions.
         III. Logarithmic Calculations.
         IV. Coördinates and Notions from Analytic Geometry.

ARTILLERY AND MACHINE GUN PROBLEMS

Chapter   V. Position Calculations. Methods of locating a fixed target: by direct observation; indirect observation involving trigonometry; map location; sound methods.
        VI. Ballistic Calculations. Initial firing data; adjustment of fire; probable errors; bracketing; effect of fire; velocity and angle of impact; penetration.
      VII. Safety Zones and Dead Areas.
     VIII. Barrage Fire.
      IX. Theoretical Ballistics. Discussion of the construction of firing tables.

ARMY ENGINEERING PROBLEMS

Chapter   X. Graphical Methods; Rates; Maxima; Work; Momentum.
      XI. Statics; Bridge Structures; Cranes; Inclined Planes.
     XII. Flexure of Beams; Suspension Cables.

AVIATION PROBLEMS

Chapter XIII. Principles of Flight. Stability; Equilibrium.
     XIV. Bombing.
     XV. Spherical Trigonometry and Navigation.

NOTE. For further information concerning the preceding course, consult Professor F. L. Griffin, Reed College, Portland, Oregon.

48(1941), 362

Adjustments in Mathematics to the Impact of War

G. B. Price

Frank B. Jewett said in a public address [1] early in 1942: "Without insinuating anything as to guilt, the chemists declare that this is a physicist's war. With about equal justice one might say that it is a mathematician's war." This note contains a report on how the mathematicians have participated in the war, the problems that have arisen, the measures that have been taken to meet them, the changes that have been made in the mathematics curriculum, and the effects that war has had on graduate work in mathematics [2]. It is based on information obtained through questionnaires [3] from forty-one of the fifty-four institutions in the United States and Canada that offer the Ph.D. degree in mathematics and from thirteen additional colleges and universities in the United States. No information was obtained from Canada. It is important to observe that the information was gathered between November 17 and December 15, 1942. Since the government has now announced (December 17, 1942) its program for the utilization of the colleges and universities during the war, this report can be considered a record of the first year, and of the first phase, of the participation of mathematics in the war.

The report can be summarized in the following conclusions.

1. There has been an enormous increase in enrollments in mathematics courses; although the most frequently reported increase was thirty per cent (enrollment for the first semester of 1942–43 compared with enrollment for the first semester of 1941–42), many institutions reported increases up to sixty and seventy per cent, and six colleges and universities reported increases ranging from 100 per cent to 300 per cent.

. . .

2. The percentage increase in enrollments in mathematics is much smaller for women than for men. Four women's colleges in the East reported increases ranging from twenty-five to forty per cent. Most coeducational institutions reported either no increase or only a small one; exceptions are one university, 62 per cent; three colleges and universities, 100 per cent; one university, 125 per cent. The total number of women studying mathematics is small. For example, the University of Michigan, after a ten per cent increase over last year, had only 337 women in undergraduate and first-year graduate courses in mathematics in the arts college; approximately two-thirds of this group is enrolled in first-year courses. Michigan apparently has one of the largest groups of women studying mathematics [4]. Of two large universities reporting 100 per cent increases in women in mathematics, one still has fewer than 150 in all undergraduate and graduate courses and the other has fewer than 200.

3. There exists a tremendous and acute shortage of mathematics instructors. This shortage is attested by such statements as "we could have used four additional instructors this fall" and "twenty per cent of the staff is in the army or navy," but perhaps even more by the emergency steps taken to secure instructors.

· · ·

4. Spherical trigonometry, in many cases combined with solid geometry, and the mathematics of navigation have been added to the curriculum almost universally.

5. Ten schools reported that they had added additional elementary courses, other than solid geometry and spherical trigonometry, in plane geometry, algebra, and trigonometry, or some combination of these, to enable students to remove high school deficiencies in mathematics or to obtain training needed for entrance into some branch of the armed forces. Some of this work, especially in algebra and trigonometry, has always been given in many institutions.

6. There has been practically no change in the content or organization of the foundation courses in mathematics (algebra, trigonometry, analytic geometry, and calculus); in continuing to teach the fundamentals in courses which have long been standard the mathematicians have carried out the recommendations of the Army, the Navy, and of their own War Preparedness Committee.

7. Fifteen institutions reported that they had dropped some of their undergraduate courses, especially advanced courses in pure mathematics but also mathematics courses for special groups of students.

· · ·

8. There has been a big increase in teaching and research in applied mathematics: fifty-seven *new courses* (other than those named in 4 and 5 above) in approximately twenty different subjects in applied mathematics are being given in the fifty-four institutions reporting. There are eight courses in mathematics of artillery fire or exterior ballistics; eight courses in cryptography and cryptanalysis; five courses in principles of mechanics; five courses in mathematics for meteorology; and many others with smaller frequencies. In addition, there was inaugurated at Brown University in the summer of 1941 a notable program of Advanced Instruction and Research in Mechanics; Brown will begin publication of *Journal of Applied Mathematics* early in 1943. The University of Wisconsin has established a four-year course leading to the degree of Bachelor in Applied Mathematics and Mechanics.

9. Mathematicians are doing a large amount of teaching in training schools for the Army and Navy, in ESMWT courses, and in refresher courses for those about to enter the army and navy.

· · ·

10. Five institutions have provided special courses involving mathematics for the training of women; in general, however, there have been no special provisions for women and no concerted efforts to enlist them in the study of mathematics. Purdue is introducing a new sequence of courses to prepare women as statistical workers. Chicago has a special course in mathematics for women students of electronics. The University of Pennsylvania is providing a special course in engineering drafting which requires mathematics. The University of California at Los Angeles has blocked out several curricula designed to prepare women as aircraft workers [6]; these include plane and solid analytic geometry, engineering mechanics, descriptive geometry, and so on. The University of New Mexico is planning a two-year specialized program for women that will involve the College of Engineer-

ing and the Departments of Mathematics and Physics. Smith College has plans under way for a summer school next summer that will likely emphasize mathematics and science. All the information indicates a strong demand for women trained in mathematics and an increasing tendency for them to enter the field. Positions on all levels are open to them. The case is cited of a young woman who graduated from a university in the mid-west with an A.B. degree in mathematics (and a Phi Beta Kappa key!) and who obtained an attractive position in the Research Laboratory of the United Aircraft Corporation at East Hartford, Connecticut.

11. Graduate work in mathematics is rapidly approaching the vanishing point. One school reported that such work had been abolished; in many others it is rapidly disappearing as a result of (a) the loss of graduate students, (b) the employment of graduate instructors on war research and for elementary instruction, and (c) the use of graduate students as instructors. One large university reports that it is offering only ten graduate courses this year instead of the usual fifteen, and that only five are planned for next year. Many institutions reported decreases in enrollment in graduate courses in mathematics ranging from fifty to seventy-five per cent. The following are the actual numbers of graduate students in mathematics reported at ten institutions: Brown, 20; Cincinnati, 5; George Washington, 20; Iowa State, 11; University of Iowa, 11; Lehigh, 5; Northwestern, 12; Ohio State, 11; St. Louis, 18; Virginia, 11. But even these figures are misleading; the information indicates that most of these *students* are actually part-time or full-time *instructors* or *research workers*. A careful analysis of the information indicates that only two of these ten institutions are giving essentially a full program of graduate courses to full-time students. M.I.T. reports that the only students attending graduate classes are (a) undergraduates taking advanced work, (b) staff members from the mathematics or other departments, and (c) men employed on war research projects.

12. One school reported that both the mathematics club and the department colloquium had been discontinued; two others reported that the departmental colloquium had ceased to function. Twelve other institutions reported either fewer meetings of their club and colloquium or more programs devoted to applied mathematics and subjects related to war. These facts emphasize further the shortage of mathematicians, the extent to which they have turned from pure to applied mathematics, and their complete devotion to winning the war.

50(1943), 31–34

A SPEED TEST QUESTION. A PROBLEM IN GEOGRAPHY

E. J. MOULTON, Columbia University

How long does it take you to get a correct answer to the following mathematical problem? You may be surprised to know that most mathematicians require at least ten minutes. You are urged to time yourself for a little amusement. Read the problem carefully, solve it mentally, write your answer down, and note the time required. (We add the remark that you are assured that this is a legitimate mathematical problem, in which the earth is assumed to be a sphere.)

THE PROBLEM. *Starting at a point on the earth's surface, a man walked due south ten miles, then due east ten miles, then due north ten miles, and found that he had returned to the starting point. Where must he have started?*

51(1944), 216

COMMENTS ON THE PROBLEM IN GEOGRAPHY (p. 216)

E. J. Moulton, Columbia University

When this problem is proposed orally to mathematicians, the usual response after a short time, varying from one second to a few minutes, is that *the man must have started at the North Pole*. When a mathematician is then told that his answer is wrong—and this answer *is* wrong—he is likely either to look for some non-mathematical explanation of his error or to become slightly belligerent in defense of his answer. However, he is assured that the problem is a bona fide question in the geometry of a sphere, and that his answer is mathematically incorrect. In spite of years of training in logical thinking, he has slipped on the word "must" in the question, "Where must he have started?" He has found one point where the traveller *may* have started, but there is another point where he may have started.

On second thought, which may last ten minutes or longer, mathematicians discover infinitely many points from which the man may have started; the usual answer then given is, *the man must have started either at the North Pole or at a point on a parallel of latitude which is ten miles north of a parallel of latitude whose circumference is ten miles*. But *this answer* also *is wrong*—the man *may* have started at any one of these points, but you should not assert that he *must* have started at one of them.

A correct (but not very satisfying) answer is that

(1) *he must have started at some point on the earth's surface.*

A more satisfactory answer, which is also correct, is that

(2) *he must have started at some point of a locus S which consists of the North Pole and the circles which are ten miles north of the parallels of latitude whose circumferences are $10/n$ miles, where n ranges over the positive integers.*

This locus S is interesting. It consists of an isolated point and an infinite set of circles which have as a limit circle the parallel of latitude which is ten miles from the South Pole; this limit circle is not a part of the locus.

It is also interesting to see what modifications are required in the preceding discussion if the number ten is replaced by a number x, and to consider the locus S as a function of x on the range $x = 0$ to $x = 12,500$ (calling the circumference of the earth 25,000 miles). I consider this function of x to be one of the most interesting which I have encountered, possessing as it does some startling discontinuities.

51(1944), 220

Applied Mathematics and the Present Crisis

R. G. D. Richardson

1. Introduction. It may be useful at this time to make certain inquiries regarding the present status of applied mathematics in America and the tendencies which should be fostered for the future. This great nation, now in the throes of war, finds startling deficiencies in its material, intellectual, and spiritual resources. There is scarcity of tin and chromium; of basic scientific knowledge in fields like aeronautics; of comprehension of what part America should play in establishing a world order. We are caught in a predicament which gives us serious food for thought in contemplating the long future. Is it not our duty as mathematicians to give some aspects of these matters most searching consideration?

Why has research in the applications not kept pace with that in pure mathematics? Is the present a strategic moment for a significant advance in that field? Is there an obligation on American Science to assume a greater share of the world's progress in this particular sector?

Other problems come to mind. In terms of educational policy after the war is over, what will result as a residue of the present participation by universities in war programs? Should mathematicians consider some more comprehensive sort of organization of their interests? The physicists have their Institute which is proving markedly effective during the war. The mathematicians are well organized for publication and research but not on the promotional side; in the present crisis, Washington found that it had to turn to physicists and others for some of the help which we would have furnished had we been more completely organized.

There is at the moment a marked trend toward increasing the group of professional mathematicians whose work lies in government agencies and industrial fields rather than in teaching. When the proper time comes, can we define the word "mathematician" in such a connection so that it has a meaning in the scientific world?

2. The deficiency in Applied Mathematics. Of those mathematicians whose names are starred in *American Men of Science*, the number who are now working in the field of applications is almost negligible. The percentage has decreased with each of the six issues and the new list to appear soon will not change this picture. Only one man (E. W. Brown) interested primarily in applied mathematics has been elected to the presidency of the American Mathematical Society since 1900, though before that time nearly all the presidents were in that field. An examination of the list of doctorates conferred in the past few years shows a very small percentage (certainly not more than 8%) in the applications. The great majority of our leading institutions make no provision for carrying students to the doctorate in applied mathematics. A scrutiny of American scientific journals will show only a small (though happily increasing) number of investigations into the fundamental reaches of this sector. To be sure, mathematical physics and statistics are well represented in America; investigations in electrodynamics and thermodynamics have achieved considerable success; but in the fields of mechanics (including such branches as fluid dynamics, elasticity, and plasticity), acoustics, and optics, the names are few and mostly of those of foreign birth.

3. A surprising situation. This deficiency is surprising in a nation that considers itself practical, above all. American engineering in many of its aspects has achieved a high level of performance. In physics and in astronomy America has built instruments and achieved results which are the marvel of the world. For these reasons, it would be hard to bring it home to the layman that America has had to rely largely, and still is relying largely, on foreign countries to build up the bases on which many of the advances in our mechanical engineering rest. But that such is the fact can easily be verified by any scientist.

It is surprising, too, because American research before 1890 was weighted toward the applications. Gibbs with his epoch-making theories which form the bases for much of modern physical chemistry; Newcomb in mathematics and astronomy—in his day the best known of American scientists internationally; G. W. Hill, in mathematical astronomy, who brought renown to this country: it would seem that these products of American scholarship might well have influenced American mathematics to proceed in quite another direction from that which it actually took.

In 1900, physics and engineering in America were essentially descriptive and there were opportunities to influence the theoretical development of these subjects. These opportunities were unfortunately neglected by the mathematicians; the physicists and engineers themselves proceeded to develop the mathematics underlying their respective disciplines and with such marked success that a large proportion of the applied mathematicians of the present day are the result of their training. Moreover, physicists have cultivated almost exclusively such sectors as wave mechanics, statistical mechanics, cosmic astrophysics; and the resulting researches have been printed in physics publications and only rarely in mathematical journals. Since many of these investigations are essentially mathematical in character and have little reference to experimentation, this development strikes one not only as unfortunate but as bizarre.

4. Causes of the deficiency. In the period since 1900, America has made unprecedented strides in mathematical research as a whole. Nevertheless, while this country is now easily the world leader in many branches of pure mathematics and stands high in some divisions of applied mathematics, it lags far behind in some other divisions. In Applied Mechanics there are a few excellent men, but they are widely scattered throughout the industries and the universities. For a quarter of a century the deficiency in this direction has been realized by the leaders of mathematical thought in the country and attempts have been made by two or three universities to build up a continuing school where young men interested in certain fields of applied mathematics can get the very best broad training. The Massachusetts Institute of Technology, the University of Wisconsin, and a few other institutions have over a long period cultivated the applications; and more recently California Institute of Technology has entered this field. Industrial concerns, in particular the Bell Telephone Laboratories, have assembled groups of men skilled in applied mathematics who have made noteworthy contributions. But for one reason or another (probably in part because of the tradition built up by the mathematicians who went early to Germany for training, in part because of lack of consistent coöperation between universities, and in part because of lack of funds), these various attempts to build up strong departments of applied mathematics in the universities have not achieved complete success and nothing outstanding as a center of instruction in this field has persisted. An able presentation of these facts was made by Thornton C. Fry of the Bell Telephone Laboratories in his 1940 report on mathematics in industry, prepared under the auspices of the *Committee on Survey of Research in Industry*. This report has had wide circulation and has evoked strong expressions of approval.

As the result of the influence of European mathematicians, including Sylvester, Klein, Kronecker, and Hilbert, the prospective leaders of American mathematics—such as Moore, White, Osgood, Bôcher, Van Vleck, and Pierpont—when they came back to America after having studied abroad during the period 1890–1910, brought with them an enthusiasm for rigor which was to galvanize instruction and to develop this country into a leader in pure mathematics. But no one with enough influence was able to transport the ideas seething in the mind of Felix Klein, who realized that if Germany was to be strong, the country would have to foster institutes of applied mathematics like that in Applied Mechanics at Göttingen established under his influence.

There are many reasons why young men have gone largely into pure rather than applied mathematics. In the first place, there has been an unhappy cleavage between the groups of leaders in the two fields, to the disadvantage of both. In the second place, it is probably fair to state that it is more difficult to do significant mathematical research in applied mathematics than it is in some of the fields of pure mathematics. Furthermore, our system of mathematics in the high schools and colleges does not prepare the minds of prospective mathematicians for interest in applications in the field of mechanics, as does, for example, the British system. In the American high schools, mathematics has been divorced from physics to the detriment of both. Attitudes of mind formed in the adolescent period are not easy to change. College teaching has fostered this divorce. The exigencies of the war are changing this attitude, but it remains to be seen if the impetus given to the fusion of these disciplines will persist.

In Germany, Britain, and France there have been at least half a dozen institutions where applied mathematics was sedulously cultivated. For example, there are great research institutes for aeronautics in Germany where even before the war many hundreds of persons (mathematicians, engineers, etc.) were concentrated. To illustrate the trend, it might be pointed out that since the beginning of the present war a new Institute has been established in Göttingen to study a particular branch of aeronautics known as Problems of Unsteady Flow and it has issued numerous important memoirs. There are other great German centers also to which engineering problems that involve mathematics can be sent and where both the theoretical and the experimental phases can be explored. These centers are adequately supported by the government and the industries. We have nothing in America of precisely the same kind. It is true that there are universities and technical schools where the highest grade of engineering instruction is carried on and where research is actively prosecuted. It is true also that there have been Research Institutes to which practical engineering problems on a somewhat lower level are assigned by industrial organizations. Furthermore, in a host of private industrial laboratories there has been highgrade research carried on. And to the extent that personnel is available, these activities are being expanded for the war.

But there has been in America almost no institution to which a young man could go for the broadest training in the advanced reaches of mathematics applied to engineering, and in which he could catch the spirit of research and learn the necessary techniques. Probably there should eventually be several institutions of this sort in this vast country; but at present there are not enough mathematicians expert in this applied field to staff any considerable number of such schools. Besides, there are decided advantages in concentrating intensive instruction in a very few institutions until strong centers are established; this seems especially true in the present emergency.

5. Effect of the emergency. The war has greatly intensified the need for remedying America's inadequacies in industrial mathematics. It has made the most striking demands upon a host of industries related to war activities and has simultaneously isolated us from communication with other centers of leadership in this field. Even the British institutions where applied mathematics is cultivated are virtually shut off from us by the enormous demands made upon them by their own country.

Thus the normal needs for orderly industrial development are greatly augmented by the present national emergency. Adequate exploitation of aerodynamics and other fields bearing directly upon defense activities awaits the basic work of mathematicians. A program must have the double purpose of serving the nation's immediate war needs and of pointing the way to a means for attacking some of the more difficult problems of organization of engineering research which lie ahead. America cannot afford to lag behind either now or in the period of reconstruction when competition is bound to be of the very keenest. We should be ready to assume our rightful place in this science as in others; South America, Asia and, to a lesser extent, Europe must look to us for leadership.

6. The growing realization of need. There are many persons influential in American science who have been for some time convinced that greatly increased attention should be given to the applications; that only as theory and practice stimulate and supplement one another can either achieve permanent strength worthy of our nation. The large influx of foreigners during the past two decades has brought to our shores some outstanding figures in the applications to Mechanics; we may cite Friedrichs, Den Hartog, Kármán, Mises, Prager, Reissner, and Timoshenko. These and others form a nucleus for instruction and research which we trust betokens a far-reaching development.

The country long since has passed the stage when it is necessary to demonstrate the vital importance of fundamental scientific research in agriculture and industry. It has been a question, however, as to whether the nation would support the rapid expansion of research in a sector such as Mechanics until it could visualize clearly the possibility of adding not only to theoretical knowledge, but also to practical application. The war is increasingly proving that such contributions can be made; mathematics is serving the spectacular developments taking place in aeronautics and other war endeavors. Problems in ballistics are enlisting groups of mathematicians at Aberdeen and elsewhere; radio research makes its calls to the Radiation Laboratory and to other centers; the Taylor Model Basin has mathematical problems in ship construction; Langley Field has enlisted a number of men to work on urgent questions in aeronautics; this list could be greatly expanded. The call for men of ability and adaptability in all of these fields greatly exceeds the supply.

. . .

7. The objectives of a new development. The United States has for years been training an enormous number of undergraduate engineers and these men prove a highly useful factor in the development of our resources. On the other hand, compared with a country like Germany, there are relatively few engineers with a training as extensive as that of men in other professions. It is fair to compare the training of engineers with that of physicians, since both professions demand a theoretical and a practical knowledge, which are obtained by a combination of university instruction and practical experience under guidance. The emergency reveals how inadequate are our reserves of personnel with three of four years of graduate training in engineering. If America is to compete on the higher levels of research in engineering, it seems clear that more of our ablest young men should be getting additional training of a graduate school character. Extensive grounding in classical physics and applied mathematics must be combined with a first-hand knowledge of some important practical problems. A prospective mathematical engineer would still have to acquire practical experience, just as a physician must serve as an interne; but he would be able to progress rapidly to problems beyond the mere routine.

Men teaching mathematics in engineering schools should have an understanding of, and sympathy with, the outlook of their colleagues in the engineering faculties. Many of them should have had their work for the doctorate primarily in applied mathematics. Moreover, every candidate for the degree of Ph.D. in mathematics should have a modicum of training in applied mathematics just as he should be prepared in analysis, in geometry, and in algebra. In a country reputed to have a genius for material development, such suggestions seem very moderate. Participation in courses of high grade in the applied field would aid in giving the proper attitude to those not concerned in delving deeply themselves into this area.

The student body for such a venture must consist of men who have a flair for the practical in science as well as for the theoretical. Talented men must be enlisted at the end of their college careers as well as at later stages of their scientific development. They should exhibit unusual aptitude for the physical sciences as well as all-round competence. One potent reason why more of the abler neophytes graduating from engineering schools have not proceeded to further study is the competition of the industries; in order to enlist such men it is necessary to have adequate funds for fellowships.

. . .

9. Activities of American mathematicians relating to the war. It will be recalled that the American Mathematical Society and the Mathematical Association of America appointed a joint Committee on War Preparedness in September, 1939, only a few days after war was declared in Europe. Much effort has been given to planning the utilization of our national mathematical resources. Advocates of the Kilgore Bill now pending in Washington claim that science is poorly organized for assistance to the war effort and that altogether too small a fraction of scientists are enlisted. On the other hand, W. L. Lawrence, writing in the New York *Times* for January 3, 1943, states that 87% of mathematicians engaged in research have turned their attention to war work; the truth of the matter probably lies between these two extremes. As individuals, mathematicians have in general responded promptly to the call for assistance in the emergency. More than one hundred have left their teaching positions to give full time to research in war problems; an equal number are in uniform, engaged in research or giving instruction to the armed forces; probably as many more are engaged part-time in active investigation for the government or industry; and the great bulk of the remainder are teaching men in college programs in connection with the armed services. All this emphasis on the practical side of mathematics cannot fail to leave an impression on future instruction. The profession needs wise leadership in determining what permanent influence this diversion of effort will leave behind. It would be folly to let the pendulum swing back to the opposite pole and abandon the present gains made in scientific prestige by our participation in practical applications. We have resources ample enough to cultivate many diverse interests, including various branches of applied, as well as of pure, mathematics.

50(1943), 415–423

PERMANENT INVESTMENTS OF THE ASSOCIATION

	Par Value	Market Value Dec. 31, 1942
U. S. Savings Bonds	$1,275.00	1,468.00
U. S. Treasury 1% Notes Ser. A 1946	3,000.00	2,979.00
U. S. Treasury $2\frac{3}{4}$% Bond 1947	1,000.00	1,042.50
U. S. Government $1\frac{3}{4}$% Bonds 1948	2,000.00	2,012.00
U. S. Treasury 2% Bonds 1950	3,000.00	3,051.00
HOLC 3% Bonds Ser. A 1944–52	3,000.00	3,096.00
Phelps Dodge Corp. $3\frac{1}{2}$% conv. deb. 1952	1,000.00	1,047.50
U. S. Savings $2\frac{1}{2}$% Bonds Ser. G 1953	3,000.00	3,000.00
U. S. Savings $2\frac{1}{2}$% Bonds Ser. G 1954	8,200.00	8,200.00
Texas Power & Light Co. 5% First Mort. Bond 1956	1,000.00	1,075.00
Amer. Tel.& Tel. Co. 3% Bonds conv. 1956	2,000.00	2,140.00
Commonwealth Edison Co. $3\frac{1}{2}$% Bonds. conv. deb. 1958	2,000.00	2,180.00
N. Y. Steam Corp. $3\frac{1}{2}$% First Mort. Bond 1963	1,000.00	1,061.25
Montana Power Co. $3\frac{3}{4}$% First Ref. Mort. Bonds 1966	3,000.00	3,112.50
Gatineau Power Co. $3\frac{3}{4}$% First Mort. Bond Ser. A 1969	1,000.00	918.75
Penn. R. R. Co. $3\frac{3}{4}$% Genl. Mort. Bonds Ser. C 1970	2,000.00	1,775.00
Cols.& So. Ohio Elec. Co. $3\frac{1}{4}$% First Mort. Bonds 1970	2,000.00	2,160.00
Shawinigan W.& P. Co. $4\frac{1}{2}$% First Mort. Bonds 1970	2,000.00	2,010.00
C.& O. Ry. Co. $3\frac{1}{2}$% Ref. Mort. Bonds Ser. D 1996	3,000.00	3,060.00
Land Trust Certif., Hotel Cleveland Site	700.00	510.00
	$45,175.00	$45,898.50

The Nature of Mathematical Proof

R. L. Wilder

1. Introduction. In presuming to come before you with such a title as "The nature of mathematical proof," let me assure you that I am not doing so with the idea of presenting any new or startling facts. I do this because I think it is good for us, as mathematical specialists of one sort or another, speaking in terminologies that frequently render us obscure even to one another, to pause and reflect now and then on just what we are doing and how we are doing it. For certainly we put a great part of our time and energy into the act of proof. We ask one another, "Do you think anyone will ever prove the Fermat Theorem?" or, "Do you think anyone will ever prove the continuum hypothesis?" A host of mathematical questions would receive answers if only we were able to find proofs for the theorems which hold the keys to their solutions.

. . .

3. Mathematical dogmatism. Now we are probably quite familiar with the fact that mathematical proof is a function of the time. History shows this conclusively —Euclid would probably have complained of the lack of rigor displayed by his predecessors; Weierstrass felt it necessary to reorganize the foundations of analysis; and so on. The present is a time when it seems appropriate to reflect on the new and still uncertain elements that have come into mathematical proof. One may berate those who make their proof methods dependent on some particular mathematical philosophy, such as intuitionism, but at the same time be taking considerable license when, in order to prove a much desired result, he resorts to methods that are both novel and uncertain. We may, if we like, hide behind the excuse that a proof which uses a hypothesis which so far as anyone knows has never led to contradiction, is at least a clue to a possible mathematical fact, and as such it may encourage us to establish the fact on firmer ground. This recalls Weyl's observation to the effect that giving a non-constructive existence proof is like informing the world that somewhere there exists buried treasure but not stating where it lies!* But more to the point is the fact that there is certainly some worth in a proof which shows that the theorem is in the same category with the not generally accepted hypothesis, as, for example, when it is shown that certain theorems are equivalent to the continuum hypothesis, or just as it was shown that the choice axiom, well-ordering theorem and comparability principles are all equivalent. Nevertheless, these facts are no ground for dogmatism.

The present division of mathematical thought into schools, insofar as it degenerates into dogmatism, is not, in my opinion, healthy. On the other hand, insofar as

---

*Weyl, H., Philosophie der Mathematik und Naturwissenschaft (Teil I), Handbuch der Philosophie, Abt. IIA, Munich and Berlin, 1926, p. 41.

it leads to dispassionate discussion of fundamental principles, it is a decidedly healthy development. I cannot refrain, in this connection, from expressing my protest against what I like to call, variously, the "mathematical dogmatist" or "the mathematical fascist." He may have no religious philosophy or affiliation in the ordinary sense, yet if you venture to doubt the validity of some one of his favorite proof-methods, you may find yourself in danger of physical violence. In his eyes, you are a mathematical free-thinker, an anarchist, and paradoxically although he may be both of these in his political beliefs, he won't tolerate them in mathematics. It is as though he were the mathematical prototype of the arch-reactionary whose political and economic emotions stem from his possession of an unusually large share of the world's goods; the mathematical reactionary is motivated by his possession of a large body of mathematical theory whose foundations are in danger of attack by the mathematical revolutionist.

In the case of some individuals, it is almost as though mathematics had become a kind of religious fanaticism, rather than a labor of love. They give one a feeling that we have in mathematics the intuitionist, formalist and logistic "Theologies." Insofar as the leaders of these systems become dogmatic, just so far do they become the mathematical analogues of the Dalai Lama or the Pope of Rome. And although I am not concerned here with dogmatism in religion, I think I am justified in protesting its presence in mathematics. Of course, analogy must not be pushed too far—we don't venture to predict the creation of an annual collection to be called "Kronecker's pence" to be presented to Brouwer. But the claim to possess mathematical infallibility, the attempt to set up formalistic or intuitionistic proof rituals—any of these acts and their like, are not, I venture to say, exemplifications of the truly mathematical spirit. On the other hand, to set forth these ideas as suggestions for the cure of mathematical ills or the improvement of proof-methods in the direction of greater rigor, is, it seems to me, to proceed in a truly scientific way.

As mathematicians, dealing with a subject that must be kept scrupulously abstract, which has no material connections except in the manner of what we call "applied mathematics," we must ever be on guard against dogmatism. It is natural for the layman to regard our works with awe, to think of us as the possessors of absolute truth, since we have been singularly successful in avoiding contradiction in applied mathematics. But we must not allow this veneration with which we are regarded to tempt us to set up our own mathematical cults or political philosophies. Rather we need to practice democracy, not sneering at new ideas as dangerous to our pet mathematical theories, but regarding them as possible improvements on the existing system. I think that the worth of new ideas can safely be left, in the long run, to the judgment of the mathematical electorate. A mathematical idea that never "takes" is probably not worth taking—although we must remember that here, as in other human activities, the long range point of view should prevail. What is unpopular today may become the fashion of tomorrow —which is only further admonition to be receptive to new ideas.

And I want to point out here what seems to me to be a heartening fact, although I realize that I may be sorely trying some of my listeners, personal philosophies: Namely, that no matter if we do use, frequently, questionable methods of proof, or even make outright errors in proof, we are usually gathering mathematical fruit. That the calculus of Newton and Leibnitz had virtually no basis at all from the viewpoint of modern standards did not invalidate the calculus as mathematics. All it needed was bolstering up. Even now we argue about such items as Duhamel's

Theorem or what we mean by the differential, but no one suggests that we throw the calculus out of our mathematical libraries, for we recognize intuitively that it is a body of acceptable mathematics—still susceptible to improvement, perhaps. It is not that we consider that the theorems of the calculus have been elegantly and conclusively proved in the theory of functions. To put the matter bluntly, it is a case of our *knowing mathematics when we see it*. And we don't set out to prove a theorem in the first place unless *we think it is worth proving*.

. . .

51(1944), 309–323

AN INTERESTING THEOREM

All real numbers are uninteresting. For the integer one is evidently uninteresting. (It has only trivial representations a sum of Mth powers of integers, and trivially divides every other number.) Now assume that there are interesting integers, and let N be the least such. Then N is greater than one. Hence $N-1$ is a *very* interesting integer; for it is the first uninteresting integer whose immediate successor is not interesting. This contradiction shows that no positive integer is interesting. But every positive rational is trivially expressible in an infinite number of ways as an ordered couple of uninteresting numbers and hence is uninteresting. Since every negative rational may be obtained from an uninteresting number by a mere change of sign, no negative rational is interesting. But zero and each of the remaining real numbers is representable as the least upper bound of a set of uninteresting numbers. . . .We conclude that no real numbers are interesting. The extension to complex numbers and Hilbert space is left as an exercise to the reader.—*Morgan Ward*.

52(1945), 540

FEDERATION OF WOMEN'S CLUBS METRIC RESOLUTION

At the annual convention of the General Federaion of Women's Clubs held April 25–28, 1944, in St. Louis, the following resolution was introduced and adopted unanimously by the delegates. This organization represents 16,500 clubs and 2,500,000 individual members.

Whereas, the irregular, numerous, unwieldy, and complicated units of weights and measures used in the United States and Great Britain are a hindrance to the teaching of arithmetic, every day commercial transactions, and world trade, and

Whereas, the metric system of weights and measures has only three units; meter, liter, and gram, interrelated and decimally divided like our dollar, and

. . .

Whereas, the full adoption of the metric system by the United States would be of great benefit to this country in post-war reconstruction, in promoting international commercial relations, particularly with the countries of Latin America. Continental Europe and Asia, therefore be it

Resolved, that the General Federation of Women's Clubs in Convention assembled, April, 1944, endorses legislation in Congress for the nation-wide adoption of the metric system of weights and measures.

52(1945), 227–228

How To Solve It. A New Aspect Of Mathematical Method. By G. Polya. Princeton University Press, 1945. 15 + 204 pages. $2.50.

The interest of this book is pedagogical. First it should be noticed that Professor Polya explicitly disavows the word "new" in the subtitle. As he observes in his preface, "heuristic...has a long past." Probably 2300 years would not be an excessive estimate. Then there is the statement on the jacket, "*A system of thinking which can help you solve any problem.*" Anyone who has been so rash as to write a book for what commercial publishers call "the trade," will know that an author is not responsible for the enthusiasms of his publishers, who must "make friends and influence people," or go out of business. Only those mathematicians who refrain from introspection and from observing their colleagues, might believe that a facility in solving problems in elementary mathematics will help them to "solve any problem." It may be suspected that Professor Polya did not write the copy for the jacket of his book.

What he did write, is an instructive exposition of the heuristic method applied to the solution of problems in elementary mathematics. Not to delude the reader into expecting more than can be offered, Professor Polya states (p. 158) that "Infallible rules of discovery leading to the solution of all possible mathematical problems would be more desirable than the philosopher's stone, vainly sought by all alchemists." And, quite bluntly: "The first rule of discovery is to have brains and good luck. The second rule of discovery is to sit tight and wait till you get a bright idea." Those of us who have little luck and less brains sometimes sit for decades. The fact seems to be, as Poincaré observed, it is the man, not the method, that solves a problem.

So far as instruction is concerned, the tactics of problem solving as expounded here are probably better known to teachers of secondary-school mathematics than they are to a majority of university professors. A generation ago, courses in the pedagogy of mathematics for prospective teachers in American secondary schools included substantially the subject-matter of this book. Such may still be the case; only those having direct contact with the teaching of teachers will know. If heuristic is no longer taught, *How To Solve It* may supply the deficiency. Every prospective teacher should read it. In particular, graduate students who are required to do some teaching, will

find it invaluable if they have not already profited by observing one of their own teachers—if they were lucky enough to have one—who knew how to teach. "The traditional mathematics professor" (p. 181) who reads a paper before one of the Mathematical Societies, might also learn something from the book: "He writes *a*, he says *b*, he means *c*; but it should be *d*."

E. T. Bell

Lucas's Tests for Mersenne Numbers

Irving Kaplansky

1. The theorem. The purpose of this note is to give a brief self-contained proof of the following theorem, which includes as special cases the famous Lucas tests for primality of Mersenne numbers.

THEOREM. *Let $q = 2^k - 1$ and suppose integers c, d are known such that d and $c^2 - d$ are quadratic non-residues of q, if q is prime. Define $W_2 = 2(c^2 + d)/(c^2 - d)$, $W_{i+1} = W_i^2 - 2$. Then q is prime if and only if* [*] $q|W_k$.

Note. The conditions of the theorem are a little peculiar in that we are called upon to announce in advance that d and $c^2 - d$ would be non-residues of q, if q were prime; we then use this information to test the primality of q. However because of the special form of $q = 2^k - 1$, we can be certain that some numbers are non-residues if q is prime, *e.g.*, $-1, -2, 3, 6$, and products of these by squares. By trial we can then find a variety of suitable values of c and d. We list below all distinct values of W_2 obtainable with c^2 and $|d|$ less than 100.

c	d	W_2
1	-2	$-2/3$
1	3	-4
3	6	10
5	-2	$46/27$
9	6	$58/25$
5	27	-52

These six values of W_2 provide us with as many "universal" tests for Mersenne numbers, $W_2 = -4$ being the one discovered by Lucas. If however the form of q is further prescribed, there may be still other valid choices of c and d. For example, if k is of the form $4n - 1$, then 5 is a non-residue of q, and we may take $d = 5, c = 1, W_2 = -3$. This is the other of Lucas's two tests and is the one he used to prove that $2^{127} - 1$ is prime.

2. Preliminary lemmas. Our proof of the above theorem is modelled on one given by D. H. Lehmer† for the case $W_2 = 4$.

Define $U_n = (a^n - b^n)/(a - b)$ with $a = c + \sqrt{d}$, $b = c - \sqrt{d}$, and let $V_n = a^n + b^n = U_{2n}/U_n$. In the following lemmas it will always be tacitly assumed that the primes under discussion do not divide $2abcd$.

---

[*]If W_k is not an integer, we mean that q divides the numerator of W_k when the latter is in its lowest terms. In a particular case we could in any event always begin by making W_2 an integer mod q.

†Journal London Math. Soc. 10, 1935, 162–165. The author is indebted to Professor Lehmer for his suggestions concerning this paper.

We begin by noting two readily verified identities:

(1) $$U_{m+n} = U_n U_{m+1} - ab U_m U_{n-1},$$

(2) $$U_n^2 - U_{n-1}U_{n+1} = (ab)^{n-1}.$$

LEMMA 1. *A prime cannot divide two successive U's.*

Proof. Take $n = 2$ in (1):

(3) $$U_{m+2} = (a+b)U_{m+1} - ab U_m.$$

It follows that a prime which divides two successive U's will divide all of them, including $U_1 = 1$.

LEMMA 2. *A prime cannot divide both U_n and V_n.*

Proof. From (1) with $m = n$ we obtain

$$V_n = U_{n+1} - ab U_{n-1}$$
$$= (a+b)U_n - 2ab U_{n-1} \quad \text{by (3)}.$$

A prime which divides both U_n and V_n will therefore also divide U_{n-1}.

LEMMA 3. *The integers r for which a prime p divides U_r are all multiples of a single integer.*

Proof. It will suffice to show that $p|U_m$ and $p|U_n$ imply that $p|U_{n \pm m} (n > m)$. That $p|U_{n+m}$ is evident from (1). Now replace n by $n - m$ in (1):

$$U_n = U_{n-m}U_{m+1} - ab U_m U_{n-m-1}.$$

Then $p|U_{n-m}U_{m+1}$. But p cannot divide both U_m and U_{m+1}; hence $p|U_{n-m}$.

The next two lemmas form the foundation on which the Lucas theory is built. For completeness we include the familiar proofs.*

LEMMA 4. *If p is prime, $U_p \equiv (d|p) \pmod{p}$.*

Proof. From the definition of U_n we readily find

$$U_p = pc^{p-1} + \binom{p}{3}c^{p-3}d + \binom{p}{5}c^{p-5}d^2 + \cdots + d^{(p-1)/2}.$$

All the binomial coefficients except the last vanish mod p. Hence

$$U_p \equiv d^{(p-1)/2} \equiv (d|p)\pmod{p}.$$

LEMMA 5. *If p is prime, $U_{p-(d|p)} \equiv 0 \pmod{p}$.*

Proof.

$$U_{p+1} = (p+1)c^p + \binom{p+1}{3}c^{p-2}d + \cdots + (p+1)cd^{(p-1)/2}.$$

Mod p, the only binomial coefficients which survive are the first and last. Also $c^p \equiv c \pmod{p}$ by Fermat's "little theorem." Hence

(4) $$U_{p+1} \equiv c(1 + d^{(p-1)/2})\pmod{p}$$

$$\equiv 0 \pmod{p} \quad \text{if and only if } (d|p) = -1.$$

---

*See for example Hardy and Wright, An Introduction to the Theory of Numbers, pp. 147–149.

Now set $n = \overset{*}{p}$ in (2). We find, using Lemma 4 and Fermat's theorem,

$$(5) \qquad\qquad U_{p-1}U_{p+1} \equiv 0 \;(\mathrm{mod}\; p).$$

Lemma 5 follows at once from (4) and (5).

LEMMA 6. *If p is prime, and $(ab|p) = (d|p) = -1$, then $p|V_t$, where $t = (p + 1)/2$.*

Proof. By Lemma 5, $p|U_{2t}$. Next take $n = t, m = t - 1$ in (1).

$$U_p = U_t^2 - abU_{t-1}^2.$$

If $p|U_t$ we have

$$U_p \equiv (d|p) \equiv -1 \equiv -abU_{t-1}^2 (\mathrm{mod}\; p),$$

in contradiction of $(ab|p) = -1$. Hence $p \nmid U_t, p|V_t = U_{2t}/U_t$.

LEMMA 7. *If $q|V_t$ with t a power of 2, and $q \leq 4t^2 - 4t$, then q is prime.*

Proof. If q is not prime, select a prime divisor $p \leq \sqrt{q}$. Then $p|V_t$ and a fortiori $p|U_{2t}$. The smallest integer r for which $p|U_r$ is then, by Lemma 3, a power of 2. But by Lemma 5, $r \leq p + 1$, and $p + 1 \leq \sqrt{q} + 1 < 2t$. Hence $r \leq t$ and p divides both U_t and V_t in contradiction of Lemma 2.

3. Proof of the theorem. The completion of the proof is now immediate. We note that $W_2 = (a^2 + b^2)/ab = V_2/ab$ and an induction readily verifies that

$$W_k = V_{2^{k-1}}/(ab)^{2^{k-2}}.$$

Now suppose $q|W_k$, then $q|V_{2^{k-1}}$ a fortiori, and by Lemma 7 q is prime. Conversely if q is prime, the hypotheses of Lemma 6 are fulfilled and $q|V_{2^{k-1}}$. Since q is prime to $ab = c^2 - d$, we have $q|W_k$.

52(1945), 188–190

REPORT OF KANSAS SECTION

The following papers were presented:

2. *Mathematics for women*, by Sister M. Helen Sullivan, Mount St. Scholastica College.

The aim of the paper is to convey a basic, educational attitude which the writer feels is missing from the present day philosophy of teaching mathematics. The speaker's thesis is this—that since the majority of women are destined to be homemakers, our approach in the teaching of mathematics in women's institutions must be entirely different from that heretofore employed. We have erred in using text books and other devices that cater to the tastes and interests of men. Mathematics has much to offer in the development of a well-rounded feminine personality. It is the task of teachers of mathematics in women's schools to employ all the forces of mathematics in the training of women.

54(1947), 67–68

Inside Back Cover, MONTHLY 40 (1933)

A MATHEMATICAL CONTEST

A contest like the one that follows has been tried several times at Wellesley College and has seemed to interest the students. Perhaps it may appeal to Mathematics Clubs in other colleges. It is obviously not suitable for a club meeting, since it involves the use of histories of mathematics, anthologies of poetry and so on. However, the officers of a club might like to sponsor such a contest and offer a prize to its members or to all students taking mathematics in the institution in question.

HELEN A. MERRILL
MARION E. STARK

"A prize is offered for the best set of answers. Any student who has taken a course in mathematics in college may compete. To win the prize a student must have at least fifteen answers correct. After each answer give an exact reference, stating where it was found."

A. Name the following:

1. A mathematician, now dead, who wrote very popular books for children.
2. The man responsible for the coödinates x and y of a point.
3. A woman mathematician who married at the age of eighteen in order to escape from Russia.
4. A Greek mathematician who was also a musician and a philosopher.
5. A mathematician whose name furnishes the title of a well-known poem by Robert Browning.
6. The man who wrote the oldest mathematical textbook still in actual use.
7. A father and daughter who were both mathematicians of note.
8. A professor of romance languages who has written an unusually interesting arithmetic.
9. The first famous woman mathematician.
10. A man deflected by a fire from mathematics to architecture as a profession.
11. An English mathematician who knew only two tunes. One of them was "God save the Queen" and the other wasn't, and he recognized the first by the fact that people stood up to sing it.
12. A mathematician who is a fine violinist.

. . .

49(1942), 191–192

William Deweese Cairns, Secretary-Treasurer of the MAA 1916–1943

A Manual for Young Teachers of Mathematics

H. E. Buchanan

1. Introduction. The novice who attempts to do his first job without the advice and counsel of older, experienced men, is not acting wisely. The young teacher is no exception. Today in America there are hundreds of young men and young women who are facing their first college classes in mathematics. Many of them do not have as much training as they should have and many have been in our armed forces or other war work so long that their mathematics is rusty from lack of use. This monograph is an effort to help them. We attempt to set forth in a somewhat orderly fashion the experience of nearly a half century of teaching, working and playing with college boys. They are the best type in America and, perhaps, anywhere else. A small percentage are brilliant, many more are bright enough to do good work and, of course, a few are poorly equipped mentally. The lazy we have always with us!

These American students take discipline well and punishment in a manly way. They are inherently honest. The opportunity to work with them is a privilege which is not to be taken lightly. The rewards to the teacher are many. They keep coming over the years and often from surprising quarters. A word of appreciation from an old student is a great part of the satisfaction one gets from living and working with youth.

Teaching is to a large extent a matter of the personality of the teacher. No fixed set of rules can be given to ensure success. Every young teacher should have a definite idea of how he is going to proceed. A good way to begin is to follow the methods of some good teacher he has known, or to read some good text on teaching and follow its suggestions. All such methods, however, are merely for a beginning. He must study his own personality and equipment as objectively as possible and gradually make his initial attempts conform to a conscious method best suited to him and probably to him alone.

2. Class discipline. The instant an instructor walks into a class room the students begin to size him up. On the part of the student this is for the most part unconscious. As a class they will soon find out all his weak points. If his weak point is discipline the class will soon become a rabble. A young teacher must be very sure that there is good behavior in his class from the first day to the last day of the term. To obtain it he should discipline himself. Let him speak distinctly and loud enough to be heard in the rear of the room. It is usually best to stand while speaking. Require attention from all the students while you are explaining even if you are talking to one student. His difficulty is likely to be the same as that of many others. No whispering should be tolerated while the instructor is talking and no newspaper reading in class.

All the students should work all the time of the class hour. If only a part of them are at the board give those seated a list of problems to work or an article to study.

It may be possible to allow a little whispering by those seated while a part are at the board but the instant the instructor begins to explain a problem all should be quiet and listen to him. There is always a chance that they may learn something.

It is a good plan to have students write their problems or questions on the board before the class begins. The students should be told to have them on the board ready for the instructor when he comes in. This saves time. The first ten or fifteen minutes may well be spent answering these questions.

It is much better to be too strict at first and perhaps ease up a bit later. A class once out of hand is very hard to control or re-discipline. Finally, make very few rules of behavior. Never make a statement as to discipline or class behavior unless you are sure you can and will enforce it. This last is a good rule to follow all your life whenever you have to organize and control a group of human beings.

3. Board work and homework. Work to be done at home and to be handed in should be assigned as often as the teacher can read it. It is not wise to assign work to be handed in unless you read it. All such work should be corrected and returned. In homework each problem should be stated at the beginning exactly as it is in the text. This is important for several reasons. One is that the student is more likely to understand what his problem or theorem means if he writes it out in full. Another reason is that the person who reads the work later will not have to guess which problem the student is trying to solve from the too often queer hieroglyphics that are on the page. If another reason is needed it is that it is good English to state first what is to be done and then proceed in an orderly fashion to carry through the work. A part of every problem is to find and apply a check to ensure accuracy.

Every piece of homework and every exercise done at the board should follow this procedure: write out what is to be done, draw and explain a figure, write out the solution, find and apply a check. Some teachers object that it takes too long to do problems this way. The answer is that it is better for a student to do one exercise carefully than to do five problems sloppily and write them up carelessly.

The symbols used in mathematics are nothing but abbreviations for words, phrases or sentences. They should be punctuated as such. For this reason it is not desirable to begin a sentence with a mathematical symbol for one cannot begin such a sentence with a capital letter without changing the notation. Mathematics is thought of by most freshmen as a mixture of x's, square roots, exponents, plenty of $+$ and $-$ signs, and with a few equal signs thrown in for variety or good measure. To them, and unfortunately to some of their high school teachers, there is nothing resembling logic in it. It is difficult to change their point of view. A program of neat, careful work such as is outlined above will at first discourage them but applied gently and persistently will teach them that mathematics is a process of thinking and the symbols are only for the purpose of helping with this thinking. Properly trained students will eventually take pride in the fact that they look for the answer in the back of the book only to see if the author got it right.

4. Quizzes and examinations. The quizzes and examinations are a necessary evil. Their chief value lies in the fact that they are the best way known to make the student take a review and get a comprehensive view of a chapter or, in the case of final examinations, of a book. They emphatically should not be used as the only means of determining a student's grade. The term grade should be made up, for the most part, from the instructor's day by day contact with the student, from oral quizzes, from blackboard work, and even from the type of questions that the student asks. Quizzes and examinations are merely a check on these. Students should know that their grades are so made up with the exception that it is probably not wise to tell them that the type of questions they ask helps the instructor to make up the grade. To tell them that would limit the number of questions they would ask and partly stifle their self expression.

A quiz should be given at a natural division of the subject matter. Usually this is at the end of a chapter. In most good texts a summary and review is given at the end of each chapter. This summary may be used for a day of review before a quiz. Sometimes the instructor may write on the board a list of ten or twelve questions and spend an hour helping the students answer them. Then a quiz may be made up from these. It is best to change the numbers in any numerical problems in this list of questions for an obvious reason.

The making up of a good quiz is an art. It is more important than most young instructors think it is. It should be constructed so that the best students may finish it in thirty or forty minutes and have time to go over it looking for faulty expression or numerical mistakes. The average student may finish most of it and even the poor student who has worked hard may finish two-thirds of it. In every quiz there should be a question or part of a question requiring some independent thinking. This serves to help pick out the brilliant or good students.

5. The instructor's preparation for class. We could begin and end this paragraph with the dogmatic statement that no instructor, old or young, should go before a class without preparation. But since the human race, from Adam down to the present, does not like to accept a blunt statement which sounds like a command it is perhaps better to amplify and clarify a bit.

One of the most embarrassing things that can happen to a young instructor is to get "stuck" on a problem or to find that he does not fully understand the text. He is making bad matters worse if he tries to pass his difficulty off by saying that the problem is not a good one or that he doesn't agree with the text. The students very quickly sense his difficulty and lose their respect for him. He is bluffing and they know it. It is better to admit lack of preparation or better still to study the text carefully. Read it as a whole to get the author's point of view, then read each assignment just before going to class. This isn't enough. A large percentage of the problems and exercises should be solved. This will save time in class and will prevent a situation like the one described above.

It is a good habit to quiz the class orally for a few minutes at the beginning of each class hour. Such a quiz can be done successfully only if the instructor knows what articles are in the day's assignment.

After a recitation which consists of an oral quiz and board work the instructor should record his opinion of those students whose answers he remembers. Some of this may be done in class if it does not interfere with class instruction. You will find that quiz grades and examination grades are usually a confirmation of your composite opinion.

6. The honor system. In a large number of our colleges and universities, particularly in the South, an honor system is in operation. This system is usually conducted by the students themselves under faculty supervision. The maintenance of an honor system is excellent training for students since it inculcates ideas of truth and honor and gives the whole student body practice in good citizenship. The existence of an honor system does not mean that the instructor writes the questions on the board and then leaves the room. He must do more than that. There are three things he should do:

1. He should explain it to the students, which implies that he reads any pamphlets that are available.

2. He should remove temptation as far as that is possible by asking them to put away all books, papers and notes and to sit on alternate seats. If the class is large they can move the chairs about so that it is not easy to see what others are doing.

3. He should explain that quiz grades and examination grades are not important enough to make it pay to cheat. Quiz grades are merely a check on the grades obtained day by day from board work and oral quizzes. He should convince them that he has complete confidence in them, that the arrangements are made not because he suspects them but to protect them against what may be unjust accusations.

Cheating is largely the fault of too large classes or of a wrong attitude on the part of the instructor. The vast majority of students are honest. A great many are young, immature and weak. They should be protected as far as possible but punished resolutely and irrevocably when found guilty.

Finally no honor system ever continues to work by itself. There must be a continual infusion of new strength into it year after year. This must be done either by a faculty member who believes in it, or from a group of students. This group of students often gets its ideals and enthusiasm from a faculty member in whom they have confidence and with whom they may associate and talk freely.

7. Conducting a class. There is no set rule for conducting a class. Every teacher has to devise his own technique. This paragraph is intended to give the young teacher something to go by till he can get on his feet and begin to develop his own methods. There are a few ideas, however, that are always good and should be followed even after he has had experience.

After the first day has passed, the text books having been announced, and an assignment made, let him begin by asking for questions on the lesson. This can be followed by an oral quiz. Perhaps only a few questions are necessary at first. They should be directed to finding out who have read the book and what was learned from the reading. A constant pressure should be kept up to make them read the text. It may well be a gentle pressure for a week or so but increase it until you can get them to reproduce some paragraphs in their own words orally or in board work. Students will get very little value from a course in Freshmen mathematics unless they learn to read a scientific book. It is quite possible that the greatest educational value they will get is the ability to read a closely written scientific text.

When the questions have been answered and the oral quiz ended send as many as possible to the board. While they are getting up and moving to their places give those who are seated a list of exercises to work. Insist that they work. It is often a good plan to assign numbers to those who are at the board, and give a different problem to the even and odd numbers. This, in part, prevents blind copying by adjoining students. Have them write the problem exactly as in the text and require neat work.

A good instructor does very little talking himself. He will explain just enough to keep his students from getting too discouraged. Students gain strength, power and skill in mathematics by doing it themselves. They can no more learn mathematics by listening to someone explain it than they can learn to play baseball by sitting in the bleachers. It is for this reason that the lecture method is a poor way to teach mathematics.

· · ·

53(1946), 371–377

The Teaching of College Mathematics

F. D. Murnaghan

$\cdots$

The subject of my talk has two main aspects: What should we teach and how should we teach it? Both of these are important, but I am firmly convinced that the importance of the second is overwhelmingly greater than the importance of the first.

$\cdots$

2. The Freshman course. What should be the content of the Freshman course? In this friendly family group I may be permitted to speak my mind frankly without thought or fear of giving offense. Where did this worship of algebra (I care not whether you call it college algebra or algebra unadorned) and of trigonometry arise?

$\cdots$

The one thing that is of mathematical value and interest in a trigonometry class is simply this: The cosine of an angle is a simpler concept than the angle itself. I have to confess that I have never been fortunate enough to meet a student who has passed a trigonometry class and who has grasped this one essential fact. When I ask such students what a right angle is they reply without exception that it is 90 degrees; and when I ask what a degree is they are always pleased to be able to tell me that it is one-ninetieth part of a right angle. There is something malevolently wrong and hypnotic about a course which permits a bright, alert young American who can detect sham in many matters to hoodwink himself in this way.

My first suggestion, then, is that you forget your complaints about the poor preparation in algebra and the lack of trigonometry in the high school. Start teaching analytic geometry in the first term of the Freshman year, but teach it by the methods of vector analysis.

$\cdots$

3. The calculus. How should calculus be taught? Here I am very definitely in the minority, but I have the conviction of the true believer: I *know* that I am right. Most of my friends, in a kindly endeavor to prevent me from wasting my time, assure me that calculus simply cannot be taught correctly. The only thing to do is to teach manipulative technique; the student who can differentiate and integrate complicated expressions is at least not scared when he sees symbols for derivatives and integrals. The few students who can clearly understand what a derivative or an integral is will find this out later either in more advanced courses or by independent study.

· · ·

Why is it that so few students grasp the fundamental concepts of calculus in a first course? It is simply because they are deluged with terms such as limit, derivative, differential, integral, and they are never told what a number is. No student can hope to know what a limit is until he understands that every number is defined by a nested sequence of intervals. It takes energy on the part of the teacher to explain this and to make clear, for example, what $3^{1/2} - 2^{1/2}$ means. But you are simply not worth your salary, and are certainly not earning it, if you have not the energy and the will to do this. It is quite unnecessary to use high-sounding phrases such as Dedekind cuts, but even poor students can be taught such a theorem as the one that says that a continuous function over a closed interval is bounded. They will make mistakes in repeating the proof, but this is no reason for lying awake at night worrying. They will have some idea of what calculus is about and will have at least a chance of understanding the Theorem of the Mean which is the central theorem of differential calculus. After all, calculus is the science of calculation, and the Theorem of the Mean is the theorem on which calculations are based. We may devoutly hope that students of calculus will in the future understand the paradox which is resolved by the Theorem of the Mean. All calculations involve a fateful leap from a safe initial point a to a final point b, and differentiation is a local process involving no such leaps. How, then, can calculations be made by means of derivatives?

· · ·

Most of the teaching of calculus is devoted to functions of a single variable, but most of the interesting applications to engineering, physics and chemistry have to do with functions of several variables. This makes it important that functions of a single variable should be treated in a manner that is readily extensible to functions of several variables. Instead of plunging at once into a discussion of the derivative, I find it advantageous to introduce the concept of a differentiable function; a differentiable function is one which may be linearized or straightened out in the neighborhood of a given point. A linear function $y = y(x)$ is one for which $\Delta y = c\Delta x$, where c is a constant, and a differentiable function is one for which $\Delta y = c\Delta x + \nu|\Delta x|$ where ν is a null-function of Δx, i.e., a function whose limit at $\Delta x = 0$ is zero. The constant function c of Δx is then simply the derivative of $y = y(x)$. Similarly a function $z = z(x, y)$ of two variables is differentiable if $\Delta z = c_1\Delta x + c_2\Delta y + \nu r$, where ν is a null-function of Δx and Δy and $r = \{(\Delta x)^2 + (\Delta y)^2\}^{1/2}$. The differentials $dy = c\Delta x$ and $dz = c_1\Delta x + c_2\Delta y$ appear then in a natural way and the student understands that we always take differentials of functions or dependent variables (the differential dx of the independent variable x being merely a name for the differential of the identity function $y = x$ of x). This emphasis on Δy rather than on the derivative $y' = \lim \Delta y/\Delta x$ makes the proof of the important chain-rule theorem much simpler.

· · ·

53(1946), 419–425

Retrospect

L. R. Ford

1. These five years. This is the last issue of the MONTHLY to appear under my editorship. As I pause and look back over the past five years and while the details are fresh in my mind, it seems appropriate to relate for the readers something of the story of this eventful period.

My editorship has spanned the war years. The Japanese attack on Pearl Harbor occurred while the first issue was being set into type. We are reading proof on the last number while treaties of peace are being debated in a broken world. The four dozen issues between them were produced amidst all sorts of difficulties engendered by the war. If the editor who took up his duties so light-heartedly five years ago could have foreseen the future, would he have recoiled from the task? I am not sure.

. . .

6. Rejected papers. After careful study by competent referees about forty per cent of the papers presented were adjudged unsuitable for publication in the MONTHLY. Why? Perhaps an attempt to analyze the reasons for rejection will be of some help to future authors. Certainly a higher rate of acceptability will lighten the tasks of my successor.

There were a few papers which had little to do with mathematics and there were some which were simply wrong. But the main causes of rejection were three in number.

(1) Some papers are too technical for our readers. They are sound mathematically but they lie in fields of advanced research and narrow specialization. They would be read by few in the collegiate field. A paper which lacks some universality of appeal probably should be sent to a research journal.

(2) Some papers are what might, in not too derogatory a sense, be called trivial. They develop at length results that might be considered as mathematical exercises. They might better form contributions to the problem departments or to Discussions and Notes.

(3) Very often papers are poorly written. Fifty per cent of the papers which are ultimately accepted have to be revised. Nobody who has not been an editor can realize how much bad writing college professors can do. Many a paper composed in a lengthy, prolix, and awkward prose might have been saved by a sprightly and fluent style. Those who write on the teaching of mathematics are still the chief sinners in this group. My advice to you would be to swallow your pride and show your paper to a friend in the English department.

7. Angle trisections and Fermat's theorem. I have not included in the previous count the long list of angle trisections and proofs of Fermat's last theorem. The diverting correspondence connected with these has done much to lighten the editorial load. The authors are blest with a boundless confidence and enthusiasm. "The new discovery is so important," says one trisector, "that I am in hopes of being invited to attend the Annual Meeting of the Association." Much of their work is copyrighted.

Sometimes the writer seeks profit, but this is exceptional. "For the disclosure of my discovery I should like to have some financial benefit" runs a letter from the West Indies. Another hopes that our publication of his results will help him "to find someone to finance the construction of my machine for interplanetary navigation."

My letters telling the authors firmly but not too bluntly that their work is erroneous have elicited varied responses. Some are angry. "I have become the dupe of bad incompetency." "I am very suspicious of your qualifications as to judging fairly the merits and demerits of my paper." Usually, however, the news is received in a spirit of resignation, born, no doubt, of oft-repeated experience. "I cannot make this any plainer and shall leave the matter in the hands of the university experts." "You know what happened to the men who first proclaimed that the earth is round." Usually we have parted friends, but I am sure that I have never convinced anybody of his errors.

53(1944), 582–590

A SLOWLY DIVERGENT SERIES

R. P. Agnew, Cornell University

It is well known, and is easily proved by the integral test for convergence and divergence of series, that the series

$$(1) \quad \sum \frac{1}{n}, \quad \sum \frac{1}{n \log n}, \quad \sum \frac{1}{n \log n \log \log n}, \cdots$$

are all divergent. These are classic examples of slowly divergent series, each one after the first diverging more slowly than its predecessor. The following theorem, which is a simple corollary of a theorem of B. Pettineo,* gives a series which diverges more slowly than any of those in (1).

THEOREM. *For each $x > 0$, let $P(x)$ denote the product of x and all of the numbers*

$$(2) \qquad \log x, \quad \log \log x, \quad \log \log \log x, \cdots$$

which are greater than 1. *Then the series*

$$(3) \qquad \sum_{n=1}^{\infty} \frac{1}{P(n)}$$

is divergent.

---

*B. Pettineo. Estensione di una classe di serie divergenti. Atti della Reale Accademia Nazionale dei Lincei, Rendiconti, Classe di Scienze Fisiche, Matematiche e Naturali. Series 8, vol 1 (1946), pp. 680–685.

54(1947), 273–274

Victor Thébault—The Man

Col. W. E. Byrne

In 1932, Victor Thébault became a member of the Mathematical Association of America. It was not long before his name appeared frequently among the contributors to the problem department. Since then many readers of this MONTHLY have wondered about the author of so many interesting problems and discussions.

Victor Michel Jean-Marie Thébault was born on March 6, 1882, at Ambrières-le-Grand (Mayenne), France. His father was a weaver. At the local primary school his teacher noted his native ability and took steps to obtain for him a scholarship at the teacher's college in Laval (Mayenne), where he remained as a student from 1898 to 1901. After graduation he became a school teacher at Pré-en-Pail (Mayenne) (1902–1905) until he was called to be a professor at the technical school of Ernée (Mayenne). In 1909, he received his certificate of capacity for a professorship (scientific) in teachers' colleges; this resulted from winning first place in competitive examination.

The modest salary of a professor was not sufficient to care for his family, now blessed with six children. Therefore Monsieur Thébault reluctantly gave up teaching to become a factory superintendent at Ernée (1910–23). This was followed by another change, to the position of Chief Insurance Inspector at Le Mans (Sarthe) (1924–40). In 1940 he retired to live at "Le Paradis" in Tennie (Sarthe).

That the farewell to teaching of mathematics did not mean any loss of interest in that field is attested by the honor conferred upon him by the French government. In 1932 he was made an Officier de l'Instruction Publique upon the recommendation of M. d'Ocagne, Member of the Institute, with the citation: "Personally I hold him in high esteem for his outstanding talent as a mathematician as shown by the numerous ingenious contributions to what is called elementary geometry, an unending source of problems whose solution requires a quite special gift of invention."

In 1935, he was made a Chevalier d l'Ordre de la Couronne of Belgium because of his activity in connection with the Scientific Society of Bruxelles, its *Annales*, and *Mathesis*.

Not only has Monsieur Thébault continued his publications (which include 15 communications to the Paris Académie des Sciences, hundreds of memoirs and articles concerning the modern geometry of the triangle and tetradhedron and the theory of numbers, and more than 1000 original problems), but also he has contributed articles to the new French *Intermédiaire des Recherches Mathematiques*, and put his own library and services at the disposition of other mathematicians. Quite typical of the man is a notice appearing in the June, 1946, *Mathesis*: "M. V. Thébault, desirous of giving a new proof of his interest in *Mathesis*, offers to foreign subscribers the means of facilitating currency operations by accepting payment of their subscriptions at his postal checking account 339-03, Rennes, France." Moreover he established in 1943 the Prix Victor Thébault to be awarded every two years by the Paris Académie des Sciences to the author, preferably a teacher of the primary or secondary system, of an original study or of an interesting work on geometry or number theory.

Those who have had an occasion, as I have had, to correspond with Monsieur Thébault, have been greatly impressed by his charm and tact as well as by his keen interest in mathematics.

54(1947), 443–444

BENJAMIN FRANKLIN FINKEL

Professor Benjamin Franklin Finkel, founder of the AMERICAN MATHEMATICAL MONTHLY, was born July 5, 1865, and died February 5, 1947. He studied in the rural schools of Ohio and taught in the schools of Ohio and Tennessee, with short periods of study at a normal school, until he became an instructor in mathematics and astronomy at Kidder Institute, Kidder, Missouri, in 1892. During these years he had become keenly aware of the poor instruction given in elementary mathematics, and he had indulged his natural bent for solving problems by contributing solutions to several mathematical and educational journals. While at Kidder Institute he published his well-known *Mathematical Solution Book*, and proposed to establish a journal "devoted solely to mathematics and suitable to the needs of teachers of mathematics in these schools" (high schools and academies).

. . .

The first number of the MONTHLY appeared in January, 1894, and Professor Finkel continued to bear the sole responsibility for the editorship and the business management of the journal until he was joined by L. E. Dickson in October, 1902, by H. E. Slaught in January, 1907, and by G. A. Miller in January, 1909. The MONTHLY was formally transferred in January, 1913, to a board of editors representing fourteen colleges and universities. Then, in January, 1916, the magazine became the official journal of the Mathematical Association of America. Professor Finkel remained one of the editors of the department of *Problems and Solutions* through 1933, and was still a member of the board of editors at the time of his death.

. . .

During the first fifteen or twenty years of the MONTHLY's existence there were great financial and technical obstacles; a picture of these difficulties was vividly presented by Professor Finkel in an address given at the meeting of the Association in Cleveland, January 1, 1931, and published in the MONTHLY for June-July of that year. American mathematics is indebted to him for his imagination and his persistent courage in establishing and promoting a journal which has become most important in the field of mathematics. His zeal, intelligently directed as it was along useful lines, has earned for him a unique place in the history of mathematical publications.

WILLIAM DeWEESE CAIRNS

54(1947), 311–312

The Teaching of Mathematics in Colleges and Universities

E. P. Vance

. . .

5. The influence of the Federal government. At the present time the presence of Federal contracts for research has had little obvious effect on the quality of science teaching in departments of mathematics. In four or five cases the granting of such contracts was reported as beneficial to the effectiveness of teaching, while it was considered harmful by only two.

With regard to policies that educational institutions should follow in the acceptance of Federal contracts for research, including a consideration of the effects of such contracts on teaching, there were several distinctly different replies. Some stated that contracts should be accepted covering fundamental research only, and only when the proposed contracts fit into existing research programs. Others felt that all contracts should be accepted when at all possible, since in the near future there would not be enough stipends available to subsidize graduate students, and research contracts would supply this deficiency. Still other schools felt that such contracts might hamper the universities unduly in the execution of their primary function of training students for teaching and fundamental research. Since this is such a fundamental matter, and one to which most mathematicians have given much thought, I quote the comments of three influential mathematicians, and chairmen of large departments, in this country:

1. I believe that educational institutions in the acceptance of Federal contracts for research should stipulate that in most cases the research men involved should be appointed on a split basis on which about half time should be spent in university teaching and only half time on the contracts. It should also be specified that so far as national safety is not involved, individuals should retain the right of publication. Under these conditions the teaching profession will be enriched by the research activity of the teachers without the customary cost for research purposes of the staff. If scientists devote their entire time to research, the total progress of the research itself may be very great but the benefits to teaching and to the students are likely to be negligible.

2. I certainly do not believe an educational institution should secure a Federal grant, then rob some other institution for men to do the work. This takes men out of teaching where they are so much needed at present. The teaching will certainly be poorer and the rate of training men will be reduced. I believe too much emphasis is being put on research *at the present time*. For a few years the emphasis should be put on teaching and training young men. Unless this is done, the situation will get worse instead of better. Taking men from the teaching staff of our universities is "killing the goose that laid the golden egg." Why doesn't the Federal government do something to help the universities obtain trained staffs in the immediate emergency of increased enrollments? These large enrollments are due largely to the action of the government in sending the veterans to college.

3. In the acceptance of Federal contracts for research, consideration should be given to the following:

(a) Is the research significant in the judgment of men who are competent in the field?

(b) Would the work be done if a Federal contract were not forthcoming?

(c) Is the university in question avoiding its proper obligations? If a university does not encourage and support research, it is no longer a university.

(d) Would the granting of such a contract accentuate the unfortunate separation of education and research?

(e) Under Federal contracts, to what extent would research be directed into specific channels, which may or may not be desirable?

It is interesting to note that several universities, requested by the Federal government to provide assistance in training scientific employees for them, did so willingly, and reported that in no case did it affect the quality of teaching. For example, graduate instruction was provided for qualified personnel at Fort Monmouth; instruction was given to members of the Ames Aeronautical Laboratory, Moffett Field; courses were set up for employees of the United States Navy at the Underwater Sound Laboratory in New London; and, of course, several schools are participating in the NROTC program, thereby offering a place for the training of "junior mathematicians."

6. The role of research. Only a small number of institutions stated that there were any significant changes in the type or amount of research now being carried on. If a change was reported, it was either an increase in applied research, or an increase in both applied and fundamental. The reasons given in case of an increased amount of research were either (1) an increase of staff, (2) an increase of university funds for research, or (3) a contract or project with some branch of the Government.

The comments immediately above were the result of the question, "Has the role of research in your department changed significantly since before the war? Are you, for example, doing more or less applied research as distinguished from fundamental research?" The idea was expressed frequently in this connection that one must not place too much emphasis on the distinction between pure and applied mathematics; mathematics, is indivisible, and one must not emphasize the diversity but rather its unity. Research must always remain an excursion into the unknown, with truth alone as its objective.

With the present overcrowded enrollments in all universities, it is gratifying to know that over three-fifths of the reports indicated that staff members had adequate time and opportunity for research and study.

· · ·

55(1948), 57–64

MATHEMATICS

C. O. OAKLEY, Haverford College

Mathematics is one component of any plan for liberal education. Mother of all the sciences, it is a builder of the imagination, a weaver of patterns of sheer thought, an intuitive dreamer, a poet. The study of mathematics cannot be replaced by any other activity that will train and develop man's purely logical faculties to the same level of rationality. Through countless dimensions, riding high the winds of intellectual adventure and filled with the zest of discovery, the mathematician tracks the heavens for harmony and eternal verity. There is not wholly unexpected surprise, but surprise nevertheless, that mathematics has direct application to the physical world about us. For mathematics, in a wilderness of tragedy and change, is a creature of the mind, born to the cry of humanity in search of an invariant reality, immutable in substance, unalterable with time.

56(1949), 19

The Scholar in a Scientific World*

C. C. MacDuffee

1. Introduction. Recorded history began with the scholar. Sometime in that great blank period from twenty to six thousand years ago, the art of writing was developed so that the scholar could record the legends and history of his nation. The scholar was then also the priest and the lawyer and the physician as he has been throughout most of recorded history.

Since earliest times scholars have lived in communities. These communities or monasteries, or musea, were the early colleges and universities. They have constituted the chief vehicle in the development of civilization. If these little bands of scholars had not existed, we should now be in the woods carrying stone axes.

A nation is just about as great as its universities. They are the ganglia in the central nervous system of the nation, whence come the nerves that stimulate its intellectual, industrial, and political life. Without the leadership of the universities, national life would probably continue for a while by inertia but would gradually slow down and succumb to the competition of rival nations.

The function of the university as a teaching institution is purely incidental. Its principal function is to keep alive the great wealth of knowledge and culture that past ages have collected, and to add to that wealth through the encouragement of scientific research and creative art. Knowledge and art can be kept alive only by the creation of scholars, and the scientific horizon can be extended only by the creation of scientists. To create scientists and scholars is the primary function of the university.

This statement of the function of the university is orthodox. It is as old as the universities themselves. It is the idea which induced Alexander to establish the Museum at Alexandria, the institution which shone out over the ancient world for the 600 years of its highest culture and where much of our mathematics was made. Alexandria was doubtless the greatest university of all time, and the burning of its library in 389 might well mark the beginning of the Dark Ages.

The Renaissance was marked by, and was in a very large measure due to, the establishment of the great universities in Padua, Paris, and Oxford, and soon in many other places. The study of the great legacy of Greece was brought into Europe by these universities, where it had been extinct for many years. Men had to learn again that it is not evil to think, and that we live in a reasonable world. The scientific developments of our generation could not have taken place if these medieval universities had not come into being.

---

*Retiring address as President of the Mathematical Association of America, Athens, Georgia, January 1, 1948.

2. Science in Modern America. American universities have progressed amazingly in a century, and so has the country. From the position of a backward nation we have come to the very front. The opportunity to establish undisputed intellectual leadership is ours. Whether we can maintain this position when the influx of German scientists stops remains to be seen.

As time goes on, and more and more is known of science, archaeology, and history, it becomes increasingly difficult to keep some branches of our culture from becoming lost. The day has long since passed when one man could have a working knowledge of all branches of science. The goal of the philosopher to know all things and to correlate them into one harmonious whole now seems very difficult of fulfillment.

The last war has destroyed many of the cultural links between our age and the past, particularly by the wanton bombing of the libraries and museums by both sides. It has also reduced the number of scholars in Europe so that some links that we had with our past are probably now gone forever. We must be willing to support scholars in all fields without everlastingly measuring their output in terms of American dollars.

Everything that I have said so far is trite and generally admitted, but we in the United States do not make it part of our practical thinking. We do not believe that the production of scholars is in itself a worthy objective, nor that the scholar is worthy of his hire. Our colleges and universities are supported by state and private funds, but the stated objective is the instruction of vast numbers of young men and women in ways and means of earning a better living than their fellows. They are taught citizenship and the American way of life. I am not always clear just what this means. Occasionally it seems to be football and the cult of the gods and goddesses of Hollywood.

We are now living in the Age of Science. One might almost say that Science has become our national religion. It has revolutionized our daily lives, our knowledge of the past, and our hopes for the future. Its miracles are commonplace and we can perform them ourselves. The researchers in medicine are the angels of mercy, and in popular imagination the Atomic Scientists are the devils.

This elevation of Science to a god-like eminence creates many difficult problems for the scientist, who is not primarily interested in becoming a high priest. First, everyone wants to be a scientist. Astrologers, numerologists, even religious groups, conjure with the name "Science." On a little higher plane, we have Political Science and the Science of Economics, when everyone knows that these subjects are not yet truly scientific. Even Psychology has renounced its parent, Philosophy, and wishes to be known as a science.

Everyone is pathetically eager to understand science, and of course most persons are unable or unwilling to pay the price in long years of study. Our book stores are flooded with books on "Relativity in Five Easy Lessons," illustrated with doodles, and "How to Make an Atomic Bomb." Even the elementary schools have courses in General Science, and eight-year-olds lisp that Einstein discovered the relativity of time and space. It is questionable whether sophistry at this level does more harm than good, but probably nothing can be done about it. It is part of the ritual in the worship of Science. A classic example is the quotation from the school boy's paper: "Gravitation was discovered by Sir Isaac Walton while he was digging for hookworms under an apple tree. It is more noticeable in the fall than in the spring."

If it were necessary or desirable to draw a fine distinction between the scholar on the one hand, and the artist or scientist on the other, it would have to be that the scholar is primarily interested in perpetuating the known, the scientist in new creation. But they are not independent. New discoveries are now rarely made by attic inventors. The creator is so steeped in the knowledge of his predecessors that he can begin where they left off. Otherwise he merely rediscovers.

There are fashions and fads in intellectual pursuits as well as in politics or clothes. At the time of Socrates in Greece, speculative philosophy was the fashion, and so was sculpture in stone. In the Middle Ages about the only outlet was theology. There were later periods in which great music was produced; others were notable for great paintings. In sixteenth century Italy there was a remarkable flowering of painting and sculpture. It must have been true that the people of this period were enthusiastic about art, and that the great painters were honored and important people. The result was that the best minds of the period were attracted into this field, and that they were stimulated by knowing that their work was appreciated.

In this period lived Leonardo da Vinci, one of the greatest potential scientists of all time. It has been said that if he had published his scientific discoveries, science would have been revolutionized. But he could not have published, for he had no sympathetic audience. In his own period he was known and honored as an artist, and it is only now in the Age of Science that his scientific thoughts are appreciated. If he had lived today, it is probable that he would have become a great scientist and that his potentialities as an artist would have remained undeveloped.

Deans of graduate schools and those who have served on university committees of award for scholarships will testify that, by and large, the undergraduate records of those who apply for fellowships in mathematics and the natural sciences excel all others. It is all very well to argue that we need the best minds in Sociology and Politics. The fact remains that the country is worshipping at the shrine of the Natural Sciences.

3. Science in the curriculum. There is now a widespread movement in the colleges and universities to incorporate some science into the liberal arts program. It seems as if all the weight of logic is in favor of this. We study and admire the gems of various older cultures, their stories, poems, songs, pictures, statues, and philosophical concepts. Included in the education of the well-rounded scholar should be the gems of thought of modern man. These gems are largely the processes of thought of the modern scientist.

· · ·

The mass production of high school graduates has had some unexpected by-products. We believe so firmly in the equality of men under the law that we assume them to be equal emotionally and mentally. At first we tried to offer everyone a college preparatory education. Many failed and were frustrated, so this system was considered unsatisfactory. The only obvious way out was to water down the contents of the courses to the capacities of the least capable, and this has been one of the aspects of "progressive education" during the past couple of decades. Since the mountain could not be persuaded to go to Mahomet, Mahomet was obliged to go to the mountain. This method may produce citizens but it does not produce scholars.

For many high school pupils, the education which they receive is quite satisfactory. It is all they can absorb, they are not frustrated, and they are kept off the streets. They learn citizenship, although it must be admitted that their courses on citizenship do not seem to prevent them from littering the streets around the school with paper, much as they did before the Social Sciences were introduced into the high school curriculum. They are indoctrinated in the "American Way of Life," which means a mild contempt for scholarship and the belief that football is the only worthy masculine activity. With the substitution of swordsmanship for football, we have the way of life of the Norman nobles of the Middle Ages.

It is quite possible that some of these lads would have been frustrated by the old fashioned grammar school where grammar was taught instead of Social Studies. They will, eventually, be frustrated by life, but they will not take it too hard. Extraverts are not easily frustrated.

The greatest failure of our secondary schools is in their unsympathetic handling of the potential scholars. These superior students will be the future leaders of America, and their education is of primary importance. Without them, our civilization will fall. But they are frequently sensitive creatures, slow to mature, and possibly somewhat unlovely in adolescence. They are frequently unsympathetically handled by extravert teachers, and many of them are definitely frustrated. All they ask is to be allowed to pursue their serious interests at their own normal speed, but this is denied them, and they are made to feel that this desire is in some way abnormal. Frequently they find no kindred spirit even among their teachers.

European schools, and the American schools of forty years ago, cultivated more ground and ploughed it deeper than most of our modern schools. They had a curriculum of solid material so that the pupil had to develop at an early age the habit of serious study. In the grades he really learned spelling and grammar and arithmetic, and with a feeling of accomplishment following effort, came a genuine liking for the subjects. In high school he learned two languages, he learned to read them and to write them, and he learned their grammar. Possibly he learned to speak them. He also learned a little English, Ancient and American History, and a dab or two of the sciences. He took Algebra and Geometry as a matter of course, and frequently he took four years of mathematics. This was a college preparatory course. A student unable to complete this course went as far as he could with it. Modern experts say that this was bad, that his entire curriculum should have been changed and watered down. I stoutly maintain that this judgment is still unproved.

The colleges of today are facing an almost impossible task in trying to make scholars in four years out of unprepared students. What should be a leisurely course in College Algebra has become a mad drive to teach high school algebra to students who have already passed it at 97% in high school. Of course, the casualties are high but the colleges are trying to hold the line. But unless relief comes soon, they will yield, and will spend two years doing what the high schools should have done.

What is the basic cause of this condition, and what is the remedy? I can do no better than to quote Dr. John Guy Fowlkes, Dean of the School of Education of the University of Wisconsin, from an address to his faculty last month: "More than ever before in the history of our country it is clear that teachers in our public schools need to be scholars."

It was an evil day when the direction of policy in our schools passed out of the hands of the scholars. When the direction of our medical schools shall pass from the hands of our doctors of medicine, and the direction of our law schools from the hands of our lawyers, they too will become ineffective.

This state of affairs is clearly due to the great increase in size which our school systems have undergone in recent years. There are not enough scholars to go around, and there is not enough money available in some local units to pay teachers adequately. This is the crux of the whole difficulty. High school teaching as a profession must be made more attractive to persons of both sexes who have a liking for the scholarly life.

· · ·

There was a time when the colleges exercised some control over the secondary schools by demanding a measure of accomplishment from entering students, *and let no one persuade you that the colleges have no right to make such demands*. Those students who were headed for college had to have adequate instruction. This frustrated the pressure groups in the schools, and became the object of a concerted and violent attack. Most legislatures were persuaded to force the state-supported colleges to allow the high school principals to decide who is prepared for college and who is not.

This seems to be the critical point at which the scholars completely lost control of the schools. Long before this they had ceased to be in the majority in their own schools, but they had support from the colleges, most of which still remain scholarly institutions. But with the loss of this last restraining influence, degeneration has followed rapidly.

6. Obligations of the Association. The Mathematical Association of America was founded in 1916 with the avowed purpose of assisting in promoting the interests of mathematics in America. During the thirty-one years of its existence it has been influential and respected. The Young Report on the reorganization of mathematics in the secondary schools was and is respected, and there have been other projects to our credit. But we must look forward rather than backward.

The Association must not be just another pressure group whose only purpose is to further the interests of its own members, and I don't think it has ever been such. Mathematics can hold its own in any fair competition. We must, however, work for a sane and realistic program of education, a return to the scholarly ideal.

With the great American experiment of mass education we should have no quarrel. But this project is being quite adequately handled, and presents no critical problem. The adequate education of scholars by scholars is not being well handled and needs our best thoughts. Call it the double-track system or what you will, there must be some way to give a genuine education to those who want it, and to persuade all those who are competent to want it. The ideal of scholarship must be made more honorable and more attractive.

· · ·

It seems clear that most of the opposition to mathematics in the schools has been because of the idea that it stood in the way of mass education. I think we have no quarrel with mass education provided it does not interfere with the training of scholars. Instead of trying to force a high mathematics requirement on all schools, I think we should demand instead that those students who are going to go to college shall be adequately trained, and that the colleges shall have the right to decide whether a student is fitted to enter.

· · ·

7. Conclusion. All of these considerations lead us back to our original text: *Scholars are important persons*. They are more important than principals, than athletic coaches, than vocational guidance directors. Scholars are worthy of public respect, and should occupy a respected social position in this country as they do in most other countries. They should be given adequate salaries so that they can live as well as the physicians and lawyers, whose equals in public esteem they should be. The schools should be under the direction and management of the scholars, and the growing gap in both prestige and salary between the principal and the teacher should be closed.

If such a conversion of the American people could take place, our school problems, and most of our political, economic, and social problems as well, would be solved. It is of course an idle dream. But scholars have always been dreamers, for only their dreams have made life endurable for them. And only those small bits of their dreams which have come to pass have made life endurable for the rest of the world.

55(1948), 129–140

Coin-Tossing Experiment

E 796 [1946, 596]. *Proposed by Henry Scheffé, University of California at Los Angeles*

Describe a coin-tossing experiment in which the probability of success is one-third.

I. *Solution by R. V. Andree, University of Wisconsin.* Following are some solutions:

1. Five uniform coins are tossed until they do *not* all show the same face. Then (a) the probability of exactly two heads is $1/3$, (b) the probability of exactly three heads is $1/3$, (c) the probability of a four to one split is $1/3$.

2. Given one normal coin and one two headed coin, one coin is selected at random and tossed. It comes down heads. Then the probability that it is the normal coin is $1/3$.

3. A game of chance is played as follows: A has two coins, B has one coin. All three coins are tossed. If all three coins agree, each player takes his own coins back and no exchange is made. Otherwise A wins if both of A's coins disagree with that of B and B wins if only one of A's coins disagrees with that of B. Then the probability that A will win is $1/3$.

4. An interesting possibility is brought up by discarding the assumption that the coin is "thin" and thus that the probability of its standing on edge is negligible. Let us consider a coin the ratio of whose diameter to height is $\sqrt{3}/1$. Then the probability that this coin will land heads is $1/3$.

55(1948), 500

Note on a Paper by L. S. Johnston

Ivan Niven

In a paper entitled *Denumerability of the Rational Number System* in the February, 1948, issue of this MONTHLY, L. S. Johnston has exhibited a one-to-one correspondence between the positive integers and the rational numbers, with a specific construction provided. The purpose of this note is to give another constructive correspondence.

For any positive integer n, let r be the number of zeros occurring in the representation of n in the binary scale of notation. These r zeros divide the representation into $r + 1$ blocks of 1's, some blocks being perhaps empty: for $j = 2, 3, \ldots, r$ let a_j denote the number of 1's between the $(j - 1)$th and the jth zeros, counted from the right; define a_1 and a_{r+1} as the number of consecutive 1's on the right and left ends of the representation. E.g., if n is 110010 in the binary scale, then $r = 3$ and $(a_1, a_2, a_3, a_4) = (0, 1, 0, 2)$. If $n = 1$, then $r = 0$, and $a_1 = 1$. Note that n determines a unique set of values $a_1, a_2, \ldots, a_{r+1}$, with positive a_{r+1}, and conversely.

Now define $b_j = -a_j/2$ for even a_j, and $b_j = (a_j + 1)/2$ otherwise: this is a one-to-one correspondence between the non-negative integers a_j and all integers b_j. Finally, if $p_1, p_2, p_3, \ldots$ are the consecutive primes $2, 3, 5, \ldots$, we define

$$(1) \qquad f(n) = \prod_{j=1}^{r+1} p_j^{b_j}.$$

Thus to any positive integer n corresponds a unique positive rational number $f(n)$. Conversely, any positive rational number f, except 1, can be written uniquely in the form (1) with $b_{r+1} \neq 0$. The exponents b_j determine a unique set a_j, with $a_{r+1} \neq 0$, and these give a unique positive integer n. E.g., if $f = 33/49$, then $(b_1, b_2, b_3, b_4, b_5) = (0, 1, 0, -2, 1)$ and $(a_1, a_2, a_3, a_4, a_5) = (0, 1, 0, 4, 1)$, so that $n = 1011110010$ in the binary scale.

We now obtain a one-to-one correspondence between the positive integers and all positive rational numbers, by letting $n + 1$ correspond to $f(n)$, written $n + 1 \leftrightarrow f(n)$, for all positive integers n, and in addition we let $1 \leftrightarrow 1$. On the other hand, if we want a one-to-one correspondence between the positive integers and all rational numbers, we can use the correspondences $1 \leftrightarrow 0, 2 \leftrightarrow 1, 3 \leftrightarrow -1, 2n + 2 \leftrightarrow f(n), 2n + 3 \leftrightarrow -f(n)$, where n ranges over all positive integers.

55(1948), 358

Earle Raymond Hedrick, First President of the MAA 1916

With, or Without, Motivation?*

G. Pólya

The following lines present the same proof twice, first briefly without motivation, then broadly with motivation. I think that the comparison of these two presentations may clarify a few not quite trivial points of class-room technique.

1. Deus ex machina. A mathematical lecture should be, first of all, correct and unambiguous. Still, we know from painful experience that a perfectly unambiguous and correct exposition can be far from satisfactory and may appear uninspiring, tiresome or disappointing, even if the subject-matter presented is interesting in itself. The most conspicuous blemish of an otherwise acceptable presentation is the "deus ex machina." Before further comments, I wish to give a concrete example.†

2. Example. I wish to present the proof of the following elementary, but not too elementary, theorem: *If the terms of the sequence $a_1, a_2, a_3, \ldots$ are nonnegative real numbers, not all equal to 0, then*

$$\sum_1^\infty (a_1 a_2 a_3 \ldots a_n)^{1/n} < e \sum_1^\infty a_n.$$

Proof. Define the numbers $c_1, c_2, c_3, \ldots$ by

$$c_1 c_2 c_3 \ldots c_n = (n+1)^n$$

for $n = 1, 2, 3, \ldots$. We use this definition, then the inequality between the arithmetic and the geometric means, and finally the fact that the sequence defining e, the general term of which is $[(k+1)/k]^k$, is increasing. We obtain

$$(1) \qquad \sum_1^\infty (a_1 a_2 \ldots a_n)^{1/n}$$

$$= \sum_1^\infty \frac{(a_1 c_1 a_2 c_2 \ldots a_n c_n)^{1/n}}{n+1} \leqq \sum_1^\infty \frac{a_1 c_1 + a_2 c_2 + \cdots + a_n c_n}{n(n+1)}$$

$$= \sum_{k=1}^\infty a_k c_k \sum_{n \geqq k} \frac{1}{n(n+1)} = \sum_{k=1}^\infty a_k c_k \sum_{n=k}^\infty \left(\frac{1}{n} - \frac{1}{n+1} \right)$$

$$= \sum_{k=1}^\infty a_k \frac{(k+1)^k}{k^{k-1}} \frac{1}{k} < e \sum_{k=1}^\infty a_k.$$

*Presented at the meeting of the Northern California Section of the Mathematical Association of America, San Francisco, January 29, 1949.

†I may be excused if I choose an example from my own work. See G. Pólya, Proof of an inequality, Proceedings of the London Mathematical Society (2) v. 24, 1925, p. LVII. The theorem proved is due to T. Carleman.

195

3. Motivation. The crucial point of the proof is the definition of the sequence $c_1, c_2, c_3, \ldots$. This point appears right at the beginning without any preparation, as a typical "deus ex machina." What is the objection to it?

"It appears as a rabbit pulled out of a hat."

"It pops up from nowhere. It looks so arbitrary. It has no visible motive or purpose."

"I hate to walk in the dark. I hate to take a step, when I cannot see any reason why it should bring me nearer to the goal."

"Perhaps the author knows the purpose of this step, but I do not and, therefore, I cannot follow him with confidence."

"Look here, I am not here just to admire you. I wish to learn how to do problems by myself. Yet I cannot see how it was humanly possible to hit upon your . . . definition. So what can I learn here? How could I find such a . . . definition by myself?"

"This step is not trivial. It seems crucial. If I could see that it has some chances of success, or see some plausible provisional justification for it, then I could also imagine how it was invented and, at any rate, I could follow the subsequent reasoning with more confidence and more understanding."

The first answers are not very explicit, the later ones are better, and the last is the best. It reveals that an intelligent reader or listener desires two things:

First, to see that the present step of the argument is correct.

Second, to see that the present step is appropriate.

A step of a mathematical argument is appropriate, if it is essentially connected with the purpose, if it brings us nearer to the goal. It is not enough, however, that a step *is* appropriate: it should *appear so* to the reader. If the step is simple, just a trivial, routine step, the reader can easily imagine how it could be connected with the aim of the argument. If the order of presentation is very carefully planned, the context may suggest the connection of the step with the aim. If, however, the step is visibly important, but its connection with the aim is not visible at all, it appears as a "deus ex machina" and the intelligent reader or listener is understandably disappointed.

In our example, the definition of c_n appears as a "deus ex machina." Yet this step is certainly appropriate. In fact, the argument based on this definition proves the proposed theorem, and proves it rather quickly and clearly. The trouble is that the step in question, although vindicated in the end, does not appear as justified from the start.

Yet how could the author justify it from the start? The complete justification takes some time; it is supplied by the following proof. What is needed is, not a complete, but an *incomplete justification*, a *plausible provisional ground*, just a hint that the step has some chances of success, in short, some heuristic *motivation*.

In many similar cases, the motivation can be given in a few words, but this is not always so. In some cases a plausible story of the discovery supplies an attractive motivation. Such stories are much more suitable for oral presentation than for print, but just for once I take the liberty of printing such a story, even if it is not quite short. It is almost unnecessary to remind the reader that the best stories are not true; they contain, however, some elements of truth.

4. Another presentation of the example. The theorem proved in section 2 is surprising in itself. We should be less surprised, if we would know, how it was discovered. We are led to it naturally in trying to prove the following: *If the series* *with positive terms*

$$a_1 + a_2 + a_3 + \cdots + a_n + \cdots$$

is convergent, the series

$$a_1 + (a_1 a_2)^{1/2} + (a_1 a_2 a_3)^{1/3} + \cdots + (a_1 a_2 a_3 \ldots a_n)^{1/n} + \cdots$$

is also convergent. I shall try to emphasize some motives which may help us to find the proof.

A suitable known theorem. It is natural to begin with the usual questions.‡

What is the hypothesis? We assume that the series Σa_n converges—that its partial sums remain bounded—that

$$a_1 + a_2 + \cdots + a_n \text{ not large.}$$

What is the conclusion? We wish to prove that the series $\Sigma (a_1, a_2 \ldots a_n)^{1/n}$ converges—that

$$(a_1 a_2 \ldots a_n)^{1/n} \text{ small.}$$

Do you know a theorem that could be useful? What we need is some relation between the sum of n positive quantities and their geometric mean. *Have you seen something of this kind before?* If you ever have heard of the inequality between the arithmetic and the geometric means, it has a good chance to occur to you at this juncture:

(A) $$\qquad (a_1 a_2 \ldots a_n)^{1/n} \leqq \frac{a_1 + a_2 + \cdots + a_n}{n}.$$

This inequality shows that $(a_1, a_2 \ldots a_n)^{1/n}$ is small when $a_1 + a_2 + \cdots + a_n$ is not large. It has so many contacts with our problem that we can hardly resist the temptation of applying it:

(2) $$\sum_{n=1}^{\infty} (a_1 a_2 \ldots a_n)^{1/n} \leqq \sum_{n=1}^{\infty} \frac{a_1 + a_2 + \cdots + a_n}{n}$$

$$= \sum_{k=1}^{\infty} a_k \sum_{n=k}^{\infty} \frac{1}{n}$$

—complete failure! The series $\Sigma 1/n$ is divergent, the last line of (2) is meaningless.

Learning from failure. It is difficult to admit that our plan was wrong. We would like to believe that at least some part of it was right. The useful questions are: *What was wrong with our plan? Which part of it could we save?*

The series $a_1 + a_2 + \cdots + a_n + \cdots$ converges. Therefore, a_n is small when n is large. Yet the two sides of the inequality (A) are different when $a_1, a_2, \ldots a_n$ are not all equal, and they may be very different when $a_1, a_2, \ldots, a_n$ are very unequal. In our case, a_1 is much larger than a_n, and so there may be a considerable gap between the two sides of (A). This is probably the reason that our application of (A) turned out to be insufficient.

Modifying the approach. The mistake was to apply the inequality (A) to the quantities

$$a_1, a_2, a_3, \ldots, a_n$$

---

‡About the rôle of such questions see the author's booklet, How to Solve It, Princeton, 5th enlarged printing, 1948.

which are too unequal. Why not apply it to some related quantities which have more chance to be equal? We could try

$$1a_1, 2a_2, 3a_3, \ldots, na_n.$$

This may be the idea! We may introduce such increasing compensating factors as $1, 2, 3, \ldots, n$. We should, however, not commit ourselves more than necessary, we should reserve ourselves some freedom of action. We should consider perhaps, more generally, the quantities

$$1^\lambda a_1, 2^\lambda a_2, 3^\lambda a_3, \ldots, n^\lambda a_n.$$

We could leave λ *indeterminate* for the moment, and choose the most advantageous value later. This plan has so many good features that it seems ripe for action:

$$(3) \qquad \sum_1^\infty (a_1 a_2 \ldots a_n)^{1/n} = \sum_1^\infty \frac{\left(a_1 1^\lambda \cdot a_2 2^\lambda \ldots a_n n^\lambda\right)^{1/n}}{(1 \cdot 2 \ldots n)^{\lambda/n}}$$

$$\leqq \sum_{n=1}^\infty \frac{a_1 1^\lambda + a_2 2^\lambda + \cdots + a_n n^\lambda}{n(n!)^{\lambda/n}}$$

$$= \sum_{k=1}^\infty a_k k^\lambda \sum_{n=k}^\infty \frac{1}{n(n!)^{\lambda/n}}.$$

We run into difficulties. We cannot evaluate the last sum. Even if we recall various relevant tricks, we are still obliged to work with "crude equations" (notation $\approx$, instead of $=$):

$$(n!)^{1/n} \approx ne^{-1},$$

$$\sum_{n=k}^\infty \frac{1}{n(n!)^{\lambda/n}} \approx e^\lambda \sum_{n=k}^\infty n^{-1-\lambda}$$

$$\approx e^\lambda \int_k^\infty x^{-1-\lambda} \, dx$$

$$= e^\lambda \lambda^{-1} k^{-\lambda}.$$

Introducing this into the last line of (3) we come very close to proving

$$(3') \qquad \sum_1^\infty (a_1 a_2 \ldots a_n)^{1/n} \leqq C \sum_1^\infty a_k$$

where C is some constant, perhaps $e^\lambda \lambda^{-1}$. Such an inequality would, of course, prove the theorem in view.

Looking back at the foregoing reasoning we are led to repeat the question: "Which value of λ is most advantageous?" Probably the λ that makes $e^\lambda \lambda^{-1}$ a minimum. We can find this value by differential calculus:

$$\lambda = 1.$$

This suggests strongly that the most obvious choice is the most advantageous: the compensating factor multiplying a_n should be $n^1 = n$, or some quantity not very different from n when n is large. This may lead to the simple value $C = e$ in (3').

More flexibility. We left λ indeterminate in our foregoing reasoning (3). This gave our plan a certain *flexibility*: the value of λ remained at our disposal. Why not give our plan still more flexibility? We could leave the compensating factor that multiplies a_n quite indeterminate; we call it c_n, and we will dispose of its value later, when we shall see more clearly what we need. We embark upon this further modification of our original approach:

$$(4) \quad \sum_1^\infty \left(a_1 a_2 \ldots a_n \right)^{1/n} = \sum_{n=1}^\infty \frac{\left(a_1 c_1 \cdot a_2 c_2 \ldots a_n c_n \right)^{1/n}}{\left(c_1 c_2 \ldots c_n \right)^{1/n}}$$

$$\leqq \sum_{n=1}^\infty \frac{a_1 c_1 + a_2 c_2 + \cdots + a_n c_n}{n \left(c_1 c_2 \ldots c_n \right)^{1/n}}$$

$$= \sum_{k=1}^\infty a_k c_k \sum_{n=k}^\infty \frac{1}{n \left(c_1 c_2 \ldots c_n \right)^{1/n}}.$$

How should we choose c_n? This is the crucial question and we can no longer postpone the answer.

First, we see easily that a factor of proportionality must remain arbitrary. In fact, the sequence $cc_1, cc_2, \ldots, cc_n, \ldots$ leads to the same consequences as $c_1, c_2, \ldots, c_n, \ldots$.

Second, our foregoing work suggests that both c_n and $(c_1, c_2 \ldots c_n)^{1/n}$ should be asymptotically proportional to n:

$$c_n \sim Kn, \qquad \left(c_1 c_2 \ldots c_n \right)^{1/n} \sim e^{-1} Kn = K'n.$$

Third, it is most desirable that we should be able to effect the summation

$$\sum_{n=k}^{n=\infty} \frac{1}{n \left(c_1 c_2 \ldots c_n \right)^{1/n}}.$$

At this point, we need whatever previous knowledge we have about simple series. If we are familiar with the series

$$\sum \frac{1}{n(n+1)} = \sum \left(\frac{1}{n} - \frac{1}{n+1} \right)$$

it has a good chance to occur to us at this juncture. This series has the property that its sum has a simple expression not only from $n = 1$ to $n = \infty$, but also from $n = k$ to $n = \infty$—a great advantage! This series suggests the choice

$$\left(c_1 c_2 \ldots c_n \right)^{1/n} = n + 1.$$

Now, visibly $n + 1 \sim n$ for large n—a good sign! What about c_n itself? As

$$c_1 c_2 \ldots c_{n-1} c_n = \left(n + 1 \right)^n, \qquad c_1 c_2 \ldots c_{n-1} = n^{n-1},$$

$$c_n = \frac{\left(n + 1 \right)^n}{n^{n-1}} = \left(1 + \frac{1}{n} \right)^n n \sim en;$$

the asymptotic proportionality with n is a good sign. And the number e arises—a very good sign!

We choose this c_n and, after this choice, we take up again the derivation (1)—with more confidence than before.

Now, we may understand how it was humanly possible to discover that definition of c_n which appeared in section 2 as a "deus ex machina." The derivation (1) became also more understandable. It appears now as the last, and the only successful, attempt in a chain of consecutive trials, (2), (3), (4) and (1). And the origin of the theorem itself is elucidated. We see now how it was possible to discover the rôle of the number e which appeared so surprising at the outset.

5. Demonstrative conclusions and heuristic motives. The two presentations, in section 2 and in section 4, are very different. The most obvious difference is that one is short and the other long. The most essential difference is that one gives proofs and the other plausibilities. One is designed to check the *demonstrative conclusions* justifying the successive steps. The other is arranged to give some insight into the *heuristic motives* of certain steps. The demonstrative presentation follows the accepted manner, usual since Euclid; the heuristic presentation is extremely unusual in print. Yet an ambitious teacher can use both manners of exposition. In fact, he should teach his students two things:

First, to distinguish a valid demonstration from an invalid attempt, a proof from a guess.

Second, to distinguish a more reasonable guess from a less reasonable guess.

The first point is generally recognized and I need not stress it. The second point is, in my opinion, even more important, but much more subtle. If my long presentation can serve this subtle second aim to some little degree, its length is amply justified.

Of course, various transitions or compromises are possible between the two manners of presentation. An alert teacher should be able to find out how much stress on motivation suits his audience, how much suits himself personally, and how much time he has for motivation.

I cannot omit a final remark on logic. Some authors distinguish two branches of logic, deductive logic and inductive logic. Yet these two branches differ widely. Deductive logic is a firmly established branch of science, and became in its latest development, as symbolic logic, practically a branch of mathematics. Inductive logic is an interesting subject of philosophical discussion, but can scarcely be regarded as an established science. Deductive logic is concerned with the validity of proofs. Inductive logic which I would prefer to call *heuristic logic*, in order to emphasize its wider scope, is concerned with plausible inference only. That deductive logic is closely connected with mathematics, is widely recognized; some modern authors think, that its proper object is the analysis of the deductive structure of mathematical theories. Now I come to my point: I think that also heuristic logic is closely connected with mathematics, but not with mathematical theories and their deductive structure, rather with mathematical problems and the invention of their solution. In fact, I think that heuristic logic could make serious progress in studying such plausible motives of the solution as were emphasized in the long presentation of our example.

57(1950), 684–691

A Symmetrical Notation for Numbers

C. E. Shannon

The possibility of representing real numbers in various scales of notation is well known. Thus, in the scale r an arbitrary positive number b may be expanded in the form,

$$b = \sum_{-\infty}^{N} a_n r^n, \qquad 0 \le a_n \le r - 1,$$

and represented in the "decimal" notation as $a_N a_{N-1} \ldots a_0 \cdot a_{-1} a_{-2} \ldots$. Negative numbers are represented by prefixing a minus sign to the representation of the corresponding positive numbers. Although it seems unlikely that the scale ten will ever be changed for ordinary work, the use of other scales and systems of notation is still of practical as well as mathematical interest. In some types of computing machines, for example, scales other than ten lend themselves more readily to mechanization.

A slight modification of the ordinary expansion gives a representation for numbers with certain computational advantages. Assuming r to be odd, it is seen easily that any positive or negative number b can be represented as

$$b = \sum_{-\infty}^{N} a_n r^n, \qquad -\frac{r-1}{2} \le a_n \le \frac{r-1}{2},$$

and we may denote b as usual by the sequence of its digits

$$b = a_N \ldots a_0 \cdot a_{-1} \ldots$$

Both positive and negative numbers are thus represented by a standard notation without a prefixed sign, the sign being implied by the digits themselves; the number is positive or negative according as the first (nonvanishing) digit is greater or less than zero. Every real number has a unique representation apart from those whose expansion ends in an infinite sequence of the digits $(r-1)/2$ or $-(r-1)/2$, each of which has two representations. If this notation were to be used, a simple notation should be invented for the negative digits which suggested their close relation to the corresponding positive digits. For typographical simplicity we shall here denote the negative digits by placing primes on the corresponding positive digits. The notation for the first nine positive and negative integers with $r =$

201

$3, 5, 7, 9$ is as follows:

r	-9	-8	-7	-6	-5	-4	-3	-2	-1
3	1′00	1′01	1′11′	1′10	1′11	1′1′	1′0	1′1	1′
5	2′1	2′2	1′2′	1′1′	1′0	1′1	1′2	2′	1′
7	1′2′	1′1′	1′0	1′1	1′2	1′3	3′	2′	1′
9	1′0	1′1	1′2	1′3	1′4	4′	3′	2′	1′

r	0	1	2	3	4	5	6	7	8	9
3	0	1	11′	10	11	11′1′	11′0	11′1	101′	100
5	0	1	2	12′	11′	10	11	12	22′	21′
7	0	1	2	3	13′	12′	11′	10	11	12
9	0	1	2	3	4	14′	13′	12′	11′	10

In general the negative of any number is found by placing a prime on each unprimed digit and taking it off each primed digit. Arithmetic operations with this system are considerably simplified. In the first place the symmetries introduced by this notation make the addition and multiplication tables much easier to learn. For the scale $r = 9$ these tables are, respectively, as follows:

+	4′	3′	2′	1′	0	1	2	3	4
4′	1′1	1′2	1′3	1′4	4′	3′	2′	1′	0
3′	1′2	1′3	1′4	4′	3′	2′	1′	0	1
2′	1′3	1′4	4′	3′	2′	1′	0	1	2
1′	1′4	4′	3′	2′	1′	0	1	2	3
0	4′	3′	2′	1′	0	1	2	3	4
1	3′	2′	1′	0	1	2	3	4	14′
2	2′	1′	0	1	2	3	4	14′	13′
3	1′	0	1	2	3	4	14′	13′	12′
4	0	1	2	3	4	14′	13′	12′	11′

	4′	3′	2′	1′	0	1	2	3	4
4′	22′	13	11′	4	0	4′	1′1	1′3′	2′2
3′	13	10	13′	3	0	3′	1′3	1′0	1′3′
2′	11′	13′	4	2	0	2′	4′	1′3	1′1
1′	4	3	2	1	0	1′	2′	3′	4′
0	0	0	0	0	0	0	0	0	0
1	4′	3′	2′	1′	0	1	2	3	4
2	1′1	1′3	4′	2′	0	2	4	13′	11′
3	1′3′	1′0	1′3	3′	0	3	13′	10	13
4	2′2	1′3′	1′1	4′	0	4	11′	13	22′

The labor in learning the tables would appear to be reduced by a factor of at least two from the corresponding $r = 9$ case in ordinary notation. There is no need to learn a "subtraction table"; to subtract, one primes all digits of the subtrahend and adds. The sign of the difference automatically comes out correct, and the clumsy device of "borrowing" is unnecessary. More generally, to add a set of numbers, some positive and some negative, all are placed in a column without regard to sign and added, *e.g.* ($r = 9$):

(1′)		(1)		carried numbers
1	3′	1′	2	
2′	3	1	4	
4′	1′	2	3	
3	2′	3′	4	
3	0	0	1′	
1′	2′	1	3′	
1′	4	1	0	

This process may be contrasted with the usual method where the positive and negative numbers must be added separately, the smaller sum subtracted from the larger and the difference given the sign of the larger, that is, three addition or subtraction processes and a sign rule, while with the symmetrical system one standard addition process covers all cases. Furthermore, in such a sum cancellation is very common and reduces considerably the size of numbers to be carried in memory in adding a column; this follows from the fact that any digit cancels its negative and these may be struck out from a column without affecting the sum. If all digits are equally likely and independent, the sum in a column will have a mean value zero, standard deviation $\sqrt{p(r^2 - 1)/12}$ where p is the number of numbers being added, while in the usual notation the mean value is $p(r/2)$ with the same standard deviation.

Multiplication and division may be carried out also by the usual processes and here again signs take care of themselves, although in these cases, of course the advantage of this is not so great.

We may note also that in the usual system of notation, when we wish to "round off" a number by replacing all digits after a certain point by zeros the digits after this point must be inspected to see whether they are greater or less than 5 in the first place following the point. In the former case the preceding digit is increased by one. With the symmetrical system one always obtains the closest approximation merely by replacing the following digits by zeros. Numbers such as $1.444\ldots = 2.4'4'4'\ldots$ with two representations are exactly half way between the two nearest rounded off approximations, and in this case we obtain the upper or lower approximation depending on which representation is rounded off. If we were using this notation, department stores would find it much more difficult to camouflage the price of goods with \$.98 labels.

We have assumed until now that the scale r is odd. If r is even, say 10, a slightly unbalanced system of digits can be used; for example, $4', 3', 2', 1', 0, 1, 2, 3, 4, 5$. The dissymmetry introduced unfortunately loses several of the advantages described above, *e.g.*, the ease of negation and hence of subtraction and also the round off property.

A more interesting possibility is that of retaining symmetry by choosing for "digits" numbers halfway between the integers. In the case $r = 10$ the possible digits would be

$$a_n = \frac{9'}{2}, \frac{7'}{2}, \frac{5'}{2}, \frac{3'}{2}, \frac{1'}{2}, \frac{1}{2}, \frac{3}{2}, \frac{5}{2}, \frac{7}{2}, \frac{9}{2},$$

and any number b can be expressed as

$$b = \sum_{-\infty}^{N} a_n r^n.$$

In this system the properties of positive-negative symmetry, automatic handling of signs, and simple round off are retained. One curious and disadvantageous feature is that the integers can only be represented as infinite decimals, and this is possible in an infinite number of different ways. For example.

$$0 = \cdot \frac{1}{2} \frac{9'}{2} \frac{9'}{2} \cdots = \frac{1}{2} \cdot \frac{9'}{2} \frac{9'}{2} \frac{9'}{2} \cdots = \frac{1'}{2} \cdot \frac{9}{2} \frac{9}{2} \frac{9}{2} \cdots \text{ etc.}$$

Symmetrical notation offers attractive possibilities for general purpose computing machines of the electronic or relay types. In these machines it is possible to perform the calculations in any desired scale and only translate to the scale ten at input and output. The use of a symmetrical notation simplifies many of the circuits required to take care of signs in addition and subtraction, and to properly round off numbers.

57(1950), 90–93

A Dart Game

E 811 [1948, 248]. *Proposed by H. D. Larson, Albion College, Michigan*

A, B, and C participate in a novel dart game, the targets consisting of three small balloons marked A, B, and C, respectively. At each turn one dart is thrown, the order of the turns being determined in advance by drawing lots. As soon as a balloon is hit and destroyed, the owner of that balloon is eliminated from the game. The balloons are placed in such a manner that there is no danger of destroying a balloon by a dart aimed at another balloon. It is known by all participants that A can hit a balloon 4 out of 5 times, B 2 out of 5 times, and C 2 out of 5 times; this knowledge is used by each player to his best advantage. What is each contestant's chance of winning the game?

Solution by Leo Moser, University of Manitoba. In a two man game, say E vs. F, with chance of hit on aim p_1 and p_2 respectively, the probability of E winning if he has first shot is

$$\sum_{i=0}^{\infty} p_1[(1 - p_1)(1 - p_2)]^i = p_1/[1 - (1 - p_1)(1 - p_2)].$$

It is clear that when all three men are in the game the best strategy for A and B is to try to eliminate each other while C does his best to miss. Thus there are only two essentially different orders of firing, A, B, C and B, A, C, and these have equal probability.

For A to win he must first win a two man game with B and then a two man game with C in which C has first shot. The case for B is similar. Taking these facts into account we find that the chances of A, B, and C winning are 1596/4807, 891/4807, 2320/4807, respectively. One observes that the poorest shot has the best chance of winning!

The Future of Mathematics*

André Weil

"At one time," says Poincaré in his Rome conference on the future of mathematics, "there were prophets of misfortune; they reiterated that all the problems had been solved, that after them there would be nothing but gleanings left . . . ." "But," he added, "the pessimists have always been compelled to retreat . . . so that I believe there are none left to-day."

Our faith in progress, our belief in the future of our civilization are no longer as strong; they have been too rudely shaken by brutal shocks. To us, it hardly seems legitimate to "extrapolate" from the past and present to the future, as Poincaré did not hesitate to do. If the mathematician is asked to express himself as to the future of his science, he has a right to raise the preliminary question: what kind of future is mankind preparing for itself? Are our modes of thought, fruits of the sustained efforts of the last four or five millennia, anything more than a vanishing flash? If, unwilling to stumble into metaphysics, one should prefer to remain on the hardly more solid ground of history, the same questions reappear, although in different guise: are we witnessing the beginning of a new eclipse of civilization? Rather than to abandon ourselves to the selfish joys of creative work, is it not our duty to put the essential elements of our culture in order, for the mere purpose of preserving it, so that at the dawn of a new Renaissance, our descendants may one day find them intact?

These questions are not purely rhetorical; upon each man's answer, or rather (for such questions do not have answers), upon the attitude which he takes in front of them, depends in large measure the trend of his intellectual efforts. It was necessary, before writing about the future of mathematics, to formulate these questions, just as the faithful cleansed themselves before consulting the oracle. Let us now interrogate destiny.

Mathematics, as we know it, appears to us as one of the necessary forms of our thought. The archaeologist and the historian have shown us civilizations from which mathematics were absent. It is indeed doubtful whether they would ever have become more than a technique, at the service of technologies, if it had not been for the Greeks; and it is possible that, under our very eyes, a type of human society is being evolved in which they will be nothing but that. But for us, whose shoulders sag under the weight of the heritage of Greek thought and who walk in the paths traced out by the heroes of the Renaissance, a civilization without mathematics is unthinkable. Like the parallel postulate, the postulate that mathe-

*Authorized translation by Arnold Dresden of the article entitled L'avenir des mathématiques in the volume Les grands courants de la pensée mathématique, edited by F. Le Lionnais. Cahiers du Sud, Marseille, 1948.

matics will survive has been stripped of its "evidence"; but, while the former is no longer necessary, we would not be able to get on without the latter.

The clinical student of ideas who limits his prognosis to the immediate future, and does not risk long-range prophecies, certainly observes more than one favorable symptom in contemporary mathematics. To begin with, while some sciences, conferring, as they now do, an almost unlimited power upon a ruthless possessor of their results, tend to become caste monopolies, treasures jealously guarded under a seal of secrecy which must of necessity become fatal to any genuine scientific activity, the real mathematician does not seem to be exposed to the temptations of power nor to the straight-jacket of state secrecy. "Mathematics" said G. H. Hardy in substance in a famous inaugural lecture, "is a useless science. By this I mean that it can contribute directly neither to the exploitation of our fellowmen, nor to their extermination."

It is certain that few men of our times are as completely free as the mathematician in the exercise of their intellectual activity. Even if some State ideologies sometimes attack his person, they have never yet presumed to judge his theorems. Every time that so-called mathematicians, to please the powers that be, have tried to subject their colleagues to the yoke of some orthodoxy, their only reward has been contempt. Let others besiege the offices of the mighty in the hope of getting the expensive apparatus, without which no Nobel prize comes within reach. Pencil and paper is all the mathematician needs; he can even sometimes get along without these. Neither are there Nobel prizes to tempt him away from slowly maturing work, towards a brilliant but ephemeral result. Mathematics is taught the world over, well here, badly there; the exiled mathematician—and who among us can to-day feel free from the danger of exile—can find everywhere the modest livelihood which allows him to pursue his work to some extent. Even in gaol one can do good mathematics if one's courage fail him not.

To these "objective conditions," or rather, as the physician would say, to these external symptoms, must be added others revealed by a more penetrating clinical examination. In recent times mathematics has demonstrated its vitality by passing through one of these periods of growing pains, to which it has been accustomed for a long time, and which are designated by the strange name of "foundation crises." It has come through it, not only without damage, but with great gain. Whenever wide domains have been added to the field of mathematical reasoning, it is necessary to inquire what techniques are allowed in the exploration of the new territory. One wants certain objects to have certain properties, one wants certain modes of reasoning to be admissible and one behaves as if they were. But the pioneer who proceeds in this way knows very well that some day the police will come to put an end to the disorder and to bring everything under the control of the general law. Thus, when the Greeks defined the ratio of two magnitudes for the first time with enough precision to raise the problem of the existence of incommensurable magnitudes, they seem to have believed and to have wanted all ratios to be rational and to have based the first sketch of their geometrical reasonings on this provisional hypothesis; some of the greatest advances in Greek mathematics are connected with the discovery of their initial error at this point. In the same way, at the beginning of the era of the theory of functions and of the infinitesimal calculus, one wished every analytical expression to define a function, and every function to have a derivative; we know to-day that these requirements were incompatible. The last crisis, which grew out of the sophistries, for which the "naive" theory of sets opened a way in its early stages, has led for us to a no less happy result, which can now be considered as permanently established. We have learned to trace our entire science back to a single source, constituted by a few signs and by a few rules for their use; this is unquestionably an unassailable stronghold, inside which we could scarcely confine ourselves without risk of famine, but to which we are always free to retire in case of uncertainty or of external danger. Only a few backward spirits still maintain the position that the mathematician must forever draw on his "intuition" for new, alogical or "prelogical" elements of reasoning. If certain branches of mathematics have not yet been axiomatized, *i.e.*, reduced to a form of exposition in which all terms are defined, and all axioms made explicit, in terms of the basic notions of set theory, this is simply because there has not yet been the time to do it. It is of course possible that some day our successors will want to introduce into set-theory modes of reasoning which we do not admit. It is even possible that the germ of a contradiction, which we do not perceive to-day, may later be discovered in the modes of reasoning we now use, although the work of the modern logicians makes this very unlikely. A general revision will then become necessary; one can feel certain even now that this will not affect the essential elements of our science.

But, if logic is the hygiene of the mathematician, it is not his source of food; the great problems furnish the daily bread on which he thrives. "A branch of science is full of life," said Hilbert, "as long as it offers an abundance of problems; a lack of problems is a sign of death." They are certainly not lacking in our mathematics; and the present time might not be ill chosen for drawing up a list, as Hilbert did in the famous lecture from which we have just quoted. Even among those of Hilbert, there are still several which stand out as distant, although not inaccessible, goals which will continue to suggest research for perhaps more than a generation; an example is furnished by his fifth problem, on Lie groups. The Riemann hypothesis, after the attempts to prove it by function-theoretic methods had been given up, appears to-day in a new light, which shows it to be closely connected with the conjecture of Artin on the L-functions, thus making these two problems two aspects of the same arithmetico-algebraic question, in which the simultaneous study of all the cyclotomic extensions of a given number field will undoubtedly play a decisive role.

· · ·

57(1950), 295–306

Benjamin Franklin Finkel (1865–1947)

5. MATHEMATICS GETS SERIOUS, 1951–1960

W. L. Duren, Jr., President of the MAA 1955

Mathematics Gets Serious

Looking back, Americans seemed particularly stern in the 1950's. Early in the decade, McCarthyism rose and finally fell when Senator Joseph McCarthy was held in contempt of Congress late in 1954. Around the world, political boundaries shifted steadily, as decolonization and revolutions moved countries in and out of the two major spheres of influence. The dollar was stable. World trade soared. Both the US and the Soviet Union developed ever more powerful nuclear weapons. Americans had a presence in the world, and they worried.

America changed. In 1951, the first transcontinental television broadcast carried an address by President Truman. In 1954, the Supreme Court outlawed racial segregation in public schools. In 1956, President Eisenhower signed the Highway Act, calling for an interstate highway system to be built in the next 20 years. In 1957, Congress passed the first civil rights bill since Reconstruction. In 1958, America (finally) launched Explorer I into orbit and discovered the Van Allen radiation belt. And in 1959, both Alaska and Hawaii were admitted into the union, changing the 48 states to 50. In 1960, John F. Kennedy was elected President. The decade had given birth to Rock and Roll, I Love Lucy, McDonalds, and strange cars with fins. America was different.

Mathematics changed too: Mathematics was more serious. The first International Congress of Mathematicians in 14 years took place in Cambridge, Massachusetts, in 1950. The plenary talks settled down to business and the titles sound specialized (except for *The Cultural Basis of Mathematics* by Raymond Wilder). Of the 22 plenary speakers, 15 were associated to American institutions. America was the leader.

The Monthly published a translation of a long, scholarly paper by Hermann Weyl called *The Future of Mathematics*. The style of papers—even short notes—became more research-like. The problems got harder, the book reviews shorter.

Support for mathematics steadily grew throughout the decade. The newly formed National Science Foundation provided funds for individual research grants, but it also sought creative ways to support the general mathematical enterprise—especially teaching. There were summer institutes for college mathematics teachers, visiting lecturer programs, and special programs for high school mathematics teachers. Everyone was concerned about the training of engineers and scientists (because of the Russians) and many advocated replacing the worn out teaching methods of the past with innovative techniques and technology. There were those who counseled caution.

The Committee on the Undergraduate Program in Mathematics (CUPM) was born in 1952 and grew throughout the decade. The School Mathematics Study Group (SMSG) was created in 1958. The Conference Board of Mathematical Sciences (CBMS) was formed in 1955. In every case, budgets for these organizations increased by orders of magnitude before the decade was over. Mathematicians organized themselves, they made their case for increased funding, they received it.

This was a decade of strange mixtures—progress and anxiety, self-assurance and self-scrutiny, keen interest in the future and contentment in the past. Just after J. D. Salinger published *Catcher in the Rye* in 1951, Eilenberg and Steenrod published *Algebraic Topology*, which was the first textbook in the subject. When Jack Kerouac was writing *On the Road* in 1956, John Milnor was writing his paper *On manifolds homeomorphic to the 7-sphere*, which led to a revolution in differential topology. And just before Kennedy won the Democratic nomination in 1960, Steve Smale announced the solution to the generalized Poincaré Conjecture. Topology had grown up in the 50's.

At the close of the decade, perhaps in reaction to 10 years of seriousness, the Monthly published a lovely short paper on the mathematics of baseball.

Lester R. Ford, Editor of the MONTHLY 1942–46

A Half-Century of Mathematics

Hermann Weyl

1. Introduction. Axiomatics. Mathematics, beside astronomy, is the oldest of all sciences. Without the concepts, methods and results found and developed by previous generations right down to Greek antiquity, one cannot understand either the aims or the achievements of mathematics in the last fifty years. Mathematics has been called the science of the infinite; indeed, the mathematician invents finite constructions by which questions are decided that by their very nature refer to the infinite. That is his glory. Kierkegaard once said religion deals with what concerns man unconditionally. In contrast (but with equal exaggeration) one may say that mathematics talks about the things which are of no concern at all to man. Mathematics has the inhuman quality of starlight, brilliant and sharp, but cold. But it seems an irony of creation that man's mind knows how to handle things the better the farther removed they are from the center of his existence. Thus we are cleverest where knowledge matters least: in mathematics, especially in number theory. There is nothing in any other science that, in subtlety and complexity, could compare even remotely with such mathematical theories as for instance that of algebraic class fields. Whereas physics in its development since the turn of the century resembles a mighty stream rushing on in one direction, mathematics is more like the Nile delta, its waters fanning out in all directions. In view of all this: dependence on a long past, other-worldliness, intricacy, and diversity, it seems an almost hopeless task to give a non-esoteric account of what mathematicians have done during the last fifty years. What I shall try to do here is, first to describe in somewhat vague terms general trends of development, and then in more precise language explain the most outstanding mathematical notions devised, and list some of the more important problems solved, in this period.

One very conspicuous aspect of twentieth century mathematics is the enormously increased role which the axiomatic approach plays. Wheras the axiomatic method was formerly used merely for the purpose of elucidating the foundations on which we build, it has now become a tool for concrete mathematical research. It is perhaps in algebra that it has scored its greatest successes. Take for instance the system of real numbers. It is like a Janus head facing in two directions: on the one side it is the field of the algebraic operations of addition and multiplication; on the other hand it is a continous manifold, the parts of which are so connected as to defy exact isolation from each other. The one is the algebraic, the other the topological face of numbers. Modern axiomatics, simple-minded as it is (in contrast to modern politics), does not like such ambiguous mixtures of peace and war, and therefore cleanly separated both aspects from each other.

In order to understand a complex mathematical situation it is often convenient to separate in a natural manner the various sides of the subject in question, make each side accessible by a relatively narrow and easily surveyable group of notions and of facts formulated in terms of these notions, and finally return to the whole by uniting the partial results in their proper specialization. The last synthetic act is purely mechanical. The art lies in the first, the analytic act of suitable separation and generalization. Our mathematics of the last decades has wallowed in generalizations and formalizations. But one misunderstands this tendency if one thinks that generality was sought merely for generality's sake. The real aim is simplicity: every natural generalization simplifies since it reduces the assumptions that have to be taken into account. It is not easy to say what constitutes a natural separation and generalization. For this there is ultimately no other criterion but fruitfulness: the success decides. In following this procedure the individual investigator is guided by more or less obvious analogies and by an instinctive discernment of the essential acquired through accumulated previous research experience. When systematized the procedure leads straight to axiomatics. Then the basic notions and facts of which we spoke are changed into undefined terms and into axioms involving them. The body of statements deduced from these hypothetical axioms is at our disposal now, not only for the instance from which the notions and axioms were abstracted, but wherever we come across an interpretation of the basic terms which turns the axioms into true statements. It is a common occurrence that there are several such interpretations with widely different subject matter.

The axiomatic approach has often revealed inner relations between, and has made for unification of methods within, domains that apparently lie far apart. This tendency of several branches of mathematics to coalesce is another conspicuous feature in the modern development of our science, and one that goes side by side with the apparently opposite tendency of axiomatization. It is as if you took a man out of a milieu in which he had lived not because it fitted him but from ingrained habits and prejudices, and then allowed him, after thus setting him free, to form associations in better accordance with his true inner nature.

In stressing the importance of the axiomatic method I do not wish to exaggerate. Without inventing new constructive processes no mathematician will get very far. It is perhaps proper to say that the strength of modern mathematics lies in the interaction between axiomatics and construction. Take algebra as a representative example. It is only in this century that algebra has come into its own by breaking away from the one universal system Ω of numbers which used to form the basis of all mathematical operations as well as all physical measurements. In its newly-acquired freedom algebra envisages an infinite variety of "number fields" each of which may serve as an operational basis; no attempt is made to embed them into the one system Ω. Axioms limit the possibilities for the number concept; constructive processes yield number fields that satisfy the axioms.

In this way algebra has made itself independent of its former master analysis and in some branches has even assumed the dominant role. This development in mathematics is paralleled in physics to a certain degree by the transition from classical to quantum physics, inasmuch as the latter ascribes to each physical structure its own system of observables or quantities. These quantities are subject to the algebraic operations of addition and multiplication; but as their multiplication is non-commutative, they are certainly not reducible to ordinary numbers.

At the International Mathematical Congress in Paris in 1900 David Hilbert, convinced that problems are the life-blood of science, formulated twenty-three unsolved problems which he expected to play an important role in the development of mathematics during the next era. How much better he predicted the future of mathematics than any politician foresaw the gifts of war and terror that the new century was about to lavish upon mankind! We mathematicians have often measured our progress by checking which of Hilbert's questions had been settled in the meantime. It would be tempting to use his list as a guide for a survey like the one attempted here. I have not done so because it would necessitate explanation of too many details. I shall have to tax the reader's patience enough anyhow.

· · ·

58(1951), 523–553

Dr. D. A. Rothrock, professor emeritus of Indiana University and a charter member of the Association, died September 2, 1949.

57(1950), 64

At Duke University the following appointments to instructorships have been made: Dr. Mary E. Estill of the University of Texas, Mr. L. M. Fulton Jr. and Dr. Walter Rudin, formerly graduate students at Duke University.

57(1950), 201

The University of Michigan announces the following: Assistant Professor Gail Young has been promoted to an associate professorship; Dr. E. E. Moise has been promoted to an assistant professorship; Dr. Raoul Bott, who has been studying at the Institute for Advanced Study, has been appointed to an instructorship; Dr. A. B. Clarke, formerly a graduate student at Brown University, has been appointed to an instructorship; Assistant Professor W. J. LeVeque is on leave of absence and is studying at the University of Manchester, England.

59(1952), 128

G. Bailey Price, President of the MAA 1957

The Problem of n Liars and Markov Chains

William Feller

1. Introduction. In the Advanced Problems section of this MONTHLY [4288, Vol. 57(1950) pp. 43–45] we find a solution of the following problem, first treated by A. S. Eddington: If A, B, C, D each speak the truth once in three times (independently), and A affirms that B denies that C declares that D is a liar, what is the probability that D was telling the truth? There are only eight different statements which A can make (or deny) and their enumeration shows that the required probability is 13/41.

John Riordan's remark that this problem applies also to switching circuits led me to look for the appropriate probability model. It turns out that the problem of n liar leads directly to the simplest Markov chain, and that natural variations of the same problem correspond to more general chains. The example may be of some didactic use since the theory of Markov chains grows in importance, but simple illustrative examples are hard to get.

2. Model for the simplest Markov chain. In the original formulation, persons C, B, and A issue successively three statements which may be true or false and which may contradict each other. However, taken at face value, each statement would imply either that D is telling the truth or that he lies. Thus, if B denies that C declares that D is a liar, he implies that D tells the truth. If A (rightly or wrongly) denies that B denied *etc.*, then he implies that D is a liar. In a continued process, an even number of denials cancel, and the implication of a statement like "A_1 asserts that A_2 denies that $A_3 \ldots$" depends on the evenness of the total number of denials.

To use a neutral terminology, we shall speak of a chance process with two possible states; at any time the observed state is 1 or 2 according as the last statement implies that D is honest or a liar. We imagine that the statements are issued at times $1, 2, 3, \ldots$. Initially, or at time 0, the "observed" state is 1 or 2 according as D tells the truth or lies. (In the original problem actually only D and C know the initial state.) Every possible sample sequence of the process is represented by a succession of the digits 1 and 2, and *vice versa*.

The fundamental assumption (to which the Markov character of the process is due) is that each person knows only the statement of the last speaker, but not the past history of the system. Moreover, at time n the observed state changes or remains unchanged according as the nth speaker tells the truth or lies. Thus we arrive at the following model which represents the simplest Markov chain:

We have a process with two possible states 1, 2. Initially (or at time 0) the probabilities of the two states are α and β, respectively $(\alpha + \beta = 1)$. Whatever the development up to time n, there is probability p that at time n the observed state does not undergo a change and probability

$q = 1 - p$ that it does. *We seek the conditional probabilities* x_n *and* y_n *that the process actually started from state* 1, *given that at time n the observed state is* 1 *or* 2, *respectively.*

In the original formulation $\alpha = p = \frac{1}{3}, n = 3$, and only x_n is required.

3. Each individual step consists in one of the four possible transitions $1 \to 1$, $1 \to 2$, $2 \to 1$, $2 \to 2$, and the corresponding transition probabilities are, by assumption,

(1) $$p_{11} = p_{22} = p, \qquad p_{12} = p_{21} = q.$$

Suppose now that at a certain time the system is in state j and let $p_{jk}^{(n)}$ be the probability (on this hypothesis) that n steps later the observed state is k. The $p_{jk}^{(n)}$ are called the n-step transition probabilities. Clearly

$$p_{jk}^{(1)} = p_{jk},$$

and

(2) $$p_{jk}^{(2)} = p_{j1}p_{1k} + p_{j2}p_{2k}.$$

More generally, we can calculate $p_{jk}^{(n)}$ from the obvious recursion formulas

(3) $$p_{jk}^{(n+1)} = p_{j1}^{(n)}p_{1k} + p_{j2}^{(n)}p_{2k}.$$

Now these are just the formulas for matrix multiplication: if the matrix (p_{jk}) is denoted by P, then the $p_{jk}^{(n)}$ are elements of P^n.

Since the initial probabilities of the states 1 and 2 are α and β, the probability $a_k^{(n)}$ of observing at time n the state k is obviously

(4) $$a_k^{(n)} = \alpha p_{1k}^{(n)} + \beta p_{2k}^{(n)}.$$

We find therefore for the two required probabilities

(5) $$x_n = \frac{\alpha p_{11}^{(n)}}{a_1^{(n)}} \quad \text{and} \quad y_n = \frac{\alpha p_{12}^{(n)}}{a_2^{(n)}}.$$

For explicit formulas we must, of course, calculate $p_{jk}^{(n)}$. The formula

(6) $$P^n = \frac{1}{2}\begin{pmatrix} 1 & 1 \\ 1 & 1 \end{pmatrix} + \frac{1}{2}(p-q)^n \begin{pmatrix} 1 & -1 \\ -1 & 1 \end{pmatrix}$$

follows directly from the canonical decomposition of matrices, and can easily be verified. Substituting from (6) and (4) into (5) we find

(7)
$$x_n = \frac{\alpha\{1 + (p-q)^n\}}{1 + (\alpha - \beta)(p-q)^n},$$

$$y_n = \frac{\alpha\{1 - (p-q)^n\}}{1 - (\alpha - \beta)(p-q)^n}.$$

For $n = 3, \alpha = p = \frac{1}{3} = 13/41$ in agreement with the solutions quoted. Moreover, $y_n = 7/20$. Clearly $x_n \to \alpha$ as $n \to \infty$. It is interesting that x_n decreases monotonically if $p > q$, while x_n is alternately larger and smaller than α if $p < q$.

4. Preferential lying. Up to now we have assumed that the chances of a person telling the truth are in no way dependent on the statement he is supposed to relay. Suppose now that each person has a preference to claim that D is honest. Then a transition $1 \to 1$ is more probable than $2 \to 2$, while $1 \to 2$ is less probable than $2 \to 1$.

We can treat this general case in the same way as before, provided we replace (1) by the general matrix of transition probabilities

$$(8) \qquad p_{11} = p, \qquad p_{12} = q, \qquad p_{21} = q', \qquad p_{22} = p',$$

where, of course, $p + q = p' + q' = 1$. All formulas of the preceding section apply, except that (6) must be replaced by

$$(9) \qquad P^n = \frac{1}{q + q'}\begin{pmatrix} q' & q \\ q' & q \end{pmatrix} + \frac{(p' - q)^n}{q + q'}\begin{pmatrix} q & -q \\ -q' & q' \end{pmatrix}.$$

The final result is now

$$(10) \qquad x_n = \frac{\alpha\{q' + q(p' - q)^n\}}{q' + (\alpha q - \beta q')(p' - q)^n}.$$

5. Generalizations. In the preceding examples we had a message capable of two forms transmitted in successive steps. It is easy to generalize this scheme to the case where the message can assume N different forms $1, 2, \ldots, N$. (For example, the message may be a digit which is repeatedly copied and is subject to copying errors.) The preceding theory applies, except that the matrix $P = (p_{jk})$ of transition probabilities is now of order N, and that more questions can be asked. We are, in this way, led to the general Markov chain with N possible states and constant (or stationary) transition probabilities. Finally, if we admit that the transition probabilities vary from step to step (variable proneness to lie) then we are led to the most general Markov chain with finitely many states.[*]

58(1951), 606–608

58(1951), 208

[*]For the general theory cf. Chapters 15 and 16 of *An Introduction to Mathematical Probability and Its Applications* by W. Feller (Wiley, New York, 1950).

BOOKS RECEIVED

1. *Essentials of College Algebra*. By J. B. Rosenbach and E. A. Whitman. New York. Ginn and Company, 1951. 10 + 322 + 30 pages. $3.00

2. *College Algebra*. By H. L. Rietz, A. R. Crathorne, and J. W. Peters. New York. Henry Holt and Company, 1951. 17 + 387 pages.

3. *College Algebra*. By H. K. Fulmer and Walter Reynolds. New York. Ginn and Company, 1951. 6 + 204 + 14 pages.

4. *Intermediate College Algebra*. By E. M. J. Pease. New York. Prentice-Hall, Inc., 1950. 7 + 420 + 36 pages. $2.85.

5. *Intermediate Algebra*. By P. K. Rees and F. W. Sparks. New York. McGraw-Hill Book Company, 1951. 8 + 328 pages. $3.25.

6. *Elements of Algebra*. By L. C. Peck. New York. McGraw-Hill Book Company, 1950. 13 + 230 pages. $2.75.

7. *Practical Mathematics, Part II, Algebra with Applications*. By C. I. Palmer and S. F. Bibb. New York. McGraw-Hill Book Company, Inc., 1950. 12 + 252 pages. $2.20.

8. *Algebra for Commerce and Liberal Arts*. By A. K. Bettinger and W. A. Dwyer. New York. Pitman Publishing Corporation, 1951. 11 + 225 pages.

Mina Rees (1902–)

A Prime-Representing Function

E. M. Wright

Recently Mills [1] proved that there is a fixed number A such that, for all positive integral values of x, the number

$$[A^{3^x}]$$

is a prime. Here $[y]$ denotes as usual the greatest integer $\leq y$. The proof depends on the result that

$$p_{n+1} - p_n < K p_n^{5/8},$$

where K is a fixed positive number and p_n is the nth prime. This result (due to Ingham [2]) is fairly deep.

Here I prove

THEOREM. *There is a number α such that, if*

$$g_0 = \alpha, \qquad g_{n+1} = 2^{g_n}, \qquad\qquad (n \geq 0),$$

then

$$[g_n] = \left[2^{2^{2^{\cdot^{\cdot^{\cdot^{\alpha}}}}}} \right], \qquad\qquad (n \geq 1),$$

is always a prime.

I require only the very elementary result that, for every $N \geq 2$, there is a prime between N and $2N$ (see [2]). By this, we can choose a sequence of primes $\{P_n\}$ such that $P_1 = 2$ or 3 and

$$2^{P_n} < P_{n+1} < P_{n+1} + 1 < 2^{P_n+1}.$$

We take all logarithms to base 2, and write,

$$u_n = \log^{(n)} P_n, \qquad v_n = \log^{(n)}(P_n + 1),$$

where $\log^{(n)}$ denotes the nth iterate of the logarithm. We have

$$P_n < \log P_{n+1} < \log(P_{n+1} + 1) < P_n + 1$$

and so

$$u_n < u_{n+1} < v_{n+1} < v_n.$$

Hence $u_n \to \alpha$ (say) as $n \to \infty$ and

$$u_n < \alpha < v_n, \qquad P_n < g_n < P_n + 1$$

and so

$$P_n = [g_n].$$

There are, of course, any number of possible values of α. For example, if

$$\alpha = 1.9287800\ldots,$$

221

we have

$$P_1 = 3, \qquad P_2 = 13, \qquad P_3 = 16381$$

and P_4 has some 5000 digits.

If we use only the almost trivial result that there is a number B such that, for every $N \geq 2$, there is a prime between N and BN (see [3]), we can prove that, for some β, $[h_n]$ is always a prime, where

$$h_0 = \beta, \qquad h_{n+1} = B^{k_n} \qquad\qquad (n \geq 0).$$

References

1. W. H. Mills, "A prime-representing function," *Bull. Amer. Math. Soc.* 53 (1947), 604.

2. E. Landau, *Handbuch der Lehre von der Verteilung der Primzahlen* I, 92 (Leipzig 1909).

3. G. H. Hardy and E. M. Wright, *Introduction to the Theory of Numbers*, second edition, 346 (Oxford, 1945).

58(1951), 616–618

1. *A theory of distribution*, by Professor Claude Chevalley, Columbia University, introduced by the Secretary.

The main object of Schwarz's theory of distributions is to put on a rigorous and uniform mathematical footing certain types of computational methods which are in frequent use by the physicists, like, for instance, the Heaviside Calculus.

One of the most frequent sources of difficulty in trying to make simple and general statements in analysis is the fact that a given function, even if it is continuous, does not necessarily have a derivative. The distributions are a class of mathematical objects wider than the class of functions, and inside which the operation of derivation is possible without any restriction. Thus, in particular, a continuous function always has derivatives of all orders; these derivatives are of course not functions in general, but distributions.

In order to define the notion of distribution, one introduces first the set D of indefinitely differentiable functions which are zero outside bounded sets (the bounded set depending on the function). This set is a vector space: the sum of two functions in D is in D, and the product by a constant of a function in D is in D. The distributions are then defined to be the linear functionals on D which are continuous relatively to a suitable notion of convergence for functions in D. A function $f(x)$ is identified with the functional which assigns to every g in D the number $\int_{-\infty}^{\infty} f(x)g(x)dx$. The derivative of a distribution S assigns to every function g in D the number $-S(dg/dx)$.

59(1952), 216

Mathematical Teaching
in Universities

André Weil

The following is the outline of a lecture once given by the author at a joint meeting of the Nancago Mathematical Society and of the Poldavian Mathematical Association. It is printed here at the editor's request, as the principles stated there seem to be of general application.

1. Improvements in the mathematical teaching in Poldavian Universities depend largely upon general improvements in the educational system in Poldavia. Mathematicians should devote themselves to the task of making such improvements as lie within their power at present, and thus contributing their share towards general reforms, which in turn will enable them to make further progress.

2. No satisfactory results can be achieved unless reforms are made both in school-teaching and in University teaching. So far as school-teaching is concerned, the efforts of mathematicians in the country should be mainly directed towards necessary changes in the curricula and towards the training of better teachers.

3. University teaching in mathematics should: (a) answer the requirements of all those who need mathematics for practical purposes; (b) train specialists in the subject; (c) give to all students that intellectual and moral training which any University, worthy of the name, has the duty to impart.

These objects are not contradictory but complementary to each other. Thus, a training for practical purposes can be made to play the same part in mathematics as experiments play in physics or chemistry. Thus again, personal and independent thinking cannot be encouraged without at the same time fostering the spirit of research.

4. The study of mathematics, as well as of any other science, consists in the acquisition of useful reflexes and in that of independent habits of thought. The acquisition of useful reflexes should never be separated from the perception of their usefulness.

It follows that problem-solving should never be practised for its own sake; and particularly tricky problems must be excluded altogether. The purpose of problems is twofold; either to drill the student in the application of some method of special importance, or to develop his originality by guiding him along some new path. Drill is essentially a school-method, and ought to become unnecessary at the final stages of University teaching.

5. Rigor is to the mathematician what morality is to man. It does not consist in proving everything, but in maintaining a sharp distinction between what is assumed and what is proved, and in endeavoring to assume as little as possible at every stage.

The student should therefore be gradually accustomed, by means of startling examples, to question the truth of every unproved proposition, until at last he is able to deduce from the ordinary axioms everything that he has learned.

6. Knowledge of a proof means the understanding of its machinery and the ability to reconstruct it. This implies: (a) perfect correctness in the definitions; (b) a faculty of connecting a given question with the general ideas underlying it; (c) a perception of the logical nature of any proof.

The teacher should therefore always follow, not the quickest nor even the most elegant method, but the method which is related to the most general principles. He should also point out everywhere the relation between the various elements of the hypothesis and the conclusion; students must be accustomed to draw a sharp distinction between premises and conclusion, between necessary and sufficient conditions, between a theorem and its converse.

7. The teaching of mathematics must be a source of intellectual excitement. This can be achieved, at the higher stages, by taking the student to the brink of the unknown; at earlier stages, by making him solve for himself questions of theoretical or practical importance.

This is the method followed in the "seminars" of the German Universities, first organized by Jacobi a century ago, and even now the most prominent feature of the German system; division of labor between students in the study of a given group of questions is a common practice in these seminars, and proves to be a powerful incentive to work.

8. Theoretical lectures should neither be a reproduction of nor a comment upon any text-book, however satisfactory. The student's notebook should be his principal text-book.

In fact, taking down notes intelligently (not under dictation) and working them out carefully at home should be considered as an essential part of the student's work; and experience shows that it is not the least useful part of it.

9. The right of any topic to form part of any curriculum is to be tested according to: (a) its importance for modern mathematics or for the applications of mathematics to modern science or technique; (b) its relations with other branches of the curriculum; (c) the intrinsic difficulty of the ideas underlying it.

This involves a revision of the present curriculum. For instance, the idea of function, the process of differentiation and integration, should appear at an early stage, because of their enormous importance both for the theory and for the most ordinary practice. Because of its practical importance, numerical calculation, and all the devices connected with it, would seem to deserve a far more prominent place in elementary teaching than they receive at present.

61(1954), 34–36

FRESHMAN MATHEMATICS
By MORRIS KLINE

. . .

2. Objections to the traditional courses. What have we been feeding the liberal arts students? The almost universal diet has been college algebra and trigonometry. I believe that these courses are a complete waste of time. What educational values are there really in exponents, radicals, logarithms, Horner's method, partial fractions, binomial theorem, the trigonometric identities, and the law of tangents, just to mention some of the conventional topics?

. . .

Suggestions to Students on Talking about Mathematics Papers*

G. E. Forsythe

The art of giving a talk is to choose and follow a dramatic technique appropriate to the occasion and to the audience. This is just as true of a mathematics talk as it is of an after-dinner speech. There are at least three types of mathematics talks which you may want to deliver: (1) a ten-minute paper before a scientific society; (2) a one-hour colloquium lecture before a department; (3) a review of a journal paper or a discussion of some problem in a working seminar.

Each of these talks has its technique, and the three techniques are different. I'd like to consider the working seminar, and give some suggestions that may help you get started. They won't fit all cases, of course, but I hope they will at least convince you of the importance of planning your talks carefully, and of looking objectively at the technique involved.

Acquiring background. As soon as you learn of a paper you will be studying, write the author for a reprint. A statement of your need may help you get it promptly. Most authors are delighted to receive requests for reprints, but don't be disappointed if the author has none left.

Allow a lot of time for study of the papers that you are to report on. You must know as much about a key paper as the author does; if you possibly can, you should understand it even better. Study where each hypothesis is used in the proof. (Could any hypothesis be weakened or eliminated? Here is often a place to start research of your own.) See how the results look in significant special cases. Actually compute numerical examples, if they are relevant, or special concrete cases where you can see what is going on. Not only are such cases illuminating, but they are also often extremely amusing.

Preparing the talk. Ask *early* for any suggestions the seminar instructor, or some older friend in your department, is able to give. When you have mastered the material, you are ready to draft a talk. Assume that your audience will know nothing about the subject except what may have been said in the seminar previously. But credit them with the intelligence to catch on quickly.

I would suggest writing down almost every word you expect to say, and everything you expect to write on the board. I would then reduce the above to an outline including the first and last sentences and the blackboard material verbatim, but leaving only key words for the rest. At this point I would throw away the

*With many suggestions from colleagues at New York University and from the editors of the MONTHLY. My students have expressed the hope that faculty members also will pay attention to these suggestions!

word-for-word write-up, as it can only harm your talk henceforth! Reading a talk is utterly ineffective.

Using your outline, have a very serious rehearsal alone with a blackboard, preferably in the room where the seminar meets. To keep an orderly array, plan where each formula will be written, and plan what will be erased and when. Write everything on the board as though your audience were present. Time yourself, and allow time for heckling from the audience later. This is a good time to check up on common sense, but often forgotten, details of speaking: posture, voice (no mumbling), writing (be sure you don't hide it, and go to the back of the room to check that your writing is readable).

At this stage you will probably have to prune your talk to save time. A second rehearsal is then in order, especially if you have changed the talk substantially. If there are one or two fellow-students who are your good friends and critics, you might consider inviting them to the second rehearsal to provide audience reaction and questions.

A suggestion which seems silly but which is apparently needed is to learn the Greek letters. You will also want to pronounce names correctly. Note, for example, that "Lewy" rhymes with "gravy," and that W. E. Milne pronounces his last name in one syllable.

Content of the talk. To gain time, I would have on the side board in advance all material—like references, tables and detailed diagrams—which takes much time to write, but which is to be read rather rapidly. On the other hand, it is deadly to write in advance anything in the main development of the subject.

Your introduction is vitally important and must be most carefully planned. You must set the stage by putting your subject in the framework of the audience's knowledge. Where do today's results fit into the general field? A historical approach is usually a good one; in this you recapitulate the main results and their discovery in chronological order. You will have to decide what definitions will first be given to make the history intelligible.

After (or before) you have traced the main developments in a general fashion, develop your basic definitions and notations carefully on the board and, if possible, leave them there for much of the talk. Lose no time here, but omit no essential detail.

A good technique is next to write down and interpret all of the theorems that you are going to cover. Include all the generality the author has contrived. Leave nothing undefined. Leave the theorems on the board (abbreviated, if necessary) for later reference.

Now discuss the implications of these results for the subjects of the seminar. Are there extensions of the theory? Do you recognize any promising unsolved problems; are there conjectures about them?

Next it is time for the key proofs. There are two good reasons for postponing them to now:

(1) you can control your time by eliminating or adding proofs, with the least harm to the total dramatic effect;

(2) you will undoubtedly lose some people in details, and you prefer to lose them as late as possible in your talk.

Select the proofs to be given on the basis of the representativeness or novelty of the methods used, the importance of the results, *etc.* When you give a proof, give it carefully. While the ten-minute talk and even the colloquium talk will ordinarily avoid details, the working seminar is expected to get to the bottom of the subject.

(If they don't, who will?) So roll up your sleeves and pitch in, and don't try to prove theorems by waving your hands. You can sometimes build a sound account of a proof around a special case. But afterwards try to tie it in with the general case. Let the audience know where each hypothesis enters the proof.

Conclude with a reiteration of the nature of the contribution here and its significance. And stop. You should welcome discussion and criticism, and one way to get it is to stop early. There are many sins that mathematical speakers fall heir to. I have tried above to show you how to avoid some of them. A common and deadly sin is to talk too long. The human animal ordinarily cannot and will not listen for more than one hour. So—when your time is up (or even a bit earlier), thank your audience for their attention and SIT DOWN.

As a final suggestion, attend all the talks you can by masters of the art, and study their techniques.

64(1957), 16–18

NEW TELEVISION SERIES

Professor Francis A. C. Sevier of the College of South Jersey is presenting a television series on WFIL-TV. The series began on January 30 and will run through May 14. It is titled "Mathematics in Modern Life". Outlines of Professor Sevier's syllabus can be obtained by writing to him at 406 Penn Street, Camden 2, New Jersey.

63(1956), 269

DU PONT FELLOWSHIPS

The University of Chicago announces six graduate fellowships of $1,920 for students who wish to prepare for teaching chemistry, mathematics or physics in secondary schools. E. I. du Pont de Nemours and Company has granted the University funds for these fellowships to encourage able college graduates of both sexes to enter high-school mathematics and science teaching.

63(1956), 269

Ivan Niven, President of the MAA 1983–1984

"If This Be Treason..."

R. P. Boas, Jr.

If I had to name one trait that more than any other is characteristic of professional mathematicians, I should say that it is their willingness, even eagerness, to admit that they are wrong. A sure way to make an impression on the mathematical community is to come forward and declare, "You are doing such-and-such all wrong and you should do it *this* way." Then everybody says, "Yes, how clever you are", and adopts your method. This of course is the way progress is made, but it leads to some curious results. Once upon a time square roots of numbers were found by successive approximations because nobody knew of a better way. Then somebody invented a systematic process and everybody learned it in school. More recently it was realized that very few people ever want to extract square roots of numbers, and besides the traditional process is not really very convenient. So now we are told to teach root extraction, if we teach it at all, by successive approximations. Once upon a time people solved systems of linear equations by elimination. Then somebody invented determinants and Cramer's rule and everybody learned that. Now determinants are regarded as old-fashioned and cumbersome, and it is considered better to solve systems of linear equations by elimination.

We are constantly being told that large parts of the conventional curriculum are both useless and out of date and so might better not be taught. Why teach computation by logarithms when everybody who has to compute uses at least a desk calculator? Why teach the law of tangents when almost nobody ever wants to solve an oblique triangle, and if he does there are more efficient ways? Why teach the conventional theory of equations, and especially why illustrate it will ill-chosen examples that can be handled more efficiently by other methods? As a professional mathematician, I am a sucker for arguments like these. Yet, sometimes I wonder.

There are a few indications that there is a reason for the survival of the traditional curriculum besides the fact that it is traditional. When I was teaching mathematics to future naval officers during the war, I was told that the Navy had found that men who had studied calculus made better line officers than men who had not studied calculus. Nothing is clearer (it was clear even to the Navy) than that a line officer never has the slightest use for calculus. At the most, his duties may require him to look up some numbers in tables and do a little arithmetic with them, or possibly substitute them into formulas. What is the explanation of the paradox?

I think that the answer is supplied by a phenomenon that everybody who teaches mathematics has observed: the students always have to be taught what they should have learned in the preceding course. (We, the teachers, were of course exceptions; it is consequently hard for us to understand the deficiencies of our students.) The average student does not really learn to add fractions in arithmetic class; but by the time he has survived a course in algebra he can add numerical fractions. He does not learn algebra in the algebra course; he learns it in calculus, when he is forced to use it. He does not learn calculus in the calculus course, either; but if he goes on to differential equations he may have a pretty good grasp of elementary calculus when he gets through. And so on through the hierarchy of courses; the most advanced course, naturally, is learned only by teaching it.

This is not just because each previous teacher did such a rotten job. It is because there is not time for enough practice on each new topic; and even if there were, it would be insufferably dull. Anybody who has really learned to interpolate in trigonometric tables can also interpolate in air navigation tables, or in tables of Bessel functions. He should learn, because interpolation is useful. But one cannot drill students on mere interpolation; not enough, anyway. So the students solve oblique triangles in order (among other things) to practice interpolation. One must not admit this to the students, but one may as well realize the facts.

Consequently, I claim that there is a place, and a use, even for nonsense like the solution of quartics by radicals, or Horner's method, or involutes and evolutes, or whatever your particular candidates for oblivion may be. Here are problems that might conceivably have to be solved; perhaps the methods are not the most practical ones; but that is not the point. The point is that in solving the problems the student gets practice in using the necessary mathematical tools, and gets it by doing something that has more motivation than mere drill. This is not the way to train mathematicians, but it is an excellent way to train mathematical technicians. Now we can understand why calculus improves the line officer. He needs to practice very simple kinds of mathematics; he gets this practice in less distasteful form by studying more advanced mathematics.

It is the fashion to deprecate puzzle problems and artificial story problems. I think that there is a place for them too. Problems about mixing chemicals or sharing work, however unrealistic, give good practice and even have a good deal of popular appeal: witness the frequency with which puzzle problems appear in newspapers, magazines, and the flyers that come with the telephone bill. There was once a story in *The Saturday Evening Post* whose plot turned on the interest aroused by a perfectly preposterous diophantine problem about sailors, coconuts, and a monkey. It is absurd to claim that only "real" applications should be used to illustrate mathematical principles. Most of the real applications are too difficult and/or involve too many side issues. One begins the study of French with simple artificial sentences, not with the philosophical writings of M. Sartre. Similarly one has to begin the study of a branch of mathematics with simple artificial problems.

We may dislike this state of affairs, but as long as it exists we must face it. It would be pleasant to teach only the new and exciting kinds of mathematics; it would be comforting to teach only the really useful kinds. The traditional topics are some of the topics that once were either new and exciting, or useful. They have persisted partly by mere inertia—and that is bad—but partly because they still serve a real purpose, even if it is not their ostensible purpose. Let us keep this in mind when we are revising the curriculum.

64(1957), 247–249

The Razor's Edge?

C. S. Ogilvy and N. G. Gunderson

Most modern calculus texts are careful to point out that the vanishing of the first partial derivatives is a necessary *but not sufficient* condition for a maximum or minimum of a function of two variables. One must examine the second partials also. Failure to do so has led more than one writer to include in his textbook an apparently plausible problem which in fact has a strange solution.* Interestingly enough the same problem, with only slight modifications, has found its way into several of the prominent texts.

In one book the problem is stated as follows: "A manufacturer produces safety razors and blades at a cost of 20¢ per razor and 10¢ per dozen blades. If he charges x cents per razor and y cents per dozen blades, he finds that he can sell $100,000/xy$ razors and $400,000/xy$ dozen blades daily. How should he fix prices so as to maximize his profit?"

The profit z is readily expressed as a function of x and y: $z = f(x, y) = 100,000(1/y + 4/x - 60/xy)$. If one equates the two first partials to zero and solves the resulting two equations simultaneously, one arrives quite happily at $x = 60, y = 15$. These apparently reasonable figures yield a profit of \$66.67. But a closer examination of the profit function discloses some disturbing facts. The ordinary three-dimensional curved surface represented by $z = f(x, y)$ has negative Gaussian curvature at the point P: $(60, 15, 6666\frac{2}{3})$; that is, the surface is saddle-shaped at P, so that although the first partial derivatives are both zero, every neighborhood of P contains points where z has a greater value and points where z has a lesser value than at P. Furthermore the surface has two horizontal rulings through P, straight lines lying in the surface, one parallel to the X-axis and the other parallel to the Y-axis. This means that so long as $x = 60$, z is independent of y, and if $y = 15$, z is independent of x. The company could sell the blades for 15¢ a dozen and the razors at a penny apiece and make the same profit.

But we have worse in store. Let the blades be sold for more than 15¢ per dozen, and it matters not how little more—say at 16¢ per dozen. Now by selling the razors at a price arbitrarily close to zero, the profits can be made to soar indefinitely! We are drawn to the reluctant conclusion that there must be something wrong with this economic dream.

The trouble, of course, lies in the original data. The immediate concern of our manufacturer is not how best to fix prices, but how most quickly to put his sales manager on the carpet and try to find out where he dredged up this formula. No law of supply and demand ever behaved that way. The price of each commodity is varying inversely as the product of both together. This says that a change in the price of only one affects the sales of both of them in the same ratio. Such a relation

---

*This was initially called to our attention by Dr. Alfred W. Jones of Bell Telephone Laboratories.

can be realistically interpreted in only one way: no one buys razors without blades or blades without razors; and each sale of one razor carries with it the automatic sale of some fixed number of blades, and vice-versa. This is another way of saying that the price of the package is to be treated as a single variable. And if indeed one lets $y = kx$, so that the profit reduces to a function of one variable, the problem begins to make sense. Even so, however, $x = 60$ and $y = 15$ is not the best solution. It is the optimum for $k = \frac{1}{4}$; but because P is a saddle point any other k will produce a maximum profit higher than \$66.67. Now where did that sales manager disappear to?

The problem can be salvaged. A corrected version appears in a revised edition of one of the texts which formerly contained the faulty problem. In the revised problem, $100,000/xy$ is replaced by $100,000/x^2y$, and $400,000/xy$ is replaced by $400,000/xy^2$. Everything else remains the same. The surface represented by the new $z = f(x, y)$ has positive curvature in the region in question and a true maximum at the point $x = 12$, $y = 24$, an elliptic point. We now have the unexpected result that the razors are being sold below cost. This is not inconsistent; that it can happen to certain items of a line is well known to all manufacturers. But there is another objection. In the revised problem both the retail prices and the daily profit are too low to be realistic. If we write 1,000,000 in place of both the 100,000 and the 400,000 in the statement of the revised problem, we end up with more reasonable figures throughout. We leave the working of this final version to the reader or his calculus class.

65(1958), 769–770

REPORT OF THE TREASURER

The Current Fund ended the year with a surplus of \$2,065. Anticipated expenditures during 1956 indicate that a rather substantial deficit will be incurred in 1956. The Finance Committee is considering whether an increase in dues to \$5.00 per year may become necessary. The Secretary-Treasurer will be glad to have the comments of members of the Association on this matter.

63(1956), 276

How To Play Baseball

Donald J. Newman

1. Introduction. This is supposed to be a mathematics paper, our preposterous title notwithstanding. Now there is very little which mathematics can do about certain aspects of the great American game; we do not expect to make any vital contributions to the axioms "Swing hard," "Pitch good," "Run fast," "Sleep lots," "Don't drink," "Play heads up," etc.

But there is one facet of baseball which does lend itself to the intrusion of the mathematical egghead, and this is the guessing game between the pitcher and batter, the game of "Should I put it in there?" and "Should I swing at the next one?"

To give a "game-theoretic" solution to this little guessing game, we must first strip to its bare essentials, and in fact we shall even idealize these bare essentials.

Thus, although in the real game the location of men on base, the score, the number of the inning, and the number of outs all should be counted in as "state variables," we will not do so. Furthermore, we also ignore the possibility of extra base hits—all this of course in order to obtain a model which is computationally feasible. To be sure, on larger computing machines more and more realistic models can be analysed, but for now we are content with the following.

2. Idealized rules.

1. *Simultaneously*, the pitcher decides whether to pitch a ball (B) or a strike (S), and the batter decides whether to swing (s) or take (t).

2. If the joint decision is, then the batter gets a

B	t	Ball
B	s	Strike
S	t	Strike
S	s	Single

3. The count is as usual

3 strikes before 4 balls	you're out
4 balls before 3 strikes	you walk

If we concentrate, for the moment, on the pitcher's strategy, we see that, depending on the count, he must make a random choice for his next pitch, random but not necessarily 50-50. Thus his decision, when the count is $m{:}n$ (m balls, n strikes $0 \leq m < 4, 0 \leq n < 3$) is given by a probability $p = p_{m,n}, 0 \leq p \leq 1$, where p is his probability of throwing a strike. The complete strategy for the pitcher, then, consists in the knowledge of the 12 numbers $p_{m,n}$.

A similar remark is valid for the batter whose strategy will be completely known when 12 numbers $q_{m,n}$ (probability of swinging when count is $m{:}n$) are found.

Our problem, then, is the computation of these numbers. Let us note that a straightforward matrix approach, which is usually tried in game theory, leads to a matrix of size approximately 128 by 128. We must seriously think of another approach, and it turns out that some ideas in dynamic programming are what click here.

3. The new approach.

Pitcher's strategy. We concentrate first on the *value* of the game rather than on the $p_{m,n}$. To this end we define $V_{m,n}$ to be the probability, given that the count is $m:n$ and that we are employing an *optimal strategy*, of striking the batter out.

Note that $V_{4,n} = 0$, $n < 3$ and $V_{m,3} = 1$, $m < 4$. These determine our "boundary values" for $V_{m,n}$. We now develop a recurrence relation.

Suppose the count is $m:n$ and that our optimal strategy calls for a strike to be pitched $p_{m,n}$ of the time. Then, if the batter swings, our new value becomes

$$A = 0 \cdot p_{m,n} + V_{m,n+1}(1 - p_{m,n})$$

whereas, if he takes, our new value becomes

$$B = V_{m,n+1}p_{m,n} + V_{m+1,n}(1 - p_{m,n}).$$

Note that the minimum of these two expressions is what we should expect as our new value! In game theory we must always assume that the opponent knows our strategy and combats it as much as possible. Furthermore, it was assumed that $p_{m,n}$ was optimal; therefore, $p_{m,n}$ maximizes $\min[A, B]$. Furthermore, $\min[A, B]$ is maximized by that value of p which forces $A = B$, that is, by

$$(1) \qquad p_{m,n} = \frac{V_{m,n+1} - V_{m+1,n}}{2V_{m,n+1} - V_{m+1,n}}$$

and so, substituting back,

$$\min[A, B] = \frac{V^2_{m,n+1}}{2V_{m,n+1} - V_{m+1,n}}.$$

But $\min[A, B] = V_{m,n}$ and

$$V_{m,n} = \frac{V^2_{m,n+1}}{2V_{m,n+1} - V_{m+1,n}}.$$

This is the required recurrence relation.

Let us now review these results. We have

$$V_{m,n} = \frac{V^2_{m,n+1}}{2V_{m,n+1} - V_{m+1,n}}, \qquad V_{4,n} = 0 \quad (n < 3), \qquad V_{m,3} = 1 \quad (m < 4),$$

which allows us to compute all the $V_{m,n}$. We have also seen, as a by-product, that (1) holds. Consequently, having all the $V_{m,n}$ we can compute all the $p_{m,n}$. This, reader will recall, gives us the entire strategy!

4. Example. Find $p_{2,2}$. From

$$p_{22} = \frac{V_{23} - V_{32}}{2V_{23} - V_{32}} = \frac{1 - V_{32}}{2 - V_{32}},$$

we see we need only find V_{32}. But,

$$V_{32} = \frac{V_{33}^2}{2V_{33} - V_{42}} = \frac{1}{2 \cdot 1 - 0} = \frac{1}{2},$$

and so $p_{22} = (1 - \frac{1}{2})/(2 - \frac{1}{2}) = \frac{1}{3}$. Thus, when the count is 2:2, pitch a strike with probability $\frac{1}{3}$.

Batter's strategy. But setting up the same conditions for the batter, we find that the equations become identical, and we conclude that $q_{m,n} = p_{m,n}$. Therefore, a knowledge of the $p_{m,n}$ is all that is required.

5. Computations. We now give the approximate values obtained for some of the $p_{m,n}$.

	0	1	2	3	m
0	.300	.347	.410	.5	
1	.262	.312	.385	.5	
2	.2	.25	.333	.5	
n					

The exact values are rather complicated fractions, e.g., $p_{20} = \frac{89}{217}$.

One further computation is of interest. We find that $V_{0,0} = .4995$, and so a one-bagger is just about as likely as an out. From this we can find the expected number of runs per inning. The result is .75 runs.

67(1960), 865–868

---

4. *The role of the mathematician in industry*, by Mr. Ross Miller, Northrop Aircraft Corporation, introduced by Professor Clifford Bell.

The profit motive is the underlying force in industry. A corporation makes money by pleasing its customers, by satisfying their wants through better products made more rapidly, more reliably, more cheaply, etc. Frequently it creates in its customers new desires which industry is in a position to satisfy. And we note in this case the opportunity for basic research.

The mathematician's goals are those of job satisfaction and interest, a feeling of challenge and responsibility. Pleasant working conditions with adequate financial rewards are also important considerations.

The goals of industry and the goals of the mathematician can be compatible and can find fulfillment in a merger of interests. The mathematician's contribution is basic and becomes ever increasing in its importance. Industry, on the other hand, stands prepared to offer the mathematician the personal rewards which are so important to the individual.

63(1956), 532

Mr. G. L. Alexanderson, Stanford University, has been appointed Instructor at the University of Santa Clara.

66(1959), 166

CALCULUS

Calculus. By Leighton (Carnegie Institute of Technology). Allyn and Bacon, 1958. 416 pages, $6.95. To be reviewed.

An Analytical Calculus, Volume IV. By Maxwell (Cambridge University). Cambridge University Press, New York, 1958. 272 pages, $4.00 Review: pp. 536–7, 1958.

Calculus, Second Edition. By Smith, Salkover and Justice (University of Cincinnati). Wiley, 1958. 520 pages, $6.50. Review: p. 377, 1958.

Elements of the Differential and Integral Calculus, New Revised Edition. By Granville, Smith and Longley (Yale University). Ginn, 1957. 556 pages, $5.75

Calculus. By Britton (University of Colorado). Rinehart, 1956. 584 pages, $6.50. Review: pp. 462–3, 1958.

Calculus. By Morrill (Johns Hopkins University). Van Nostrand, 1956. 537 pages, $6.00.

67(1960), 315

TEACHER TRAINING

By Gail Young

The nature of the mathematics recommendations. I can summarize rather quickly the reasons why there should be changes not only in the amount but in the nature of the training of the secondary teacher of mathematics. First, improvements in grades 1 through 8 through such projects as the SMSG program will mean better-prepared high school students, capable of more difficult work. Second, pressure from engineering schools and industry—and from some, though apparently not all, mathematicians—will push down into the high schools more and more work that is now usually done in college. Third, the current changes in the high school curriculum call for a differnt training. Thus the teacher of the next decade will be taking better-prepared students through new material well into college work. Many people would guess that in a decade 15% of our high school seniors will be taking calculus, a figure not unreasonable if compared with European experience.

67(1960), 795

6. MATHEMATICS EXPANDS, 1961–1970

R. L. Wilder, President of the MAA 1965

Mathematics Expands

It was the decade of tumult. President Kennedy announced we would put men on the moon by the end of the decade, and he established Advisory boards to find how it could be done. In 1961, the Bay of Pigs invasion strained international relations. In 1962, military advisors in Vietnam became actively involved in the war and the Cuban missile crisis steered the world closer to war—and then away again. In November of 1963, President Kennedy was assassinated. During the summer of 1967, riots broke out in cities across the country. In 1968, after the Tet Offensive showed a desperate situation in Vietnam, both Martin Luther King and Robert Kennedy were assassinated. Later that year, at the Democratic National Convention in Chicago, protests and rioting police were seen on television screens across the nation. Finally, near the close of the decade, Neil Armstrong became the first man to walk on the moon on July 20, 1969. A scant eight years before, the first American had been placed in suborbital flight for only a few minutes.

America was in turmoil, and mathematics reflected society. Mathematics grew and changed. In 1960, there were exactly 303 Ph.D.'s in mathematics; in 1970, there were over 1300 Ph.D.'s in mathematics—and nearly 2,000 in the mathematical sciences. Indeed, the President's Science Advisory Committee produced the Gilliland Report in 1962, calling for dramatic expansion in the production of Ph.D.'s in engineering, mathematics, and physical science. They set a goal of 7 times the 1960 production by the end of the decade; mathematicians scoffed. Graduate programs expanded and were born. Such programs required new faculty, which created more need for graduate students. Every university sought talented young mathematicians. These were the good times of summer; the fall was soon to come.

Good times? What happened to salaries during the decade? In 1960–61 the median salary for Assistant Professors in Group I universities was about $8,000. For Professors, it was about $12,000. By 1969–70, the median salary for Assistant Professors had grown to $11,500 and the median salary for Professors was now about $22,000. What about prices? In 1961, hotel rooms near the University of Chicago ranged from $5.50 to (a luxurious) $12. In 1970, hotel rooms ranged from $12 to $20. Milk was about 50¢ per gallon in 1961; it was on sale for 99¢ a decade later. Some things stayed the same: Bread was on special sale for 17¢ a loaf throughout the decade, and new cars (the cheapest Chevrolets) went from just under $2,000 to just over. In 1964, the MAA increased its dues by 20%, from five dollars to six. (Dues had been $4 from 1921 until 1957.)

Mathematics became more sophisticated as it grew. The Monthly published articles that assumed readers knew about measure theory and vector bundles. It was a golden age of clever, concise proofs of well known results. Mathematics was research.

It was also the age of curriculum reform, which had begun in the previous decade. Schools and universities knew what SMSG and CUPM stood for. Every school child (and parent) learned about the "new math" (whether it was new or not). And mathematicians discovered that balancing teaching and research in universities was not always easy.

Mathematics gave birth to Computer Science. What had been pioneering mathematics in the 40's became an important new area in the 50's. Computer science blended with the engineering of computers and with scientific computation. Computer scientists were engineers and chemists and (occasionally) mathematicians. Mathematics departments accommodated computer scientists—uneasily in most cases. By the end of the 60's, Computer Science was separate from mathematics, and growing fast. Mathematicians were perplexed, skeptical, and anxious.

It was an age of triumphs and tragedies. Just before Kennedy was assassinated, Paul Cohen was proving the independence of the continuum hypothesis while Feit and Thompson published their famous paper on the solvability of groups of odd order. In the middle of the decade, as people despaired at riots across the nation, Atiyah and Singer (and Segal) published their famous three papers on Elliptic Operators. At the close of the decade, Matijasevic published his solution to Hilbert's tenth problem (on the solution of Diophantine equations), while the press was filled with accounts of the Mylai massacre.

Mathematics had grown, but expansion was over. By 1970, it was clear that the market was saturated with Ph.D.'s, universities were saturated with tenured faculty, and America was becoming suspicious of technology (and mathematics). There were not enough talented teachers to implement educational reform; mathematicians took the blame. Research funds were reallocated to help ease the glut, sometimes creating even more hardship. Stories of talented young mathematicians driving taxi cabs were common (even if the cabbies themselves were not).

By the end of the decade, America looked to the recent past with sorrow and looked to the future with anxiety. Rachel Carson published *A Silent Spring* in 1962; the first Earth Day was in 1970. In that same year, the war expanded into Cambodia and the postal reform act established an independent and improved Postal Service. The International Congress was held in Nice, where all but one of the plenary addresses were given in English.

American mathematics had grown up, and was slightly weary.

The Recurrence Theorem

Fred B. Wright

The purpose of this note is to give a completely self-contained account of the recurrence theorem of ergodic theory. This theorem was first proved by Poincaré, and, in modern form, asserts that if T is a measure-preserving transformation on a finite measure space and if E is any measurable set, then almost every point of E returns to E infinitely often under application of T. More generally, if all of the transformations $T, T^2, T^3, \ldots,$ are incompressible, the same conclusion holds for any measure space. In 1947, Halmos [1] showed that if T is incompressible, one-one, and if T^{-1} is measurable, then all powers of T are incompressible. In 1959, Taam [2] succeeded in removing the restrictions on T, by carefully analyzing the already quite involved combinatorial proof of Halmos. Independently, in 1959 the author of this note, in the course of an investigation of the properties of endomorphisms of Boolean algebras, discovered a very simple proof of the recurrence theorem which circumvents the necessity of proving that powers are incompressible. This proof appears midway in a paper [3] which introduces considerable machinery whose function is more general than a proof of the recurrence theorem, and consequently it appears more difficult than it is, in fact. We show here that this theorem is completely on the surface.

Throughout, let X be a set, let $\mathscr{S}$ be a σ-algebra of subsets of X, and let $\mathscr{I}$ be a σ-ideal of $\mathscr{S}$. The triple $(X, \mathscr{S}, \mathscr{I})$ will be called a *measurability space*. A mapping T of X into itself will be called a *measurable* transformation if $E \in \mathscr{S}$ implies $T^{-1}E \in \mathscr{S}$. If T is a measurable transformation and if $E \in \mathscr{S}$, the set $E - \bigcup_{j=1}^{\infty} T^{-j}E$ is the set of all those points in E which never return to E, and the set $E - \bigcap_{m=0}^{\infty} \bigcup_{j=m+1}^{\infty} T^{-j}E$ is the set of those points of E which do not return infinitely often to E. The measurable transformation T is called *recurrent* if $E - \bigcup_{j=1}^{\infty} T^{-j}E \in \mathscr{I}$ for all $E \in \mathscr{S}$, and is called *infinitely recurrent* if $E - \bigcap_{m=0}^{\infty} \bigcup_{j=m+1}^{\infty} T^{-j}E \in \mathscr{I}$ for all $E \in \mathscr{S}$.

A measurable transformation T is called *incompressible* if $E \subset T^{-1}E$ implies $T^{-1}E - E \in \mathscr{I}$. We remark that T is incompressible if and only if $T^{-1}E \subset E$ implies $E - T^{-1}E \in \mathscr{I}$. For, let $E \in \mathscr{S}$ and let $F = X - E$. Then $E \subset T^{-1}E$ if and only if $T^{-1}F \subset F$, and $E - T^{-1}E = T^{-1}F - F, T^{-1}E - E = F - T^{-1}F$.

A set $E \in \mathscr{S}$ is called a *wandering set* if the sequence $E, T^{-1}E, T^{-2}E, \ldots,$ is disjoint. A measurable transformation is called *conservative* if every wandering set is in $\mathscr{I}$.

THE RECURRENCE THEOREM. *Let T be a measurable transformation of a measurability space $(X, \mathscr{S}, \mathscr{I})$. Then the following are equivalent*: (1) T *is incompressible*; (2) T *is conservative*; (3) T *is recurrent*; (4) T *is infinitely recurrent*.

Proof. For any $E \in \mathscr{S}$, let $E^* = \bigcup_{j=1}^{\infty} T^{-j}E$. We first prove that (1) implies (3). Let $E \in \mathscr{S}$, and let $A = E \cup E^*$. Then $T^{-1}A = E^* \subset A$, and since T is incompressible, $A - T^{-1}A \in \mathscr{I}$. But $A - T^{-1}A = E - E^*$.

Next, (3) implies (2). Suppose E is a wandering set; then E and E^* are disjoint, so that $E - E^* = E$. If T is recurrent, then $E = E - E^* \in \mathscr{I}$, so that T is conservative.

Now we show that (2) implies (1). If E is any set in $\mathscr{I}$, then $E - E^*$ is a wandering set. If T is conservative, $E - E^* \in \mathscr{I}$. But then if $T^{-1}E \subset E$, we have $E^* = T^{-1}E$, and hence $E - T^{-1}E \in \mathscr{I}$. Thus T is incompressible.

Finally, we prove that (4) and (3) are equivalent. First observe that if $B = \bigcap_{m=0}^{\infty} T^{-m}E^*$, then $B \subset E^*$ and $E - E^* \subset E - B$. Thus if T is infinitely recurrent, it is recurrent. Conversely, assume (3). By what has been shown, we know that (1) holds; we use both (1) and (3). It is clear that $E - B = (E - E^*) \cup [E \cap \bigcup_{m=1}^{\infty}(T^{-m}E^* - T^{-(m+1)}E^*)]$. Since T is recurrent, $E - E^* \in \mathscr{I}$. Since $T^{-1}(T^{-m}E^*) = T^{-(m+1)}E^* \subset T^{-m}E^*$ for each $m \geq 0$, and since T is incompressible, $T^{-m}E^* - T^{-(m+1)}E^* \in \mathscr{I}$. Since $\mathscr{I}$ is a σ-ideal, then $E - B \in \mathscr{I}$. This completes the proof.

We must remark that in the proof of the equivalence of (1), (2), and (3), *no use was made of any property whatsoever of the family $\mathscr{I}$.*

References

1. P. R. Halmos, Invariant measures, Ann. of Math., vol. 48, 1947, pp. 735–754.
2. C. T. Taam, Amer. Math. Soc. Notices, vol. 6, 1959, p. 813.
3. F. B. Wright, Recurrence theorems and operators on Boolean algebras, Proc. London Math. Soc. (to appear).

68(1961), 247–248

BOOK REVIEW

The First Six Million Prime Numbers. By C. L. Baker and F. J. Gruenberger. The Microcard Foundation, Madison, Wis., 1959. $35.00.

This is an interesting and somewhat unique presentation of a large volume of tabular material. Using an IBM 704, the first six million prime numbers have been computed and tabulated on 4,800 pages with 1,250 numbers to a page. On each line of the page the first prime is displayed explicitly, the remaining 24 show only their terminal 3 digits. 39 pages of tabulation are then reduced photographically and placed on one side of a 3 × 5 card with another 39 tabulation pages on the other side. In this fashion the entire 4,800 pages are reduced to 62 cards which are legible using a 5 power magnifying glass. Since a table of large prime numbers is not apt to be consulted too frequently, this technique may well be the answer to the librarian's storage problem. Presumably, if the price could be dropped to one third of its present level, many mathematicians would be interested in securing copies for their own libraries.

67(1960), 1048–1049

π Is Not Algebraic of Degree One or Two

J. D. Dixon

In [1] I. Niven gives a simple proof that π is irrational. The following is a modification of his proof to show that π does not satisfy an equation

$$(1) \qquad P(x) = ax^2 + bx + c = 0,$$

where a, b and c are rational integers, not all zero.

The following identity is given in [1]. Let $f(x)$ be a polynomial with successive derivatives $f^{(1)}(x), f^{(2)}(x), \ldots$ and let

$$F(x) = f(x) - f^{(2)}(x) + f^{(4)}(x) - \cdots.$$

Then $(d/dx)\{F'(x)\sin x - F(x)\cos x\} = f(x)\sin x$ and so

$$(2) \qquad \int_0^\pi f(x)\sin x\,dx = F(\pi) + F(0).$$

Now suppose π satisfied equation (1). To prove the theorem we obtain a contradiction by putting

$$f(x) = \frac{1}{n!}\left[P(x)\{P(x) - P(0)\}\right]^{2n}$$

for some integer n and showing that the right side of (2) is then an integer while the left side lies strictly between 0 and 1 for sufficiently large n.

Since for each integer $r \geq 0$

$$\{P(x)^r\}'' = r(r-1)P(x)^{r-2}P'(x)^2 + rP(x)^{r-1}P''(x)$$

$$= r(r-1)P(x)^{r-2}\{4aP(x) + (b^2 - 4ac)\} + 2arP(x)^{r-1},$$

the successive even order derivatives $f(x), f^{(2)}(x), \ldots$ must all be polynomials in $P(x)$ with rational coefficients. Since the factors $P(x)$ and $P(x) - P(0)$ each occur $2n$ times in $f(x)$, those expressions in $f^{(2k)}(x)$ which are nonzero for $x = 0$ or $x = \pi$ must have a factor of $(2n)!$ in the numerator. This cancels the $n!$ in the denominator and $f^{(2k)}(0)$ and $f^{(2k)}(\pi)$ are integers for $k = 0, 1, 2, \ldots$. Hence $F(\pi) + F(0)$ is an integer.

On the other hand if C is an upper bound of $[P(x)\{P(x) - P(0)\}]^2$ for $0 \leq x \leq \pi$ then $|f(x)| < C^n/n! \to 0$ for $n \to \infty$. There $\int_0^\pi f(x)\sin x\,dx < 1$ for sufficiently large

n. At the same time $f(x)\sin x \geqq 0$ for $0 \leqq x \leqq \pi$, and since $f(x)\sin x$ is continuous and not zero everywhere in the interval we have $\int_0^\pi f(x)\sin x\,dx > 0$. Thus the integral lies strictly between 0 and 1 and we have the required contradiction in (2).

Reference

1. I. Niven, A simple proof that π is irrational, Bull. Amer. Math. Soc. **53**(1947) 509.

69(1962), 636

EXCERPTS FROM AN ARTICLE BY MINA REESE

What is the prospect for providing the teachers to handle the crowds of new students? Except during the war the number of mathematics Ph.D.'s has constituted an almost constant percentage of all science and mathematics Ph.D.'s (a little under 5%) and also of all Ph.D.'s in all fields (2.8%). Thus if the total number of Ph.D.'s in all fields does double within the next decade along with the doubling of the college population, as has been predicted, we may hope to produce around 600 Ph.D.'s in mathematics in 1970. But a little arithmetic will show a terrifying imbalance even if, as seems most unlikely, the drift to industry and government were stemmed.

. . .

Why is it that the yield of mathematical doctorates is so small a proportion of all doctorates in science and mathematics together? Why is it that so few mathematics majors go on to the Ph.D.? Must a student be a genius to receive a Ph.D. in mathematics? Some of our students seriously think the answer to this question is "Yes." In physics, a B student at college can do a very good job in his Ph.D. research;

but a B student in mathematics will rarely be accepted as a candidate for a doctorate in mathematics. We shall certainly need some of our B students as teachers, particularly in our two-year colleges if these continue to spring into being as they have been doing recently. In several states, the master plan for the development of higher educational facilities to take care of the doubling of collegiate enrollment within the next decade calls for the establishment of many new community and junior colleges within a few years.

. . .

How are we to take account of the vastly expanded need for new mathematicians, particularly new teachers of mathematics? The situation is so desperate that very serious consideration should be given to the proposal from the Committee on the Undergraduate Program to the Mathematical Association of America and the American Mathematical Society that a new doctoral degree be created.

68(1961), 374–375

Yet Another Proof of the Fundamental Theorem of Algebra

R. P. Boas, Jr.

The following proof of the fundamental theorem of algebra by contour integration is similar to Ankeny's [1], but is simpler because it uses integration around the unit circle (which is usually the first application of contour integration) instead of integration along the real axis; thus there is no need to discuss the asymptotic behavior of any integrals.

Let $P(z)$ be a nonconstant polynomial; we are to show that $P(z) = 0$ for some z. We may suppose $P(z)$ real for real z. (Indeed, otherwise let $\bar{P}(z)$ be the polynomial whose coefficients are the conjugates of those of $P(z)$ and consider $P(z)\bar{P}(z)$.) Suppose then that $P(z)$ is real for real z and is never 0; we deduce a contradiction. Since $P(z)$ does not either vanish or change sign for real z, we have

$$(1) \qquad \int_0^{2\pi} \frac{d\theta}{P(2\cos\theta)} \neq 0.$$

But this integral is equal to the contour integral

$$(2) \qquad \frac{1}{i}\int_{|z|=1} \frac{dz}{zP(z+z^{-1})} = \frac{1}{i}\int_{|z|=1} \frac{z^{n-1}dz}{Q(z)},$$

where $Q(z) = z^n P(z + z^{-1})$ is a polynomial. For $z \neq 0$, $Q(z) \neq 0$; in addition, if a_n is the leading coefficient in $P(z)$, we have $Q(0) = a_n \neq 0$. Since $Q(z)$ is never zero, the integrand in (2) is analytic and hence the integral is zero by Cauchy's theorem, contradicting (1).

Reference

1. N. C. Ankeny, One more proof of the fundamental theorem of algebra, this MONTHLY, 54 (1947) 464.

71(1964), 180

A New Method of Catching a Lion

In this note a definitive procedure will be provided for catching a lion in a desert (see [1]).

Let Q be the operator that encloses a word in quotation marks. Its square Q^2 encloses a word in double quotes. The operaor clearly satisfies the law of indices, $Q^m Q^n = Q^{m+n}$. Write down the word 'lion,' without quotation marks. Apply to it the operator Q^{-1}. Then a lion will appear on the page. It is advisable to enclose the page in a cage before applying the operator.

I. J. Good

1. H. Petard, A contribution to the mathematical theory of big game hunting, this MONTHLY 45 (1938) 446–447.

72(1965), 436

R. H. Bing, President of the MAA 1963

Recent Developments in Mathematics

Jean Dieudonné

About a year ago, I rather lightheartedly accepted the invitation to give this talk, [at a conference to dedicate Van Vleck Hall at the University of Wisconsin] but, as the time drew near, I began to realize the pitfalls ahead of me, and my recklessness. When so many different topics are touched upon, for every statement of mine there will necessarily be in the audience someone who knows the subject much better than I do, and who therefore will with good reason take exception to my superficial remarks. But I am afraid I will be the target of far more stringent criticism for the selection I have had to make: it is quite clear that *some* choice was imperative in such a short review, and I had therefore to decide on what was important and what was not; no objective criterion being available, I must admit that I have merely followed my own tastes. I tried, however, not to imitate certain of our colleagues, who are so entranced by the beauty of their tiny nook in some highly specialized field, that for them it is the only and unique important thing in the whole world; and to counterbalance the danger of subjectivity, I have taken into account the opinions I have heard expressed by some of the best mathematicians of our time. I confess that this is not a very democratic procedure, but I am afraid I don't believe very much in democracy, at least in scientific matters. To make things clearer, I have selected, as prominent landmarks of present-day mathematics, the solutions of some outstanding problems bequeathed to us by previous generations of mathematicians. I readily admit that there may be some point in claiming, as some will do, that more ingenuity can be spent in unraveling, say, the structure of some fancy nonassociative algebra, than in solving Hilbert's fifth problem or disproving Burnside's conjecture; but I am not very receptive to such arguments, and on these matters I will rather follow C. L. Siegel or A. Weil.

Mathematics progresses essentially in two different ways. The mathematicians whom I might call the tacticians pounce head on at a problem, using only old and well-tested tools, and they merely rely on their cleverness to give some new twist to traditional arguments, and thus reach the solution which had eluded previous attempts. The strategists, on the other hand, will never be satisfied until the concepts involved in a problem have been so thoroughly analyzed, and their connections put in such a clear light, that the final solution almost appears as a triviality; but of course this may demand lengthy and tedious developments of seemingly unrelated very general theories, which some people will deem out of proportion with the initial question.

I believe, however, that both approaches are essential to the well-being of mathematics. Excessive reliance on individual prowess, without a corresponding renewal of methods and outlook, may well end in sterility, through intensive concentration on tiny aspects of a theory, unduly magnified; and on the other hand, an exclusive lover of generality will too often lose sight of the proper motivations,

247

and indulge in endless churning out of more and more empty theories. In the better men the two tendencies fortunately blend into a harmoniously balanced and fruitful combination, of which Hilbert is perhaps the perfect example. Let it therefore be quite clear that in following (somewhat loosely) the preceding classification, I do not intend in the least to assign higher value to one type of results over the other.

. . .

I would like to conclude with some general remarks. The first one, and the one I would like to emphasize most, is that, despite the tremendous surge and somewhat bewildering diversity of unpredictable developments during the last 20 years, mathematics is now more *unified* than it ever was. Of course, striking examples of the deep kinship of various parts of mathematics have not been unknown to classical times, from the use of Dirichlet's series in number theory to Riemann's introduction of topology in function theory. But we now have reached a point where it is practically impossible to apply to a large part of contemporary mathematical literature any one of the old labels of "algebra," "analysis" or "geometry." Algebraic geometry and "analytic geometry" already behave like identical twins, any advance in one being almost invariably matched by the corresponding theorem in the other within a short time; on the other hand, the merger of commutative algebra and algebraic geometry is all but complete, and the theory of algebraic numbers is confidently expected to fall in line within a few years. Some of the most remarkable theorems derive from the successful confrontation of two seemingly unrelated theories: it was by translating the "Riemann hypothesis" for curves over a finite field in purely geometric terms that A. Weil realized that what was needed was the "abstract" counterpart of Lefschetz's fixed point formula in topology, finally succeeded in forging in that way his famous proof, and was further led to the formulation of his conjectures, the proof of which is expected to give us at last general methods of attack in diophantine analysis. Still more revealing is the history of the "Riemann-Roch-Hirzebruch-Grothendieck" theorem. Around 1950, Kodaira understood that the Riemann-Roch theorem for classical algebraic varieties of dimension 2 and 3 could be formulated as an *equality* between topological invariants of the variety instead of an inequality as with Zariski and the Italian geometers; but he still lacked the machinery to extend these results to higher dimensional cases. A little later, he realized that the topological invariants he needed were linked to the properties of vector bundles over complex manifolds (Chern classes, Pontrjagin classes); as soon as Serre and Cartan started using coherent sheaves in that theory, Kodaira saw that this gave him one of his essential tools and in a few months he had (in partial collaboration with Spencer, and independently of Serre and Cartan) obtained far reaching results in that theory, using in particular theorems from the theory of elliptic partial differential equations. In 1952, Hirzebruch became interested in the problem; by ingenious algebraic devices, he linked the Chern classes of vector bundles on algebraic varieties to earlier invariants introduced in algebraic geometry by Eger and Todd, and a little later Serre was able to guess what the formulation of the general Riemann-Roch theorem should be. But a proof was still lacking, and it was obtained only in 1954 by Hirzebruch *via* the use of yet another set of ideas, this time Thom's cobordism theory, which had just appeared and gave just the information on Pontrjagin classes needed to fill in the gap. Such a bewildering maze of arguments was not very satisfactory, and Hirzebruch himself was aware that simpler and better proofs could probably be obtained. This was done in 1957 by Grothendieck, who kept the essential algebraic and homological ideas of his predecessors, but could dispense with all the machinery coming from harmonic forms or differential topology; this enabled him, not only to extend the formula to abstract algebraic geometry, but also to show that it was a special case of a still more general and simpler one. To do that, however, he had in particular to introduce new tools in homological algebra, what are now called the Grothendieck groups and rings, whose latent potentialities immediately attracted attention. The first step in that direction has been made by Hirzebruch and Atiyah, in an extraordinary combination where this time the new ingredient is, of all things; Bott's periodicity theorem for the homotopy groups of simple Lie groups. This has enabled them to give an extension of the Riemann-Roch theorem to differentiable manifolds; it has also provided Adams with the necessary tools for his solution of the problem of vector fields on spheres. Still more surprising is the way in which Atiyah has used these concepts in the theory of representations of finite groups, and H. Bass is now exploring with success the application of similar methods to projective modules and linear groups over Dedekind rings. This is where we stand today, and we certainly are still very far from the end of the story; but I hope this is enough to show you how complex and fruitful the interaction now is of mathematical ideas coming from every conceivable quarter.

It is sometimes feared (even by young graduates) that these powerful tendencies towards complete unification of the various branches of mathematics may in the end be self-defeating, through sheer mental impossibility to get a firm and competent grasp of so many different ideas and theories at the same time. Fortunately it seems that, as it has been the case in similar "Sturm und Drang" periods in the history of our science, the bewildering diversity of the new concepts naturally produces by reaction a search for a simplification of the unwieldy mess. This time it seems that our salvation will come from the new concepts of categories and functors, introduced in the early 1940's by Eilenberg and MacLane; they have already demonstrated their versatility and usefulness in the work of men like Eckmann, Hilton, Kan and Grothendieck, and many younger mathematicians are now engaged in this work of concentration and simplification, which on a new level repeats the story of algebra and topology of 40 years ago. Of course, as always, the price to be paid is in more "abstraction"; but it is now a well-established phenomenon that what is highly abstract for a generation of mathematicians is just commonplace for the next one, and the cries of anguish one still hears from time to time usually come from older men visibly afraid of being unable to catch up with the younger set. One tends, however, to be a little more impatient towards this manifestation of human frailty than, say, 30 years ago and the traditional jokes about "hard" and "soft" mathematics have now become a little stale. It is of course very easy, from the heap of axiomatic trash dumped every year by would-be mathematicians on the unhappy public, to select some particularly nonsensical paper and exhibit it as the typical product of modern mathematics; I leave it to you to judge whether such an attitude is compatible with even a minimum of intellectual honesty and whether those who indulge in it should not have the decency to remain silent until they have made the effort of getting more accurate information.

As a final remark, I would like to stress how little recent history has been willing to conform to the pious platitudes of the prophets of doom, who regularly warn us of the dire consequences mathematics is bound to incur by cutting itself off from the applications to other sciences. I do not intend to say that close contact with other fields, such as theoretical physics, is not beneficial to all parties concerned; but it is perfectly clear that of all the striking progress I have been talking about, not a single one, with the possible exception of distribution theory, had anything to do with physical applications; and even in the theory of partial differential equations, the emphasis is now much more on "internal" and structural problems than on questions having a direct physical significance. Even if mathematics were to be forcibly separated from all other channels of human endeavour, there would remain food for centuries of thought in the big problems we still have to solve within our own science.

Indeed our wealth is now so great that even a cursory inventory of it could not possibly be made within the span of a single talk, and I woefully realize how much valuable work I had to leave reluctantly aside in such fields as complex multiplication (A. Weil, Shimura, Taniyama), nonlinear partial differential equations, infinite Lie groups (Chern, Kuranishi, Sternberg), Riemann surfaces (Teichmüller, Ahlfors, L. Bers), potential theory (Deny, Beurling, Hunt, Choquet), harmonic analysis (Kahane, Katznelson, Rudin, Helson, Malliavin and many others), to say nothing of mathematical logic or probability theory, where my ignorance prevents me from entering at all. I hope at least to have given you some faint idea of the tremendous progress accomplished during the last 20 years; no other comparable period of our history has been so rich in new ideas and results, and we have every reason to confident that the future will ever more fulfill Hilbert's motto: "We must know and we shall know."

71(1964), 239–248

REMUNERATION FOR AUTHORS OF EXPOSITORY WRITING

At its meeting on January 24, 1962, in Cincinnati, Ohio, the Board of Governors authorized remuneration for authors of all expository writing, felt to be of sufficient quality to justify remuneration, at the rate of $6.00 per printed page for new articles and $3.00 for revised articles and reprints. Expository writing includes Carus Monographs, Slaught Papers and Miscellaneous Publications, MAA Studies, etc. At its meeting on August 26, 1962, in Vancouver, B.C., the Board authorized the Treasurer to make such payments when so directed by the Sub-committee on Expository Writing of the Committee on Publications. In accordance with these actions, payments at the rate of $6.00 per printed page were made to authors of eight papers in the MONTHLY and six papers in the MATHEMATICS MAGAZINE during 1962. The total amount expended was $858.00.

HENRY L. ALDER, *Secretary*

70(1963), 1045

The William Lowell Putnam Mathematical Competition: Early History

G. Birkhoff

The basic idea underlying the Putnam Competition was expressed by William Lowell Putnam in a three-page article in the Harvard Graduates' Magazine of December, 1921 (reprinted here as an Appendix) from which the following excerpt is taken: "...it is a curious fact that no effort has ever been made to organize contesting teams in regular college studies. All rewards for scholarship are strictly individual and are given in money, or in prizes or in honorable mention. No opportunity is offered a student by diligence and high marks in examinations to win or help in winning honor for his college. All that is offered to him is the chance of personal reward. Little appeal is made to high ideals or to unselfish motives.

Is not this one of the reasons why the effort to interest the great bulk of the undergraduate body in their studies is such an uphill task?"

These words expressed Mr. Putnam's profound conviction in the value of intellectual competition, especially by *teams* whose members took pride in the achievements of their team as a whole and the standing of the institution which it represented. This conviction was shared by Mrs. Putnam, and by her brother Abbott Lawrence Lowell, then President of Harvard. In a letter about the competition some years later, Mr. Lowell referred to the importance of "the promotion of scholarship by contest, as athletics have been," and of "the spirit of intercollegiate emulation in scholarship."

To give substance to the ideas expressed above, Mrs. Putnam established a $125,000 trust in her will, written in 1927. She directed that the trustees "shall have always in mind the purpose of this trust, as set out in my husband's own words," and incorporated a copy of his article in her will.

It was some years after Mrs. Putnam wrote her will, however, before the Putnam Competition assumed its present form. In the meantime, various short-lived experiments were tried.

The first experiment was a competition in 1928 in the field of English between Harvard and Yale. The Yale and Harvard teams each contained ten senior concentrators in English. One member of the Harvard team was Nathan Marsh Pusey, now president of that institution. It was won by Harvard, a prize of $5,000 going to the winning institution. Both Yale and Princeton declined to repeat the contest, and an offer to compete in economics with Cambridge University was also declined, partly because of a fear of undesirable publicity.

The second experiment was a similar mathematical competition in 1933 between Harvard and West Point juniors. The examination stressed ability to solve problems in the calculus, analytic geometry, and elementary differential equations.

This close competition was won by the somewhat more mature and (according to local tradition) better trained West Point cadets. Again, the competition was not repeated, though all ten West Point participants expressed enthuasiasm in personal letters to Mrs. Putnam.

Mr. Lowell having retired as President of Harvard in 1933, no further contest was held until after the death of Mrs. Putnam in 1935. In her will, she appointed her sons, George Putnam and August Lowell Putnam, as trustees for the William Lowell Putnam Memorial Fund. These two brothers then consulted George David Birkhoff at Harvard as to the best use to be made of the bequest, a question which the latter had often discussed with Mrs. Putnam. It seemed natural to have the competition in mathematics, both because mathematical ability can better be tested by a set examination than ability in most subjects, and because of the strong mathematical traditions of the Putnam family.

The possibility of discovering outstanding mathematical ability at an early age by set examinations had, indeed, been demonstrated conclusively in Hungary, where a competitive examination (named after the physicist Roland Eötvös) had been given annually (since 1894) to young people entering the University. It certainly contributed to the development of Hungarian mathematics; among the winners were such internationally known names as Leopold Fejér, Theodore von Kármán, Dénes König, Alfred Haar, Marcel Riesz, Gabor Szegö, Tibor Radó, and Edward Teller.

Somewhat similarly, the Cambridge University mathematical examinations were used in the nineteenth century, as a means of detecting both mathematical and legal ability, the Senior Wrangler being considered a man "most likely to succeed" in either profession.

The mathematical interests of the Putnam family were unusual, to say the least. Mrs. Putnam's brother, former President Abbott Lawrence Lowell of Harvard, received in 1877 his A.B. *summa cum laude* in Mathematics at Harvard, and his undergraduate thesis was published in volume 13 (1877), of the Proceedings of the American Academy of Arts and Sciences, pages 222–250. Mr. William Lowell Putnam received in 1882 an A.B. *magna cum laude* in Mathematics at Harvard, and was for many years Chairman of the Visiting Committee of the Mathematics Department there. George Putnam, a trustee of the Putnam Fund, had also majored in mathematics and was active in his turn on the same Visiting Committee, while the present form of the Putnam Competition was being planned. Roger Lowell Putnam also majored in mathematics at Harvard, graduating *magna cum laude* in 1915, while William Lowell Putnam's grandson, McGeorge Bundy (later Dean of the Faculty at Harvard), and his brother Harvey majored in mathematics at Yale.

George David Birkhoff saw in these various facts an important opportunity to stimulate interest in mathematics, to establish national standards of undergraduate mathematical achievement, and to discover and encourage mathematical talent at a time when good fellowships for first-year graduate students were few and far between. With these objectives in mind he proposed, in essence, the principles governing the present Putnam Competition.

The first principle was that the competition should be open to three-man teams selected by the faculty of, *and* to individuals from, *any* American or Canadian college. This was to encourage participation by smaller colleges, who might have only one or two unusually able mathematical concentrators in any given year. The second principle was to have the competition administered by the Mathematical

Association of America, which is the professional organization representing college mathematics teachers. The third principle was to distribute prizes and honorable mention to several teams and individuals, so that distinguished performance would be broadly recognized. (To be one of the top 25 or even 50 competitors in any one year, out of 1500 competitors selected from many thousands of mathematics concentrators, is no mean achievement.) The fourth principle was to offer a graduate fellowship at Harvard to *one* of the *five* top competitors. This choice was allowed because Birkhoff recognized that other evidence of talent—especially evidence of creative originality—should be weighed in evaluating men having nearly equal examination scores.

Birkhoff took a very active part in making up the first examination, which was made up by the Harvard Mathematics Department in 1938. So as to avoid favoring the few schools where advanced mathematical courses were then open to able undergraduates, a deliberate effort was made to stress questions covered in standard courses in the calculus, analytic geometry, and elementary mechanics. The emphasis was correspondingly on thoroughness, accuracy, and a clear command of detail.

The first competition was won by the University of Toronto, which was asked to set the next examination and disqualify itself from competing the following year. A similar procedure was followed for about five years, the winning teams being alternately from Toronto and Brooklyn College. These victories were attributed partly to the greater maturity and more intensive concentration program of Canadian mathematical concentrators, partly to the British tradition of problem-solving, and partly to the ability and devotion of Professor Harris MacNeish of Brooklyn in training students how to diagnose and solve set problems.

During the years 1943–5 the Putnam competition was not held because of war conditions. After the war, it was renewed with a few minor modifications. Thoroughness and facility in handling standard course material was given less emphasis, ingenuity in devising and using algorithms and in logical analysis being emphasized in their place. To construct an examination testing these qualities, a special committee was appointed consisting of the distinguished Hungarian-born mathematicians George Pólya and Tibor Radó, and the first Putnam Scholar, Irving Kaplansky. Since that time, membership in the committee making out the examination has rotated, and institutions have been free to win the competition in successive years whenever the ability of their teams made this possible!

72(1965), 469–471

Professor Emil Artin, Mathematisches Seminar, Hamburg, Germany, died on December 20, 1962. He was a member of the Association for 25 years.

70(1963), 522

TABLE I
Baccalaureate Origins of Ph.D. Mathematicians who
Received their Ph.D. During 1952–1962.

Rank	Institution	Number	Cumulative percent
1	University of Chicago	93	
2	City College of the City University of New York	92	7.1
3	Massachusetts Institute of Technology	82	
4	Harvard University	76	
5	University of California (Berkeley)	74	16.0
6 and 7	Brooklyn College	57	
	New York University	57	
8	University of Michigan	52	
9	Columbia University	43	
10	University of California (Los Angeles)	41	
11	Cornell University	39	
12 and 13	University of Minnesota	38	
	University of Illinois	38	
14	University of Texas	34	
15	California Institute of Technology	33	32.6
16 and 17	Yale University	28	
	Stanford University	28	
18	University of Wisconsin (Madison and Milwaukee)	27	
19	Princeton University	24	
20	Swarthmore College	23	
21 and 22	Purdue University	22	
	University of Pennsylvania	22	
23	Iowa State University of Science and Technology	21	
24, 25 and	Northwestern University	20	
26	Illinois Institute of Technology	20	
	Carnegie Institute of Technology	20	42.2
27	University of Washington (Washington)	19	
28	State University of Iowa	18	
29, 30, 31	Oberlin College	17	
and 32	Ohio State University	17	
	University of Oklahoma	17	
	University of Rochester	17	
33, 34	Brown University	16	
and 35	University of Florida	16	
	University of Oregon	16	
36, 37 and	Case Institute of Technology	15	
38	Reed College	15	
	Oklahoma University of Agricultural and Applied Science	15	

A Brief Dictionary of Phrases Used in Mathematical Writing

H. Pétard

Since authors seldom, if ever, say what they mean; the following glossary is offered to neophytes in mathematical research to help them understand the language that surrounds the formulas. Since mathematical writing, like mathematics, involves many undefined concepts, it seems best to illustrate the usage by interpretation of examples rather than to attempt definition.

ANALOGUE. This is an a. of: I have to have *some* excuse for publishing it.

APPLICATIONS. This is of interest in a.: I have to have *some* excuse for publishing it.

COMPLETE. The proof is now c: I can't finish it.

DETAILS. I cannot follow the d. of X's proof: It's wrong. We omit the d.: I can't do it.

DIFFICULT. This problem is d.: I don't know the answer. (Cf. Trivial.)

GENERALITY. Without loss of g.: I have done an easy special case.

IDEAS. To fix the i.: To consider the only case I can do.

INGENIOUS. X's proof is i.: I understand it.

INTEREST. It may be of i.: I have to have *some* excuse for publishing it.

INTERESTING. X's paper is i.: I don't understand it.

KNOWN. This is a k. result but I reproduce the proof for the convenience of the reader: My paper isn't long enough.

Langage. Par abus de l.: In the terminology used by other authors. (Cf. Notation.)

NATURAL. It is n. to begin with the following considerations: We have to start somewhere.

NEW. This was proved by X but the following n. proof may present points of interest: I can't understand X.

NOTATION. To simplify the n.: It is too much trouble to change now.

OBSERVED. It will be o. that: I hope you have not noticed that.

OBVIOUS. It is o.: I can't prove it.

READER. The details may be left to the r.: I can't do it.

REFEREE. I wish to thank the r. for his suggestions: I loused it up.

STRAIGHTFORWARD. By a s. computation: I lost my notes.

TRIVIAL. This problem is t.: I know the answer. (Cf. Difficult.)

WELL-KNOWN. This result is w.: I can't find the reference.

Exercises for the student: Interpret the following.
1. I am indebted to Professor X for stimulating discussions.
2. However, as we have seen.
3. In general.
4. It is easily shown.
5. To be continued.

This article was prepared with the opposition of the National Silence Foundation.

73(1966), 196–197

A COMMENT ON MEASURES OF PRODUCTIVITY OF MATHEMATICS DEPARTMENTS

K. O. May, Carleton College

R. B. Siebring's note "Institutional Influences in the Undergraduate Training of Ph.D. Mathematicians" (this Monthly, January, 1965) illustrates the pitfalls of educational statistics. It has been pointed out before that institutional rankings in various studies of recent years can be explained more easily by known differences in student quality than by assumptions about teaching effectiveness. Siebring's proportion of majors who later earn the doctorate in mathematics is subject to this caution, and, in addition, is largely influenced by the dgree to which a department ignores the majority who are destined for mathematical careers in teaching, government, and business without the Ph.D. This is evident if one observes that a department could earn a high score by limiting its major to students with evident ability and motivation to go on to the Ph.D. and by doing nothing for future teachers, those interested in mathematical careers not involving research, or students for whom a mathematics major is the center of a general liberal arts training or the basis for graduate work in

some field other than pure mathematics. Conversely, the great technical schools could quickly lower their ratings by offering teacher training, pre-actuarial programs, etc. Siebring's ratio tells us something about the specialization of departments on one aspect of mathematical education but very little about the quality of teaching. For such information it is necessary to discount by student aptitude and motivation, as well as to consider the variety of responsibilities shouldered by most departments.

72(1965), 664

MATHEMATICAL EDUCATION NOTES

A Dilemma in Mathematical Education

T. L. Saaty

Traditionally no field of knowledge equals mathematics in the fascination and excitement which it evokes in its practitioners. Once under its magical spell an individual cannot leave it without strong feeling of desertion—perhaps even a sense of diminishing stature. This rank of mathematics is accomplished through the work of people. They are the masters and innovators, individuals who have put themselves in harmony with the field by creative work in it. Some are discoverers, some are educators, and even some are critics and connoisseurs.

Since both the status and development of mathematics require an *esprit de corps* and a strong camp of followers who are mindful of each other in a constructive spirit, it is valuable to take a look at an aspect of mathematical education which might improve this solidarity both in method and in end result. Most mathematicians are potentially fine in human values and equally willing to give of themselves to see others obtain fulfillment by being mathematically creative. Occasionally one is inclined to feel that he must make progress all by himself. For some educators there is perhaps a passing feeling of discouragement with the available raw material.

A distinguished colleague, in an offguarded witty utterance confided this discouragement by saying: "Universities are a wonderful place to be, but the worst thing about them is that they have students." I seriously doubt that he genuinely felt this way.

In science and even more in mathematics, the rigors and high standards required of students take so much of the energy of their mentors to enforce, that little time is left for consideration of personal growth. A clear perspective of the total person is needed.

I have known many a would-be great scholar and creator of ideas who has fallen short of the attainment of his goals, not because of lack of great intelligence or talent or even self discipline, but mainly because of the absence of needed personal warmth to temper an imagined or existing nonconcern or perhaps harshness in the environment. This is particularly characteristic of some of the larger departments. The United States has always been distinguished by the generous spirit of its people towards their fellowmen. It is surprising that this spirit has not been adequately and maturely interwoven in the fabric of higher education in science. It is usually taken for granted that for producing worthy scholars, a good mind is most of what is needed and personal development is secondary. However, there are enough examples of brilliant individuals who failed because of lack of personal development that we need to re-examine this philosophy in order to decrease what I consider a substantial waste of talent, and a nontrivial loss.

Many sensitive and talented individuals learn from early childhood to live in a world of their own intellectual making. In spite of depth of a model of the natural world, the component having to do with human relations may be a naive one. Consequently, the task of growing up to assume responsibility and being effective in a real world of people is painful and for some the desire for philosophical and personal accommodation occurs at the time that a student is receiving his graduate education. At that time he is approaching a peak of his awareness of the surrounding world. Thus he faces two problems—one revolves about excellence in demonstrating his creative mental acumen and the other in adapting his "world picture" to a framework in which he can exist harmoniously. These are both strong forces. The intelligent man in this category is sure of his ability to cope with academia, but is more apprehensive about how his picture fits the world. Failure at school may be due to apathy resulting from the fact that he attaches greater significance and exerts greater effort in sorting out his relations in the world than in the fulfillment of his academic obligations to a point where his interests in the pursuit of his creative talents may cease, a result of confusion of objectives.

Assuming that this diagnosis is sufficiently valid, let us examine briefly a possible external remedy. First, it would be helpful if his teachers, who exert perhaps the strongest influence on his total development, were intellectual in outlook and had a broader awareness of life than the distinguished colleague of our introductory remark.

Of course, there are many individuals who are successful scientists who may never have faced the problem of resolving a complicated mental picture of the world. By example, they serve to integrate and inspire a student who is anxious for interpretations and encouraging suggestions.

There are also those teachers who, almost like intellectual giants, have found an accommodation for both their academic excellence and the world about them. They are the ones who must be singled out by various imaginative means to help the student through this difficult period. Every school and every department must have examples of them. To the type of student under discussion they are living examples of what he might make of himself. Without them his education may make him technically proficient, but still he must look elsewhere for a gift of human values and for a better realization of himself. A feeling of urgency in this self realization and absence of the catalytic encouragement of experienced teachers may precipitate a crisis before his academic career is finished. To be an effective teacher one must care about the student.

Some individuals who are absorbed in research are by nature such that they do not enjoy being frequently bothered but may allow questions along their line of specialization. They may be ones who have learned when to exercise mental curiosity about the world and when to block it in pursuit of academic excellence. For some of their students whose curiosity may be strong, blocking would not work. Therefore, such teachers would not be in a favorable position to understand the problem and render help at a period of concentration and these periods are known to last a long time.

Perhaps another area of improvement rests with a better exploration of a student's talents. Some students arrive at a creative apex of their field by different roads, but others are lost because at their school ordinarily only one road is available, a certain agreed upon road which simplifies the life of the educator, but not of the student. To be sure brilliance and depth do not lack with a large number of those who do not make the mark, but they just don't have the knack of walking in the footsteps of their elders or in absorbing themselves along traditional lines.

I know a number of examples of very capable students who left school or were asked to leave because they could not *adapt* themselves to the regime or learn at the hands of personally colorless "masters" or immature specialists. In many cases, perhaps especially in the smaller schools, a course seemed to mean that the expert must again go through the chore of regurgitating old stuff or unloading on the blackboard, with brilliance and with the touch of a Chef of French Cuisine, that which they have memorized from many previous presentations. The style is so perfect that interruptions with questions which might beset the student are regarded as boorish and lacking in the manner of high scholarship. Thus, the student assumes the role of a copier from the board to the paper and gains little insight from the teacher's experience. His education is mostly achieved at home. In fact since the teacher copies his material from books and published papers, it would be better training for the student to look things up for himself and spend the time in class in discussions which might illuminate points of special difficulty. He may even find out that there are things in the course which the teacher himself does not know, thus giving him the idea that one need not be perfect to be creative.

It is well known that fashions in science are changeable and that in any case at the current rapid pace of growth in 50 or 100 years what is most likely to survive is the spirit scientists convey as carriers of the torch. This indicates that rather than spending much time in perfecting the old, one perhaps should acquire and communicate the zest of the new.

The foregoing discussion indicates that to produce effective teachers, more than specialization in a field is required. Knowledge and practice in teaching methods, courage and constructiveness in human relations are essential ingredients. They provide the fine distinction between assembly line production of talent and a respect for the higher and deeper value of young people.

73(1966), 398–400

THEREFORE BE IT RESOLVED that the Board of Governors of the Mathematical Association of America
   strongly urges the Congress of the United States and the officers of the National Science Foundation, even in the face of the present extraordinary pressures on the National Budget, to consider carefully the importance to our national welfare of continued and even increased support of science education,
   strongly urges the Congress to continue directing this support of science education through the National Science Foundation rather than through other agencies which are not similarly parts of the scientific community, and
   strongly urges the officers of the National Science Foundation, even in the face of the growing need for funds in support of research, to consider carefully the importance to science in our nation of continued and even increased support of science education by the National Science Foundation.

74(1967), 474

New textbooks for the "New Mathematics," RICHARD P. FEYNMAN, *Engineering and Science*, March, No. 6, 28 (1965) 9–15.

Professor Feynman is Richard Chace Tolman Professor of Theoretical Physics at the California Institute of Technology and last year was a member of the California State Curriculum Commission, during his tenure on which he spent considerable time on the selection of textbooks for a modified arithmetic course for grades 1 to 8. The article is the result of this experience.

The article is generally quite critical of the text materials available for grades 1 to 8. For example, "Many of the books go into considerable detail on subjects that are only of interest to pure mathematicians." Or again, "It is possible to give an illusion of knowledge by teaching the technical words which someone uses in a field (which sound unusual to ordinary ears) without at the same time teaching any ideas or facts using these words." It is clear, however, that Professor Feynman's aim in the indicated grades in the "new" mathematics is quite close to those of an informed mathematician or mathematical educator, since "in the 'new' mathematics, then, first *there must be freedom of thought*; second, *we do not want to teach just words*; and third, *subjects should not be introduced without explaining the purpose or reason*." Much of the criticism in the article probably stems from looking at the first eight grades in isolation from the full school program, for much of the terminology introduced in the earlier grades is put to use in later grades.

The article is provocative, and well worth reading, in spite of the fact that one may disagree with much of what is said. It is unfortunate that some of the statements are either the result of misinformation or over-simplification; for example, "one of the best ways to solve complex algebraic equations is by trial and error." Or again, "all the elaborate notation for sets that is given in these books, almost never appear in any writings in theoretical physics, in engineering, in business arithmetic, computer design, or other places where mathematics is being used."

The journal, *Engineering and Science*, is a publication of the California Institute of Technology; individual copies of the journal are available at 50 cents each, by writing to the offices of the journal at 1201 East California Blvd., Pasadena, California 91104.

72(1965), 1018

An Easy Proof of the Fundamental Theorem of Algebra

Charles Fefferman

THEOREM. *Let $P(z) = a_0 + a_1 z + \cdots + a_n z^n$ be a complex polynomial. Then P has a zero.*

Proof. We shall show first that $|P(z)|$ attains a minimum as z varies over the entire complex plane, and next that if $|P(z_0)|$ is the minimum of $|P(z)|$, then $P(z_0) = 0$.

Since $|P(z)| = |z|^n |a_n + a_{n-1}/z + \cdots + a_0/z^n| (z \neq 0)$, we can find an $M > 0$ so large that

$$(1) \qquad |P(z)| \geq |a_0| \qquad (|z| > M).$$

Now, the continuous function $|P(z)|$ attains a minimum as z varies over the compact disc $\{|z| | |z| \leq M\}$. Suppose, then, that

$$(2) \qquad |P(z_0)| \leq |P(z)| \qquad (|z| \leq M).$$

In particular, $|P(z_0)| \leq P(0) = |a_0|$ so that, by (1), $|P(z_0)| \leq |P(z)|(|z| > M)$. Comparing with (2), we have

$$(3) \qquad |P(z_0)| \leq |P(z)| \qquad (\text{all complex } z).$$

Since $P(z) = P((z - z_0) + z_0)$, we can write $P(z)$ as a sum of powers of $z - z_0$, so that for some complex polynomial Q,

$$(4) \qquad P(z) = Q(z - z_0).$$

By (3) and (4),

$$(5) \qquad |Q(0)| \leq |Q(z)| \qquad (\text{all complex } z).$$

We shall show that $Q(0) = 0$. This will establish the theorem since, by (4), $P(z_0) = Q(0)$.

Let j be the smallest nonzero exponent for which z^j has a nonzero coefficient in Q. Then we can write $Q(z) = c_0 + c_j z^j + \cdots + c_n z^n (c_j \neq 0)$. Factoring z^{j+1} from the higher terms of this expression, we have

$$(6) \qquad Q(z) = c_0 + c_j z^j + z^{j+1} R(z),$$

where $c_j \neq 0$ and R is a complex polynomial.

If we set $-c_0/c_j = re^{i\theta}$, then the constant $z_1 = r^{1/j} e^{i\theta/j}$ satisfies

$$(7) \qquad c_j z_1^j = -c_0.$$

Let $\epsilon > 0$ be arbitrary. Then, by (6),

$$(8) \qquad Q(\epsilon z_1) = c_0 + c_j \epsilon^j z_1^j + \epsilon^{j+1} z_1^{j+1} R(\epsilon z_1).$$

Since polynomials are bounded on finite discs, we can find an $N > 0$ so large that,

261

for $0 < \epsilon < 1, |R(\epsilon z_1)| \leq N$. Then, by (7) and (8) we have, for $0 < \epsilon < 1$,

$$|Q(\epsilon z_1)| \leq |c_0 + c_j \epsilon^j z_1^j| + \epsilon^{j+1}|z_1|^{j+1}|R(\epsilon z_1)|$$

$$\leq |c_0 + \epsilon^j\left(c_j z_1^j\right)| + \epsilon^{j+1}\left(|z_1|^{j+1}N\right)$$

(9) $$= |c_0 + \epsilon^j(-c_0)| + \epsilon^{j+1}\left(|z_1|^{j+1}N\right)$$

$$= (1 - \epsilon^j)|c_0| + \epsilon^{j+1}\left(|z_1|^{j+1}N\right)$$

$$= |c_0| - \epsilon^j|c_0| + \epsilon^{j+1}\left(|z_1|^{j+1}N\right).$$

If $|c_0| \neq 0$, then we can take ϵ so small that $\epsilon^{j+1}(|z_1|^{j+1}N) < \epsilon^j|c_0|$. In that case, by (9)

$$|Q(\epsilon z_1)| \leq |c_0| - \epsilon^j|c_0| + \epsilon^{j+1}\left(|z_1|^{j+1}N\right) < |c_0| = |Q(0)|,$$

contradicting (5). So $|c_0| = 0$, and therefore $Q(0) = c_0 = 0$.

74(1967), 854–855

Quick Calculus. By Daniel Kleppner and Norman Ramsey. Wiley, New York, 1965. iii + 294 pp. $2.95.

Reviewing the work before us is like passing on the merits of the latest Howard Johnson's. A paperback by two Harvard physicists, it is more or less equivalent to the first few chapters of those texts which open with a bird's eye view of calculus. It is "programmed" for self-study in an elementary way—no branching, just skipping (over the solution, or the next problem, if you got the first one right). The "quickness" is achieved by omitting hypotheses and proofs, and working far fewer problems than the average text. A long appendix offers a summary, proofs, and more problems. Instant calculus, anyone?

I don't care if physicists—or chimpanzees, for that matter—write mathematics books, but they ought at least to bring an honest point of view to the subject. Apparently all functions have definite integrals: fine, if that's their experience. But they must be kidding then in the beginning when they define and discuss continuous functions, solemnly going through the whole δ-ϵ bit on limits. There is hardly a mention of the standard heuristic use of differentials by physicists (and mathematicians). It's as if they were afraid some mathematician were going to look at their book. In case one does, I recommend their treatment of negative area and backwards integration as being particularly humorous.

ARTHUR MATTUCK, Massachusetts Institute of Technology

75(1968), 316–317

What to Do Till the Computer Scientist Comes*

George E. Forsythe

Computer science departments. What is computer science anyway? This is a favorite topic in computer science department meetings. Just as with definitions of mathematics, there is less than total agreement and—moreover—you must know a good deal about the subject before any definition makes sense. Perhaps the tersest answer is given by Newell, Perlis, and Simon [8]: just as zoology is the study of animals, so computer science is the study of computers. They explain that it includes the hardware, the software, and the useful algorithms computers perform. I believe they would also include the study of computers that *might* be built, given sufficient demand and sufficient development in the technology. In an earlier paper [4], the author defines computer science as the art and science of representing and processing information. Some persons [10] extend the subject to include a study of the structure of information in nature (e.g., the genetic code).

Computer scientists work in three distinguishable areas: (1) design of hardware components and especially total systems; (2) design of basic languages and software broadly useful in applications, including monitors, compilers, time-sharing systems, etc.; (3) methodology of problem solving with computers. The accent here is on the principles of problem solving—those techniques that are common to solving broad classes of problems, as opposed to the preparation of individual programs to solve single problems. Because computers are used for such a diversity of problems (see below), the methods differ widely. Being new, the subject is not well understood, and considerable energy now goes into experimental solution of individual problems, in order to acquire experience from which principles are later distilled. But in the long run the solution of problems in field X on a computer should belong to field X, and computer science should concentrate on finding and explaining the principles of problem solving.

One example of methodological research in computer science is the design and operation of "interactive systems," in which a man and a computer are appropriately coupled by keyboards and console displays (perhaps within a time-sharing system) for the solution of scientific problems.

Because of our emphasis on methodology, Professor William Miller likens the algorithmic and heuristic aspects of problem solving in computer science to the methodology of problem solving in mathematics so ably discussed by Professor Pólya in several books [9]. In computer science there is great stress on the dynamic

---

*Expanded version of a presentation to a panel session before the Mathematical Association of America, Toronto, 30 August 1967. The author is grateful to Professors T. E. Hull, William Miller and Allen Newell for various ideas used in the paper.

action of computation, rather than the static presentation of logical structure. It tends to attract men of action, rather than contemplative men. Our students want to *do* something from the first day.

Computer science is at once abstract and pragmatic. The focus on actual computers introduces the pragmatic component: our central questions are economic ones like the relations among speed, accuracy, and cost of a proposed computation, and the hardware and software organization required. The (often) better understood questions of existence and theoretical computability—however fundamental—remain in the background. On the other hand, the medium of computer science—information—is an abstract one. The meaning of symbols and numbers may change from application to application, either in mathematics or in computer science. Like mathematics, one goal of computer science is to create a basic structure in terms of inherently defined concepts that is independent of any particular application.

Computer science has hardly started on the creation of such a basic structure, and in our present development stage computer scientists are largely concerned with exploring what computers can and cannot economically do. Let me emphasize the *variety*· of fields in which computing has become an important tool. One of these is applied mathematics, as Professor Lax emphasizes, but this is merely one. Others include experimental physics, business data processing, economic planning, library work, the design of almost anything (including computers), education, inventory management, police operations, medicine, air traffic control, national population inventories, space science, musical performance, content analyses of documents, and many others. I must emphasize that the amount of computing done for applied mathematics is an almost invisible fraction of the total amount of computing today.

There is frequent discussion of whether computer science is part of mathematics —i.e., applied mathematics or "mathematical science." In a purely intellectual sense such jurisdictional questions are sterile and a waste of time. On the other hand, they have great importance within the framework of institutionalized science —e.g., the organization of universities and of the granting arms of foundations and the Federal Government.

I am told that the preponderant opinion among administrators in Washington is that computer science is part of applied mathematics. I believe the majority of university computer scientists would say it is not; cf. [8]. I would have to ask you how mathematicians feel about the matter. COSRIMS (Committee on the Support of Research in the Mathematical Sciences, appointed by the National Academy of Science—National Research Council) has taken the position that computer science is a mathematical science, but many of the discussions emphasize differences between mathematics and computer science.

In spite of the infancy of our subject, there are approximately 40 computer science departments in the United States and Canada today. There is no longer any doubt that computer science will have a separate university organization for several coming decades. I believe that the creation of these separate departments is a correct university response to the computer revolution, for I do not think computers would be well studied in an environment dominated by either mathematicians or engineers. However, finding suitable faculty members is very difficult today.

What are these computer science departments doing? Answer: Roughly the same things that mathematics departments are doing: education, research, and service. We teach computer science to three types of students: to our majors at the

B.S., M.S., and Ph.D. levels, to technical students who need computing as a tool, and to any students who wish to become acquainted with computing as an important ingredient of our civilization. We do research in our several specialties: e.g., numerical analysis, programming languages and systems, heuristic methods of problem solving, graphical data representation and processing, time-sharing systems, logical design, business data processing, etc. We perform an unusually large amount of community service in helping our colleagues with their computing problems, both individually and by advising or managing the university computation center.

· · ·

What can you do now? And now follow my answers to the question of the title. *First*, you can get a little acquainted with computing. This involves two steps:

Step A: Learn to program some automatic digital computer in some language—e.g., Fortran Algol, PL/1—and actually use the computer enough to find out some of the fascination and frustrations of the computerman's world.

Step B: Read some books from the list at the end of this paper. Since computer science is not yet very deep and mathematicians are very smart people, this should not be onerous.

Second, you can study how computing intersects mathematics. Applied mathematics is no longer the same subject, now that you have a magnificent experimental tool at hand. Moreover, there are several undergraduate courses that owe their large enrollments largely to their wide applications in technology and science: e.g., linear algebra, and ordinary differential equations. I think both of these courses should be substantially influenced by computers.

In a linear algebra course, along with concepts like rank, determinant, eigenvalues, linear systems, and so on, ought to go some constructive computational methods suitable for automatic computers. There is plenty of literature now, and I think some of it should be worked into courses in linear algebra. If not, then an instructor should loudly confess that he is ignoring these topics, and furnish some reading lists for his students.

The same goes for ordinary differential equations. Here the situation is slightly different, in that textbooks in this field usually do say something about numerical methods. The trouble is that it usually dates from before the days of computers. It should be expunged and replaced with at least an equivalent amount of orientation in today's useful numerical methods for computers. See [7] for Professor Hull's suggestions.

I think also that the calculus courses should be influenced by an awareness of computing, but I do not expect this to be a very large fraction of the courses. See [6] for some ideas.

The alternative to weaving computational material into various mathematics courses is to teach computational mathematics in separate courses, in either the department of mathematics or the computer science department. This alternative is the accepted method at present, but many have felt it should be only a temporary expedient. If computational mathematics is taught in the computer science department, what effective mechanism can there be to reunite the theoretical and the computational aspects of mathematics?

There is a good deal of interest nowadays in *computer-aided instruction*. I don't expect this to have a very large application to university mathematics teaching. However, I should like to call your attention to the usefulness of a computer-controlled cathode-ray-tube display and "light pen" in giving vivid graphical representations of sophisticated concepts. In one of these, developed by Professor William McKeeman and Mr. William Rousseau at Stanford University, the scope shows both the complex z plane and the plane of $f(z)$, for any simple elementary function f typed at the console. When the light pen traces any curve in the z-plane, a dot of light traces the curve $f(z)$. Many of the elementary theorems of analytic function theory receive an impressive illustration in this way. Professor Marvin Minsky has used similar displays in dealing with non-linear ordinary differential equations.

At a more fundamental level, the emergence of computer science has added one more applier of mathematics. Along with operations research, economics, and other more recently mathematized subjects, computer science is relatively more interested in *discrete mathematics* (e.g., combinatorics, logic, graph and flow theory, automata theory, probability, number theory, etc.; see [1]), than in *continuum* mathematics (e.g., calculus, differential equations, complex variables, etc.). Hence the mathematics department (in my view) should devote much thought to organizing its curriculum suitably from the standpoint of consumers of discrete mathematics. I feel that currently common curricula are inherited from the days when continuum mathematics was more in demand (from physics, mechanical engineering, etc.).

Third, you can help the computer scientist find his way to your campus, and make him feel welcome. Above all, please don't judge him as a mathematician, for he isn't one and isn't supposed to be one—his values are different. The difference in values between mathematics and numerical analysis is the subject of a provocative paper [5].

When the computer scientist does arrive on campus, be prepared for a rather large impact. He is tied to a rampant field of rapidly growing interest to students and scholars everywhere. He will need many colleagues and new buildings. He may take some of the heat off mathematics faculties by providing a partial substitute for mathematics as a research tool. This vast energy may have some undesirable side effects on your sense of importance and even your budget.

Fourth, if you are really enthusiastic, I recommend tackling some research problems of a mathematical nature that would help computer science (and your own publication list). There are serious and important mathematical questions at almost every turn, and most computer scientists aren't very good at mathematics. I will leave to Professor Lax the important area of experimental mathematics. One area of computer science with a probable payoff is the automation of algebra and analysis. So far, most actual computing consists of automated *arithmetic*. A Fortran program, for example, asks a computer to carry out addition, subtraction, multiplication and division of (simulated) real or complex numbers, in a sequence which is dynamically determined by the course of the computation. There is little else. It is clear that computers are capable of automated *algebra*, and there have been experimental systems for this since about 1961. They are still primitive. Some of the roadblocks to further development occur at surprising places. One is the question of *simplification* (e.g., of rational polynomial expressions in n variables). What do we mean by simplification? How shall we do it? See Brown [2] for one indication of the depth of the problem.

Proposed by Dr. R. W. Hamming, but still largely in the future, is the partial automation of *analysis*. Faced with an initial-value problem for an ordinary differential equation, for example, a computer should be able to put the problem into some sort of normal form (using automated algebra, of course). Then the computer should inspect the normal form to see whether it is a recognized standard equation. If it is, then a solution formula should be obtained from a table, and then transformed (by automated algebra) back into the variables originally presented. Of course, the user may want a table of values. The computer then must decide whether to use the solution formula (if one exists), or to compute a numerical solution. In the latter case, a numerical integration formula must be automatically selected (or devised), and then used (by automatic arithmetic) to produce a table of answers and error bounds (more automated analysis). There are many unsolved problems in this program, and mathematicians are uniquely qualified to define the problems and start their solution.

Most computation to date has been *serial* in nature, with only one computation or decision being made at a time within the central processor. Soon to arrive will be *parallel* computers, in which from two to perhaps several hundred operations can be formed simultaneously. The general pattern of serial computation has been well understood since the work of Babbage, Aiken, von Neumann, and others. There are good research problems in analyzing parallel computation and identifying the important features. See [**3**] for a recent contribution.

There are good research problems in the theoretical aspects of the design of algorithms. Initiated by Post, Turing, and others, there is an important theory that tells us that some functions are computable on a "Turing machine," and some are not. (Turing machines differ in theoretical capability from existing computers only in having infinite storage capacity.) This theory has been extended to state that some problems can be solved on a Turing machine with a suitable algorithm, but for some problems no such algorithm can exist.

It is essential to know that a problem is solvable, but this is only the beginning. What is needed next is information about how much computer storage is required for the program and data, and how long the algorithm will run. In other words, we need theoretical information on the complexity of solvability. There are some results by Kolmogorov and others on the complexity of a computable function, but much more research is needed.

Other research problems lie in areas further removed from mathematics. One such area is computer graphics—the uses of computers for dealing directly with information in the form of structures. (Examples: representing graphs of mathematical trees, design of networks, recognition of three-dimensional block structures from photographs, automatic reading of bubble chamber pictures.) In this area there are problems of representing information, both visually and inside a computer store, and of processing the information. Most algorithms are bing created by persons with only a modest knowledge of mathematics, and it seems likely that an interested mathematician could both help solve some computing problems and find worth-while mathematical problems.

In summary, here are my four answers to the question of the title:

(1) Learn a little about computer science.
(2) Consider how mathematics curricula should be affected by computer science.
(3) Help the computer scientist find his way, but expect a big blast after he gets there.
(4) Think of computer science as a possible source of mathematical research problems.

. . .

77(1970), 454–462

Top: George Pólya, late 1930's
Bottom: George Pólya, 1960's

Erratum: Contrary to the information we received some months ago and which was published in the October issue of this MONTHLY, we have just been advised that Professor Emeritus G. H. Hunt is alive and well: our deepest apologies and our very best wishes to him for a long life ahead.

Hardy's "A Mathematician's Apology"

L. J. Mordell

A reprint of this most interesting book appeared in 1967 with a foreword by C.P. Snow, Hardy's friend of long standing.

It has often been reviewed and highly praised, but there are, however, some opinions expressed by Hardy which, perhaps, have not been adequately dealt with by other reviewers. Furthermore, Snow's foreword calls for some comment, especially his references to Ramanujan (1887–1920). He writes that after Hardy and Littlewood read the manuscript sent by Ramanujan to Hardy (probably Jan. 16, 1913), "they knew, and knew for certain" that he "was a man of genius.... It was only later that Hardy decided that Ramanujan was, in terms of *natural* mathematical genius, in the class of Gauss and Euler; but that he could not expect, because of the defects of his education, and because he had come on the scene too late in the line of mathematical history, to make a contribution on the same scale." While one would readily accept that Ramanujan was a man of genius, the comparison with Gauss and Euler is very farfetched. I have some difficulty in believing that Hardy made such a statement, or at any rate made it in this form. What does natural mathematical genius mean? Undoubtedly Ramanujan was outstanding in some aspects of mathematics and had great potentialities. But this is not enough. What really matters is what he did, and one cannot accept such a comparison with Euler and Gauss, whose many-sided contributions were of fundamental importance and changed the face of mathematics. In fact in 1940, in the book on Ramanujan, Hardy said "I cannot imagine anybody saying with any confidence, even now, just how great a mathematician he was and still less how great a mathematician he might have been."

Snow says that Ramanujan, as is commonly believed, was the first Indian to be elected (2 May, 1918) a fellow of the Royal Society. He was the second. The first was Ardaseer Cursetjee (1808–1877), shipbuilder and engineer, F.R.S. 27 May, 1841. Snow notes that Ramanujan was elected a fellow of Trinity four years after his arrival in England and continues, "it was a triumph of academic uprightness that they should have elected Hardy's protégé Ramanujan at a time when Hardy was only just on speaking terms with some of the electors and not at all with others." It is well that the merits of a fellowship candidate are judged by the quality of his original work and not by the political views of his sponsors.

Let us examine some of the views expressed by Hardy. They are sometimes stated too categorically, regardless of exceptions and limitations. A number of them had their origin in what he says, most gloomily, in the very first section of the Apology: "It is a melancholy experience for a professional mathematician to find himself writing about mathematics. The function of a mathematician is to do something, to prove new theorems, to add to mathematics, and not to talk about what he or other mathematicians have done."

His practice many years ago does not conform with this statement. He recalls in Section 6 that he did talk about mathematics in his 1920 Oxford inaugural lecture, which actually contains an apology for mathematics. Further in 1921, he gave an address on Goldbach's theorem to the Mathematical Society of Copenhagen. In this, he did talk about what he and other mathematicians had done. Such talks render a real service to mathematics and many have found great pleasure and inspiration in listening to or reading such expositions. Hardy had followed the practice of many eminent mathematicians in giving them. These have contributed to the richness and vividness of mathematics and make it a living entity. Without them, mathematics would be much the poorer.

No mathematician can always be producing new results. There must inevitably be fallow periods during which he may study and perhaps gather ideas and energy for new work. In the interval, there is no reason why he should not occupy himself with various aspects of mathematical activity, and every reason why he should. The real function of a mathematician is the advancement of mathematics. Undoubtedly the production of new results is the most important thing he can do, but there are many other activities which he can initiate or participate in. Hardy had his full share of these. He took a leading part in the reform of the mathematical tripos some sixty years ago. Before then, it was looked upon as a sporting event, reminding one of the Derby, and was out of touch with continental mathematics. A mathematician can engage in the many administrative aspects of mathematics. Hardy was twice secretary and president of the London Mathematical Society and, while so occupied, must have done an enormous amount of unproductive work. He served on many committees dealing with mathematics and mathematicians. He wrote a great many obituary notices. He was well aware that a professor of mathematics is a representative of his subject in his University. This entails many duties which cannot be called doing mathematics.

His reference to a melancholy experience shows how much he took to heart and suffered from the loss of his creative powers. The result is, as Snow says, that the *Apology* is a book of haunting sadness.

Further in this first section, he says despairingly, "if then I find myself writing not mathematics but 'about' mathematics, it is a confession of weakness, for which I may rightly be scorned or pitied by younger and more vigorous mathematicians. I write about mathematics because, like any other mathematician who has passed sixty, I have no longer the freshness of mind, the energy, or the patience to carry on effectively with my proper job." He had been for many years a most active mathematician and his collected works now being published will consist of seven volumes. It seems almost nonsense to say that anyone would scorn or pity him, and the use of the term 'rightly' is even more nonsensical.

We all know only too well that with advancing age we are no longer in our prime, and that our powers are dimmed and are not what they once were. Most of us, but not Hardy, accept the inevitable. There are still many consolations. We can perhaps find pleasure in thinking about some of our past work. We can read what others are doing, but this may not be easy since many new techniques have been evolved, sometimes completely changing the exposition of classical mathematics. Various reviews, however, may give one some idea of what has been done. (We can still be of service to younger mathematicians.)

His statement about a mathematician who has passed sixty is far too sweeping and any number of instances to the contrary can be mentioned, even among much older people. One need only note some recent Cambridge and Oxford professors.

Great activity among octogenarians is shown by Littlewood, his lifelong collabora-tor, Sydney Chapman, his former pupil and collaborator, and myself. There is also Besicovitch in the seventies. Davenport, who had passed sixty, was as active and creative as ever, and his recent death is a very great loss to mathematics since he could have been expected to continue to produce beautiful and important work.

The question of age was ever present in Hardy's mind. In Section 4, he says, "No mathematician should ever allow himself to forget that mathematics, more than any other art or science, is a young man's game." It seems that he could not reconcile himself to growing old. For further on, he says, "I do not know an instance of a major mathematical advance initiated by a man past fifty." This may be so, but much depends on the definition of the *advance*. But there is no need to be troubled about it. Much important work has been done by men after the age of fifty.

A number of Hardy's statements must be qualified. In Section 2, he says, that "good work is not done by 'humble' men. It is one of the first duties of a professor, for example, in any subject, to exaggerate a little both the importance of his subject and his own importance in it. A man who is always asking, 'Is what I do worthwhile?' and 'Am I the right person to do it?' will always be ineffective himself and a discouragement to others."

Though one may naturally have a better opinion of one's work than others have, there are many exceptions to his statement. I never knew Davenport to exaggerate or emphasize the importance of his work, but he was a most effective mathemati-cian and a very successful supervisor of research. Prof. Frechet told me a few years ago, that when Norbert Wiener was working with him a long time ago, he was always asking, "Is my work worthwhile?", "Am I slipping?", etc. S. Chowla is as modest and humble a mathematician as I know of, but he inspires many research students.

We comment on some more of Hardy's statements about mathematics. One of the most surprising is in Section 29, "I do not remember having felt, as a boy, any *passion* for mathematics, and such notions as I may have had of the career of a mathematician were far from noble. I thought of mathematics in terms of examinations and scholarships; I wanted to beat other boys, and this seemed to me to be the way in which I could do so most decisively."

It has often been said that mathematicians are born and not made. Most great mathematicians developed their keenness for mathematics in their school days. Their ability revealed itself by comparison with the performances of their school-mates. Their ambition was to continue the study of mathematics and to take up a mathematical career. Probably no other motive played any part in the decision of most of them.

In Section 3, he considers the case of a man who sets out to justify his existence and his activities. I see no need for justification any more than a poet or painter or sculptor does. As Trevelyan says, disinterested intellectual curiosity is the life blood of real civilization. It is curiosity that makes a mathematician tick. When Fourier reproached Jacobi for trifling with pure mathematics, Jacobi replied that a scientist of Fourier's calibre should know that the end of mathematics is the great glory of the human mind. Most mathematicians do mathematics for the very good reason that they like and enjoy doing it. Davenport told me that he found it "exciting" to do mathematics.

Hardy says that the justifier has to distinguish two different questions. The first is whether the work which he does is worth doing and the second is why he does it, whatever its value may be. He says to the first question: The answer of most people, if they are honest, will usually take one or the other of two forms; and the second form is merely a humbler version of the first, which we need to consider seriously. "I do what I do because it is the one and only thing I can do at all well." It suffices to say that the mathematician felt no need to do anything else.

Hardy is very appreciative of the beauty and aesthetic appeal of mathematics. "A mathematician," he says in Section 10, "like a painter or poet, is a maker of patterns...," and these "...must be beautiful. The ideas...must fit together in a harmonious way. Beauty is the first test: there is no permanent place in the world for ugly mathematics.... It may be very hard to *define* mathematical beauty..." but one can recognize it. He discusses the aesthetic appeal of theorems by Pythagoras on the irrationality of $\sqrt{2}$, and Euclid on the existence of an infinity of prime numbers. He says in Section 18, "there is a very high degree of *unexpectedness*, combined with *inevitability* and *economy*. The arguments take so odd and surprising a form; the weapons used seem so childishly simple when compared with the far-reaching results."

I might suggest among other attributes of beauty, first of all, simplicity of enunciation. The meaning of the result and its significance should be grasped immediately by the reader, and these in themselves may make one think, what a pretty result this is. It is, however, the proof which counts. This should preferably be short, involve little detail and a minimum of calculations. It leaves the reader impressed with a sense of elegance and wondering how it is possible that so much can be done with so little.

Somehow, I do not think that Hardy's work is characterized by beauty. It is distinguished more by his insight, his generality, and the power he displays in carrying out his ideas. Many of the results that he obtains are very important indeed, but the proofs are often long and require concentrated attention, and this may blunt one's feelings even if the ideas are beautiful.

Hardy does not define ugly mathematics. Among such, I would mention those involving considerable calculations to produce results of no particular interest or importance; those involving such a multiplicity of variables, constants, and indices, upper, lower, right, and left, making it very difficult to gather the import of the result; and undue generalization apparently for its own sake and producing results with little novelty. I might also mention work which places a heavy burden on the reader in the way of comprehension and verification unless the results are of great importance.

Hardy had previously said that he could "quote any number of fine theorems from the theory of numbers whose meaning anyone can understand, but whose proofs, though not difficult, may be found tedious." It often happens that there are significant results apparently of some depth, the proof of which can be grasped by those with a minimum of mathematical knowledge. Perhaps I may be pardoned if I give one of my own. The theorem of Pythagoras suggests the problem of finding the integer solutions of the equation $x^2 + y^2 = z^2$. This was done some 1000 years ago and is not difficult. But suppose we consider the more general equation $ax^2 + by^2 = cz^2$. This is a real problem in the theory of numbers. Legendre at the end of the eighteenth century gave necessary and sufficient conditions for its solvability. Then when the equation is taken in the normal form, i.e. abc is square-free and $a > 0$, $b > 0$, $c > 0$, Holzer showed in 1953 that a solution existed with $|z| < \sqrt{ab}$, from which it follows that $|x| \leqq \sqrt{bc}, |y| \leqq \sqrt{ca}$. I recently found a proof of this result that no one would call tedious by showing that if a solution (x_1, y_1, z_1)

existed with $|z_1| > \sqrt{ab}$, then there was another with $|z_2| < |z_1|$. This arose by taking an appropriate line through the point (x_1, y_1, z_1) to meet the conic $ax^2 + by^2 = cz^2$ in the point (x_2, y_2, z_2). I call this a schoolboy proof, because the only advanced result required is that the equation $lx + my = n$ has an integer solution if l and m are co-prime. A proof of the theorem could have been found by a schoolboy.

We conclude by examining Hardy's views about the utility or usefulness of mathematics. He seems to denigrate the usefulness of 'real' mathematics. In Section 21, he says, "The 'real' mathematics of the 'real' mathematicians, the mathematics of Fermat and Euler and Gauss and Abel and Riemann is almost wholly 'useless.'" This statement is easily refuted. A ton of ore contains an almost infinitesimal amount of gold, yet its extraction proves worthwhile. So if only a microscopic part of pure mathematics proves useful, its production would be justified. Any number of instances of this come to mind, starting with the investigation of the properties of the conic sections by the Greeks and their application many years later to the orbits of the planets. Gauss' investigations in number theory led him to the study of complex numbers. This is the beginning of abstract algebra, which has proved so useful for theoretical physics and applied mathematics. Riemann's work on differential geometry proved of invaluable service to Einstein for his relativity theory. Fourier's work on Fourier series has been most useful in physical investigations. Finally one of the most useful and striking applications of pure mathematics is to wireless telegraphy which had its origin in Maxwell's solution of a differential equation. Many new disciplines are making use of more and more pure mathematics, e.g., the biological sciences, economics, game theory, and communication theory, which requires the solution of some difficult Diophantine equations. It has been truly said that advances in science are most rapid when their problems are expressed in mathematical form. These in time may lead to advances in pure mathematics.

These remarks may serve as a reply to Hardy's statement that the great bulk of higher mathematics is useless.

It is suggested that one purpose mathematics may serve in war is that a mathematician may find in mathematics an incomparable anodyne. Bertrand Russell says that in mathematics, "one at least of our nobler impulses can best escape from the dreary exile of the actual world." Hardy's comment on this reveals his depressed spirits. "It is a pity," he says, "that it should be necessary to make one very serious reservation—he must not be too old. Mathematics is not a contemplative but a creative subject; no one can draw much consolation from it when he has lost the power or desire to create; and that is apt to happen to mathematicians rather soon." What does he mean when he says mathematics is not a contemplative subject? Many people can derive a great deal of pleasure from the contemplation of mathematics, e.g., from the beauty of its proofs, the importance of its results, and the history of its development. But alas, apparently not Hardy.

A mathematician's apology by G. H. Hardy with a foreword by C. P. Snow, University Press, Cambridge 1967 (18s.). A paperback edition is available at 8s.

77(1970), 831–836

Various Indices of Shortages of College Teachers by Fields

	Biological Sciences	Chemistry	Physics	Physical Sciences	Mathematics
Percentage of new teachers with Doctor's degrees, colleges and universities, 1964–65	50.2			59.1	28.2
Percentage of new teachers with Master's degrees or less, colleges and universities, 1964–65	29.3			27.2	51.1
Percentageof new teachers with less than a Master's degree, junior colleges 1963–64 and 1964–65	12.1	14.9	19.9		17.3
Number of colleges reporting critical shortage of qualified teachers in 1963–64 and 1964–65	167	191	406		597
Number of colleges foreseeing more acute shortage of qualified teachers	120	195	362		508
Number of budgeted positions unfilled in colleges and universities, 1963–64 or 1964–65	50	51	110		166
Number of junior colleges reporting a shortage of qualified teachers in 1963–64 and 1964–65	30	105	132		159
Number of junior colleges foreseeing a future shortage of qualified teachers	30	92	120		178

Source: *Teacher Supply and Demand in Universities, Colleges, and Junior Colleges, 1963–64 and 1964–65,* National Education Association

74(1967), 475

EDITORIAL

Entering the last quarter of its first century, this MONTHLY has a new cover. Few remember that it was ever anything but blue on blue, as introduced by Editor Lester R. Ford, Sr. in 1942. We hope the MONTHLY will be better than ever. If so, this will be due to the excellence of our authors' contributions and the hard work of the Associate and Collaborating Editors and the referees.

The Editor invites criticisms, complaints, and suggestions.

Backlog: Main articles, 11 months; Mathematical Notes, 10 months; Research Problems, 4 months; Classroom Notes, 4 months; Mathematical Education, 4 months.

HARLEY FLANDERS, *Editor*

76(1969), 2

After the Deluge

D. A. Moran

The purpose of this note is to provide a somewhat simpler proof than that given by Professor Marston Morse of his elementary theorem about pits, peaks, and passes on the sphere [1]. It should be recalled that Professor Morse views a positive, real-valued, bounded, differentiable function on the sphere as an altitude, measured from the center of some hypothetical spherical planet. Critical points of the function then correspond to pits (= minima), peaks (= maxima), and passes (= saddle points of index −1) on the planet. It is assumed that no critical points more complicated than these three types ever occur, and that no two of these singularities occur at precisely the same altitude. Our viewpoint is essentially the same as this, but we start with the following slightly different

ADDITIONAL HYPOTHESIS. *No pass is as high as a peak, or as low as a pit.*

This hypothesis is easily fulfilled, if we agree to drill deep holes at the bottom of each pit, and raise tall flagpoles atop each peak. Let N_0, N_1, and N_2 be the number of pits, passes, and peaks.

Now, let rain begin to fall on the planet which represents the sphere. Immediately N_0 lakes are created. As the water level rises to the altitude of a pass, a lake can merge with itself, creating an island, or else a lake can merge with another lake, resulting in a net decrease by 1 in the number of lakes. When the water level has risen to inundate every pass, but is not yet as high as any peak, the number of islands will be

$$1 + \text{number of island-increasing passes},$$

and the number of lakes will be

$$N_0 - \text{number of lake-decreasing passes}.$$

The number of lakes less the number of islands is therefore

$$N_0 - N_1 - 1.$$

On the other hand, at this point in time there is clearly one lake and N_2 islands, so

$$N_0 - N_1 - 1 = 1 - N_2,$$

proving Morse's generalization of the theorem of Euler.

Written while the author was partially supported by NSF grant GP 8962.

References

1. Marston Morse, *Pits, peaks, and passes* (motion picture film), Modern Learning Aids #3462, New York.

2. George Polya, Induction and Analogy in Mathematics, Princeton Univ. Press, 1954, pp. 163–165.

77(1970), 1096

Ralph Boas, Editor of the MONTHLY 1977–1981

7. MODERN TIMES, 1971–1993

Al Wilcox, Executive Director of the MAA 1968–1989

Modern Times

President Nixon visited China in 1972, establishing relations that would indirectly affect mathematics graduate programs throughout America. In that same year, five men were arrested for breaking into the Watergate complex. Nixon resigned in 1974 at almost precisely the time when Fields Medals were being awarded to Enrico Bombieri and David Mumford in Vancouver. The United States celebrated its bicentennial in 1976, when Quillen and Suslin were publishing their (independent) proofs of the Serre Conjecture. In 1979, a major accident occurred at the Three Mile Island nuclear reactor and 63 Americans were taken hostage at the American embassy in Teheran; some months later, Daniel Gorenstein and Michael Aschbacher announced the classification of the finite simple groups. After two years of political turmoil in Poland, the International Congress of Mathematicians scheduled for Warsaw was postponed from 1982 until 1983. The first woman President of the AMS, Julia Robinson, died in 1985; six months later, the space shuttle Challenger exploded.

Mathematics has changed remarkably little during the past quarter century. There have been dramatic triumphs—the Four Color Theorem, the Bieberbach Conjecture, the Classification of 4-manifolds, the Mordell Conjecture, and (most recently) the Fermat Conjecture. Some fields have blossomed, especially those associated with computation. But the profession itself has changed only slightly. After its dramatic expansion in the 60's, mathematics found itself with a surplus of Ph.D.'s in the 70's, and times were tough for young mathematicians. Things improved in the 80's, and times were good. The expansion was followed by contraction, and times are bad again. The profession is adjusting.

What is different now? Computers and calculators are, of course, but even in the early 70's mathematicians saw the possibilities for computers—for symbolic manipulation, for experimental research, for scientific calculation, and for classroom instruction. The Monthly was full of articles about computer calculus, and even articles about using handheld calculators in the classroom.

Computers have affected mathematics the most in a way that would have surprised people in 1970—because of one man (Donald Knuth) and one program (TeX).

The role of women has been different. Women were always part of the American mathematical community; in 1912, about 8% of the AMS membership were women. When the MAA was founded in 1916, 12% of the founding members were women. There were famous women mathematicians—for example, Charlotte Angus Scott, Grace Chisolm Young, Anna Pell Wheeler, Emmy Noether, and Mina Rees. But the great expansion of mathematics following the war, which accelerated through the 60's, did not include women—at least not at first. Women discovered they were an ever smaller fraction of the mathematical community, and they took action. The Association for Women in Mathematics was founded in 1971. The percentage of women among new Ph.D.'s has risen steadily beyond the previous maximum (which was achieved in the 1930's). Three women have been presidents of the MAA; one of the AMS.

Our fads have been different—Catastrophe Theory, Fractals, Chaos—but a century of the Monthly shows that mathematics, like society, is susceptible to fashion.

Our steady preoccupation with federal funding marks the last quarter of this century as unusual. American mathematics has worried a great deal about its own needs.

Some things remain the same. For one hundred years, mathematicians have bemoaned the state of mathematics education, and they have proposed solutions. (Some of them worked.) During that same time, people tried to balance research and teaching, and the two cultures of American mathematics gently sparred with one another. Applied mathematicians pointed out (at periodic intervals) that pure mathematics had wandered from its origins. Pure mathematicians complained that applied mathematics was dull. Young mathematicians believed that old mathematicians were overpaid. Old mathematicians pointed out that young mathematicians knew nothing of the history of the subject. Young and old mathematicians complained about the lack of good mathematical exposition.

For one hundred years, readers of the Monthly have read about major advances, curious proofs, and history. They have submitted articles and solved problems. They have served as editors and referees. And while details of the Monthly have changed in those one hundred years, the philosophy has stayed the same. The Monthly is a journal that binds mathematicians together.

In 1993, as the Monthly published its 100th volume, the nation slowly recovered from recession. Bill Clinton became the 42nd President of the United States. The Soviet Union had become 15 separate nations. War smoldered in the former Yugoslavia, and famine stalked Somalia. Floods swept the midwest. AIDS continued to spread across the world. Mathematicians admired the largest known prime $(2^{756,839} - 1)$, and they studied the proof of the Fermat Conjecture recently announced by Andrew Wiles.

MATHEMATICAL EDUCATION

Female Mathematicians, Where Are You?

Violet H. Larney

At the persuasive urging of the U.S. Department of Health, Education and Welfare, the Academic Dean at College X has asked his (not *her*) department chairmen to hire more women faculty. (Let X be almost any institution of higher education in the United States in 1972.) Understandably, the College Administration looks with a collectively skeptical eye at the Chairman of the Mathematics Department when he says that he would be happy to hire a well-qualified female professor, but that he hasn't been able to find one! Meanwhile, inquiries about job openings keep pouring in from mathematicians of the wrong sex!

Are Mathematics Department chairmen really the male chauvinist pigs (whatever that is!) that their hiring practices indicate they are? Let's examine the facts. For the sake of simplicity (and generosity), a *qualified female mathematician* will be defined as a female who possesses an earned doctorate in mathematics. Assume that 25 is the minimum age at which a woman earns her Ph.D. (the median age in 1966 was 29.6), and assume that she retires when she is 65. Then the female mathematicians qualified to hold academic appointments in 1970–71 would have earned their doctorates some time during the preceding forty years. A few reference books and a desk calculator yield the figure of 816 as the total number of women who received doctorates in mathematics from the academic year 1930–31 through 1969–70. (At the time of writing, no figures were yet available for 1970–71.) But in 1970 there were 1181 U.S. colleges and universities that awarded degrees in mathematics, either at the baccalaureate level or above. Hence, the upper bound for the number of available women Ph.D.'s is too small to average even one woman at each institution! When one eliminates from the set of 816 women those who are retired, are deceased, have non-academic jobs, or who are not employed (by choice or for other reasons), one might safely conjecture that in 1970 there was available only one female with a Ph.D. in mathematics for every two degree-granting institutions in the United States.

During the past four decades, when 816 doctorates were awarded to women, there were 10,742 degrees awarded to men, thus giving women 7% of the degrees earned during that period. We might ask whether the situation is beginning to improve for women, now that we have left the Victorian Era behind and are being exposed to the Liberated Woman. Interestingly enough, records for the last half-century show that women earned a record high of 40% of the 15 mathematics doctorates awarded in 1921, and they hit their low point in 1952, earning only 4% of the 205 degrees conferred that year! In the sixties they began to stage a slow

TABLE 1
Doctorates in Mathematics Earned at U.S. Universities from 1931 through 1970,
by Five-year Periods

Period	Percent of Women	No. of Women	No. of Men	Total
1931–35	15.7	63	338	401
1936–40	12.2	50	360	410
1941–45	13.5	40	257	297
1946–50	8.0	49	561	610
1951–55	4.5	51	1074	1125
1956–60	5.0	66	1247	1313
1961–65	6.6	165	2343	2508
1966–70	6.8	332	4562	4894
Total	7.1	816	10,742	11,558

comeback. Table 1 shows the Ph.D. production in mathematics over the past 40 years.

Incidentally, Table 1 also gives striking evidence of why young mathematicians are desperately searching for openings these days. It was found that 53% of all mathematicians produced in the entire forty years were produced in the last seven years! The year 1970 alone turned out more new mathematicians for the job market than were educated in the entire 15-year period from 1931 through 1945!

The two-year colleges also are being pressured into hiring more women. Up until very recently most community colleges have had to hire those who possessed a master's degree only. The proportion of master's degrees earned by women during the past two decades is revealed in Table 2. The figures indicate that a small two-year institution which has a Mathematics Department consisting of four men and one woman has its share of the national supply of females, and is not as sexually unbalanced as its president might think!

To gain a proper perspective, one must find out how the situation in mathematics compares, sexwise, with that in other fields. Table 3 gives a broad answer to that question. During the five-year period, 1965–70, 41% of all baccalaureates awarded in the U.S. in all disciplines went to women, while 36% of the undergraduate degrees in mathematics were earned by women. Although the female drop-out rate (after the baccalaureate) is high in most areas of study, it is significantly higher in mathematics than it is in general. We see that while women earned 13% of all doctorates in 1965–70, women earned only 7% of the doctorates awarded in mathematics.

TABLE 2
Masters Degrees in Mathematics Earned at U.S. Universities
from 1951 through 1970, by Five-year Periods

Period	Percent of Women	No. of Women	No. of Men	Total
1951–55	17.4	706	3349	4055
1956–60	19.7	1255	5106	6361
1961–65	19.5	3120	12,872	15,992
1966–70	25.2	6783	20,177	26,960
Total	22.2	11,864	41,504	53,368

TABLE 3
Number and Percent of Degrees Earned by Women in Mathematics and in All
Disciplines for the Five-year Period 1966–70

	Bachelors	Masters	Doctors
In Mathematics	43,329 (36%)	6,783 (25%)	332 (7%)
In All Disciplines	1,411,937 (41%)	321,476 (39%)	14,897 (13%)

We conclude with a few questions that need to be answered. If college administrators are specifying female faculty quotas for their schools and departments, are they first determining the present available supply of woman-power to fill these quotas? To correct the imbalances, are graduate schools making a real effort to admit more women, and to award more assistantships to women? Do the little girls who loved arithmetic in grade school automatically lose interest in advanced mathematics, or did a sexist attitude somewhere along the line convince them that it is unladylike to enjoy electric trains and abstract algebra? In general, do females have less aptitude for doing mathematical research than do males? (A man wouldn't dare even to ask that one!) In any event, female mathematicians, where are you?

References

Doctorate Production in United States Universities, 1920–1962, National Academy of Sciences—National Research Council Publication 1142, 1963.

Higher Education, *Earned Degrees Conferred*, National Center for Educational Statistics, 1966 and 1970.

Statistical Abstract of the United States, 1950 through 1971.

80(1973), 310–313

THE IMPACT OF COMPUTERS

By GARRETT BIRKHOFF

1. High-school preparation. Although the mention of 1984 suggests a mood of Orwellian gloom for our panel discussion, I thnk there are grounds for cautious optimism if mathematicians are willing to show restraint, to plan honestly and carefully, and to follow up their plans with effective and proportionate action. Now that our post-Sputnik euphoria has faded, and the post-war baby bulge and prosperity have had their impact on college enrollment, it seems clear that the mathematical community is faced with a return to a more normal balance between supply and demand. As a result, mathematicians as individuals seem likely to be less affluent in 1984; as a partial compensation, they should be able to receive and provide better education!

If our present rate of production of Ph.D.'s continues, which seems to me likely, we shall have about 15,000 new Ph.D.'s in mathematics by 1984. This great reservoir of highly skilled talent can greatly improve the quality of mathematical education, *provided* that Ph.D. training is broadened* and geared more realistically to our national needs. In particular, it can greatly improve the quality of undergraduate mathematical education, especially if incoming students and their teachers have clearer ideas about the power *and* limitations of mathematics. Since the number of Americans of college age will be smaller than now, students should be able to get considerably more faculty attention.

---

*Esoteric research might well be de-emphasized; see I. Herstein, this MONTHLY, 76(1969) 818–824.

EDITORIAL: BOOK PRICES

A dramatic increase in book prices during the last two decades is evident to everyone. This has brought windfall profits on books already produced in an era of lower costs. But on the whole, book prices seem to have increased pretty much in keeping with the general inflation. A tabulation of Telegraphic Reviews for 1967–1969 shows that prices have not increased appreciably during this three year period. Indeed the average list price per page has remained 1.6¢ for paperbacks, 2.4¢ for elementary (first two years) books, and 3.6¢ for advanced books. This is about twice the levels of fifteen years ago.

The variations in prices among publishers are dramatic, and exasperating. For example, paperbacks sell from as little as 0.6¢ per page to as high as 13.2¢. Elementary texts (first two college years) vary from 1.2 to 7.5¢ and advanced books from 0.5 to 8.7¢. The variations are by no means always related to quality (often photo-offset of type script is priced the highest) or sales. The tremendous variation for paperbacks is due to the recent tendency to publish very advanced treatises in paperback. Since the cost of the hard cover is negligible, the lower price of paperbacks is due to the expected large market rather than to the nature of the binding. Accordingly if a book with a small market is published in paperback, it still has to be priced high (though probably not above 5¢ even for sales of only a few hundred) in order to avoid loss. The variations in prices among publishers are to some extent orrelated with the type of book published, and therefore it is not fair to single out particular publishers for praise or blame on the basis of their average prices per page. However, as a general rule it appears that the well-established companies producing the books of highest quality are not those charging the highest prices. Indeed there appears to be no correlation whatever between price and quality of content or production.

Popular paperbacks and elementary texts sell at prices determined by competition, but the wide variation in prices of more advanced books arise from varying estimates of sales and from the amount of overhead that is charged to the book. Extremely high prices appear most commonly to be due to an attempt to make a killing from a unique book that will be a "must" purchase for libraries and by charging to the book expenses of general company promotion and development. The practice of paying the cost of a book by library sales is encouraged by the fact that many libraries order automatically or at least without regard to price. Irresponsible pricing might be discouraged if those who order books for courses, personal use, or for the library would question any book whose price was substantially above the average. Sometimes a high price is justified or the book really is a "must." More often neither is the case.

KENNETH O. MAY

77(1970), 1120–1121

Computer Science and its Relation to Mathematics

Donald E. Knuth

A new discipline called Computer Science has recently arrived on the scene at most of the world's universities. The present article gives a personal view of how this subject interacts with Mathematics, by discussing the similarities and differences between the two fields, and by examining some of the ways in which they help each other. A typical nontrivial problem is worked out in order to illustrate these interactions.

1. What is Computer Science? Since Computer Science is relatively new, I must begin by explaining what it is all about. At least, my wife tells me that she has to explain it whenever anyone asks her what I do, and I suppose most people today have a somewhat different perception of the field than mine. In fact, no two computer scientists will probably give the same definition; this is not surprising, since it is just as hard to find two mathematicians who give the same definition of Mathematics. Fortunately it has been fashionable in recent years to have an "identity crisis," so computer scientists have been right in style.

My favorite way to describe computer science is to say that it is the study of *algorithms*. An algorithm is a precisely-defined sequence of rules telling how to produce specified output information from given input information in a finite number of steps. A particular representation of an algorithm is called a program, just as we use the word "data" to stand for a particular representation of "information" [**14**]. Perhaps the most significant discovery generated by the advent of computers will turn out to be that algorithms, as objects of study, are extraordinarily rich in interesting properties; and furthermore, that an algorithmic point of view is a useful way to organize knowledge in general. G. E. Forsythe has observed that "the question 'What can be automated?' is one of the most inspiring philosophical and practical questions of contemporary civilization" [**8**].

From these remarks we might conclude that Computer Science should have existed long before the advent of computers. In a sense, it did; the subject is deeply rooted in history. For example, I recently found it interesting to study ancient manuscripts, learning to what extent the Babylonians of 3500 years ago were computer scientists [**16**]. But computers are really necessary before we can learn much about the general properties of algorithms; human beings are not precise enough nor fast enough to carry out any but the simplest procedures. Therefore the potential richness of algorithmic studies was not fully realized until general-purpose computing machines became available.

I should point out that computing machines (and algorithms) do not only compute with *numbers*; they can deal with information of any kind, once it is represented in a precise way. We use to say that a sequence of symbols, such as a name, is represented inside a computer as if it were a number; but it is really more correct to say that a number is represented inside a computer as a sequence of symbols.

The French word for computer science is *Informatique*; the German is *Informatik*; and in Danish, the word is *Datalogi* [21]. All of these terms wisely imply that computer science deals with many things besides the solution to numerical equations. However, these names emphasize the "stuff" that algorithms manipulate (the information or data), instead of the algorithms themselves. The Norwegians at the University of Oslo have chosen a somewhat more appropriate designation for computer science, namely *Databehandling*; its English equivalent, "Data Processing" has unfortunately been used in America only in connection with business applications, while "Information Processing" tends to connote library applications. Several people have suggested the term "Computing Science" as superior to "Computer Science."

Of course, the search for a perfect name is somewhat pointless, since the underlying concepts are much more important than the name. It is perhaps significant, however, that these other names for computer science all de-emphasize the role of computing machines themselves, apparently in order to make the field more "legitimate" and respectable. Many people's opinion of a computing machine is, at best, that it is a necessary evil: a difficult tool to be used if other methods fail. Why should we give so much emphasis to teaching how to use computers, if they are merely valuable tools like (say) electron microscopes?

Computer scientists, knowing that computers are more than this, instinctively underplay the machine aspect when they are defending their new discipline. However, it is not necessary to be so self-conscious about machines; this has been aptly pointed out by Newell, Perlis, and Simon [22], who define computer science simply as the study of computers, just as botany is the study of plants, astronomy the study of stars, and so on. The phenomena surrounding computers are immensely varied and complex, requiring description and explanation; and, like electricity, these phenomena belong both to engineering and to science.

When I say that computer science is the study of algorithms, I am singling out only one of the "phenomena surrounding computers," so computer science actually includes more. I have emphasized algorithms because they are really the central core of the subject, the common denominator which underlies and unifies the different branches. It might happen that technology someday settles down, so that in say 25 years computing machines will be changing very little. There are no indications of such a stable technology in the near future, quite the contrary, but I believe that the study of algorithms will remain challenging and important even if the other phenomena of computers might someday be fully explored.

2. Is Computer Science Part of Mathematics? Certainly there are phenomena about computers which are now being actively studied by computer scientists, and which are hardly mathematical. But if we restrict our attention to the study of algorithms, isn't this merely a branch of mathematics? After all, algorithms were studied primarily by mathematicians, if by anyone, before the days of computer science. Therefore one could argue that this central aspect of computer science is really part of mathematics.

However, I believe that a similar argument can be made for the proposition that mathematics is a part of computer science! Thus, by the definition of set equality, the subjects would be proved equal; or at least, by the Schröder-Bernstein theorem, they would be equipotent.

My own feeling is that neither of these set inclusions is valid. It is always difficult to establish precise boundary lines between disciplines (compare, for example, the subjects of "physical chemistry" and "chemical physics"); but it is possible to distinguish essentially different points of view between mathematics and computer science.

The following true story is perhaps the best way to explain the distinction I have in mind. Some years ago I had just learned a mathematical theorem which implied that any two $n \times n$ matrices A and B of integers have a "greatest common right divisor" D. This means that D is a right divisor of A and of B, i.e., $A = A'D$ and $B = B'D$ for some integer matrices A' and B'; and that every common right divisor of A and B is a right divisor of D. So I wondered how to calculate the greatest common right divisor of two given matrices. A few days later I happened to be attending a conference where I met the mathematician H. B. Mann, and I felt that he would know how to solve this problem. I asked him, and he did indeed know the correct answer; but it was a mathematician's answer, not a computer scientist's answer! He said, "Let $\mathscr{R}$ be the ring of $n \times n$ integer matrices; in this ring, the sum of two principal left ideals is principal, so let D be such that

$$\mathscr{R}A + \mathscr{R}B = \mathscr{R}D.$$

Then D is the greatest common right divisor of A and B." This formula is certainly the simplest possible one, we need only eight symbols to write it down; and it relies on rigorously-proved theorems of mathematical algebra. But from the standpoint of a computer scientist, it is worthless, since it involves constructing the infinite sets $\mathscr{R}A$ and $\mathscr{R}B$, taking their sum, then searching through infinitely many matrices D such that this sum matches the infinite set $\mathscr{R}D$. I could not determine the greatest common divisor of $\left(\begin{smallmatrix} 1 & 2 \\ 3 & 4 \end{smallmatrix}\right)$ and $\left(\begin{smallmatrix} 4 & 3 \\ 2 & 1 \end{smallmatrix}\right)$ by doing such infinite operations. (Incidentally, a computer scientist's answer to this question was later supplied by my student Michael Fredman; see [**15**, p. 380].)

One of my mathematician friends told me he would be willing to recognize computer science as a worthwhile field of study, as soon as it contains 1000 deep theorems. This criterion should obviously be changed to include algorithms as well as theorems, say 500 deep theorems and 500 deep algorithms. But even so it is clear that computer science today does not measure up to such a test, if "deep" means that a brilliant person would need many months to discover the theorem or the algorithm. Computer science is still too young for this; I can claim youth as a handicap. We still do not know the best way to describe algorithms, to understand them or to prove them correct, to invent them, or to analyze their behavior, although considerable progress is being made on all these fronts. The potential for "1000 deep results" is there, but only perhaps 50 have been discovered so far.

In order to describe the mutual impact of computer science and mathematics on each other, and their relative roles, I am therefore looking somewhat to the future, to the time when computer science is a bit more mature and sure of itself. Recent trends have made it possible to envision a day when computer science and mathematics will both exist as respected disciplines, serving analogous but different roles in a person's education. To quote George Forsythe again, "The most valuable acquisitions in a scientific or technical education are the general-purpose mental tools which remain serviceable for a lifetime. I rate natural language and mathematics as the most important of these tools, and computer science as a third" [9].

Like mathematics, computer science will be a subject which is considered basic to a general education. Like mathematics and other sciences, computer science will continue to be vaguely divided into two areas, which might be called "theoretical" and "applied." Like mathematics, computer science will be somewhat different from the other sciences, in that it deals with man-made laws which can be proved, instead of natural laws which are never known with certainty. Thus, the two subjects will be like each other in many ways. The difference is in the subject matter and approach—mathematics dealing more or less with theorems, infinite processes, static relationships, and computer science dealing more or less with algorithms, finitary constructions, dynamic relationships.

· · ·

81(1974), 323–326

FROM REVIEW OF THE ART OF COMPUTER PROGRAMMING (second page).

There is no doubt that this series of volumes is indispensable for the specialist in computer science and for instructors in various fields. It shows how much mathematical background a serious computer scientist ought to have and that this background must include concrete mathematics in addition to abstract mathematics.

OLGA TAUSSKY

77(1970), 900–901

The Secretary then reported briefly on the Association's financial position: the Association last year had a deficit of $9,125, after having operated in the black the previous year with an excess of income over expenditures in the amount of $3,114. The budget for this year indicates a probable deficit of about $4,660.

In spite of this, the Association will not increase dues for 1973. They will remain at $12.50 for individual members.

79(1972), 1158

PROGRESS REPORTS

Edited by P. R. Halmos

The Serre Conjecture

W. H. Gustafson, P. R. Halmos, J. M. Zelmanowitz

What does the first row of an invertible matrix look like? That depends: where are the entries? If the entries are real numbers (or, for that matter, elements of an arbitrary field), the first row of an invertible matrix can be anything except $(0, 0, \ldots, 0)$. (If the given row of length n has a non-zero entry, assume, with no loss of generality, that that entry is in the first position, and construct the square matrix of size n with that first row, the rest of the first column all 0's, and the lower right corner equal to the identity matrix of size $n - 1$.)

If the matrices about which the question is being asked are restricted to have integer entries, the answer is less obvious. Thus, for instance, $(2, 4, 6)$ cannot be the first row of an invertible integer matrix, because the determinant of any such matrix must be ± 1, whereas the determinant of an integer matrix whose first row is $(2, 4, 6)$ is necessarily even. There is a clue here: if the entries of a prescribed row have a non-trivial common divisor, then that row cannot occur in an invertible matrix. It turns out that the necessary condition thus discovered (that the greatest common divisor of the prescribed row be 1) is sufficient as well: any row satisfying it *can* occur in an invertible matrix.

The last assertion is true for the ring of integers, or, for that matter, for any principal ideal domain: the condition that a row $(a_1, a_2, \ldots, a_n)$ must satisfy is the existence of a related column $(t_1, t_2, \ldots, t_n)$ such that $a_1 t_1 + a_2 t_2 + \cdots + a_n t_n = 1$. Such rows are called unimodular rows. (The t's exist if and only if the greatest common divisor of the a's is 1. Equivalently: the t's exist if and only if the ideal generated by the a's is the entire domain. Note that in case the domain is a field the conditions are equivalent to $(a_1, a_2, \ldots, a_n) \neq (0, 0, \ldots, 0)$.)

There are various approaches to the proof of sufficiency; one that works smoothly for Euclidean domains goes as follows. Regard $(a_1, a_2, \ldots, a_n)$ as a matrix with one row and n columns, and use the existence of the t's to infer the possibility of performing a sequence of elementary column operations on it so as to convert it to $(1, 0, \ldots, 0)$. The performance of an elementary column operation has the same effect as multiplying the given matrix on the right by certain elementary matrices. If the product of all the multipliers is U, so that

$$(a_1, a_2, \ldots, a_n) \cdot U = (1, 0, \ldots, 0),$$

then

$$(a_1, a_2, \ldots, a_n) = (1, 0, \ldots, 0) \cdot U^{-1},$$

so that U^{-1} is an invertible matrix with first row $(a_1, a_2, \ldots, a_n)$.

It is now tempting to jump to a general algebraic conclusion: if R is a commutative ring with unit, and if $(a_1, a_2, \ldots, a_n)$ is a unimodular row over R, then $(a_1, a_2, \ldots, a_n)$ is fit to be the first row of an invertible matrix over R. For $n = 1$ the conclusion is trivial, and for $n = 2$ it is almost equally trivial: given (a_1, a_2) and (t_1, t_2) with $a_1 t_1 + a_2 t_2 = 1$, the matrix

$$\begin{pmatrix} a_1 & a_2 \\ -t_2 & t_1 \end{pmatrix}$$

does the trick.

For $n \geq 3$, however, the conclusion is false; the standard counter-example makes contact with a well-known part of elementary topology. Let R be the ring of all real-valued continuous functions defined on the 2-sphere, i.e., on the locus of the equation $x^2 + y^2 + z^2 = 1$ in $\mathbf{R}^3$. If $a_1(x, y, z) = x$, $a_2(x, y, z) = y$, and $a_3(x, y, z) = z$, then (a_1, a_2, a_3) is a unimodular row (because $x^2 + y^2 + z^2 = 1$). The row (a_1, a_2, a_3) cannot, however, occur in an invertible matrix over R. Reason: the second row (b_1, b_2, b_3) of such a matrix would be, at each point (x, y, z), linearly independent of the first, and, therefore, would have a non-zero projection in the plane tangent to the sphere at (x, y, z). This is impossible: there is no non-singular continuous tangent vector field on the sphere (or you can't comb a porcupine).

Serre (1955) made the conjecture that for certain important special rings the general conclusion is true; the rings are the polynomial rings $R = k[x_1, \ldots, x_m]$ in m variables, with coefficients in a field k. The original formulation of Serre's conjecture had to do with modules over these R's. Recall that a module over R is "a vector space with respect to scalars from R". A module over R is of special importance if it has a finite basis (equivalently, if it is isomorphic to R^m for some m). If a module over R has a basis, and is the direct sum of two submodules, do they too necessarily have bases? A module with a basis is called *free*, and a direct summand of a free module is called *projective*; the original formulation of Serre's conjecture was that every projective module over $k[x_1, \ldots, x_m]$ is free.

The subject makes surprising contact with topology. A *vector bundle* over a base space X is a generalization of the concept of the Cartesian product of X with a vector space. Intuitively, a bundle is a "twisted" Cartesian product, in the sense in which a Möbius strip is a twisted cylinder. The Cartesian product itself is the easiest vector bundle (called the *trivial* bundle); the generalized ones retain the projection map onto X, just as if they were Cartesian products, but are like Cartesian products only locally.

A *section* of a vector bundle is a continuous function s from X to the bundle, such that s followed by the projection onto X is the identity mapping on X. The vector structure of the bundle makes possible a natural definition of addition of sections. If, moreover, R is the ring of scalar-valued continuous functions on X, then the vector structure of the bundle makes possible a natural definition of multiplication of a section by an element of R. In other words, the set of all sections is a module over R. If the base space is at all decent, it turns out that this module is always projective; the module is free if and only if the bundle is trivial. There is, thus, a correspondence between projective modules and vector bundles, and, in that correspondence, the free modules correspond to trivial bundles.

What is known in the topological context is that a bundle over m-dimensional Euclidean space $\mathbf{R}^m$ is in a natural sense always (isomorphic to) the trivial bundle. How much of that conclusion continues to make sense and remains true in an algebraic context? If, in other words, k is an arbitrary field, does it make sense to speak of vector bundles over k^m? The answer is yes; a topology that gives the phrase its sense (the Zariski topology) is definable in purely algebraic terms. (A closed set in k^m is the locus of common zeros of a set of polynomials in m variables.) A vector bundle over k^m, with "fiber" given by the vector space k^n, is a "twisted" version of $k^m \times k^n$. The role of the ring of scalar-valued continuous functions on the base space k^m is played by the ring $R = k[x_1, \ldots, x_m]$ of polynomials. The sections form a projective module over R; Serre's conjecture is that such a module is necessarily free, i.e., that such a bundle is necessarily trivial.

The conjecture remained open for more than 20 years. The first non-trivial step was Seshadri's (1958); he proved the conjecture for $m = 2$. A later and important step was taken by Horrocks (1964) who proved an analogous result for local rings (rings with only one maximal ideal). The final step was taken simultaneously and independently by Quillen and Suslin (1976).

Quillen in effect reduced the general case to the one treated by Horrocks. The essence of Quillen's method is induction on the number m of variables. The step from $m - 1$ to m is similar to the procedure of complexifying a real vector space. It involves showing that every projective module over $k[x_1, \ldots, x_m]$ is a tensor product of a projective module over $k[x_1, \ldots, x_{m-1}]$ with $k[x_1, \ldots, x_m]$. The reason the induction step is successful is that the property of having a basis survives the construction.

References

1. D. Quillen, Projective modules over polynomial rings, Invent. Math., 36 (1976) 167–171.

2. A. A. Suslin, Projective modules over a polynomial ring are free, Soviet Math. Dokl., 17 (1976) 1160–1164.

3. T.-Y. Lam. Serre's conjecture, Springer-Verlag, Berlin, 1978.

85(1978), 357–359

PROGRAMMABLE CALCULATORS

Introduction. A number of studies and projects have demonstrated that the use of a computer in conjunction with the teaching of calculus can be an effective means for improving student achievement. In 1973 a study was conducted at SUCC to investigate the effect of the incorporation of an electronic programmable calculator (mini-computer) in a first course in calculus. The rationale for the study was that the computational requirements for such a course are far below the capabilities of a computer (but well within the range of a mini-computer) and it is not reasonable to engage the computer to perform in a manner far beneath its potential (nor to have undergraduates compete for computer time with a faculty member or researcher whose studies require the capacity of the computer); the absence of a need for a formal programming language; the mini-computer is mobile and can be used in both the classroom and the mathematics laboratory; and over a period of a few years the cost of a mini-computer is less than the rental of a terminal.

83(1976), 281

R. A. Rosenbaum (1967–68)

Harley Flanders (1969–73)

Alex Rosenberg (1974–77)

Ralph Boas (1977–1981)

Paul R. Halmos (1982–86)

Herbert S. Wilf (1987–91)

Recent Editors of the MONTHLY

The Retrial of the Lower Slobbovian Counterfeiters

Michael Hendy

Lower Slobbovia is a poor country, unable to mint its local currency, the Rasbucknik [1]. Hence nine (N) coiners, $C_1, C_2, \ldots, C_9$ were engaged to produce coins to government specifications. However, it was suspected that some of the coiners were counterfeiting by introducing some base metal into the alloy. Any pair of counterfeit coins weighed the same, but differed slightly in weight from good coins. Each coiner produced either all good coins or all counterfeits. A procedure was called for to determine in three weighings which (if any) of the coiners were dishonest, using a beam balance with a set of infinitely refinable weights, as many coins from each coiner as may be needed, and one good coin.

The Court of Lower Slobbovia was supplied with the following solution [2]. In the first weighing, the weight Wg of the good coin is determined. In the second weighing, a single coin from each coiner is selected and these nine (N) coins are weighed, giving a total weight T. If the discrepancy $D = T - NWg$ is zero then all the coiners are honest. In the third weighing a sample of 2^{i-1} coins are selected from C_i, $i = 1, \ldots, 9$, and weighed, giving a total weight T'. The discrepancy here is $D' = T' - (2^N - 1)Wg$.

Now the integer S such that $D'/D = S/\beta(S)$ is determined where $\beta(S)$ is the number of ones in the binary representation of S, i.e., $\beta(S) = \sum_{i=1}^{N} B_i$ where $S = \sum_{i=1}^{N} B_i 2^{i-1}$. For such an S, coiner C_i is a counterfeiter if and only if $B_i = 1$, and $\beta(S)$ is the number of counterfeiters.

This procedure was followed at the trial of the coiners in 1954 and the fateful ratio $D'/D = 23$ was determined. After further computation it was discovered that $S = 69 = 2^0 + 2^2 + 2^6$, with $\beta(69) = 3$, satisfied the ratio and accordingly coiners C_1, C_3, and C_7 were arrested, to be detained indefinitely at the Archduke's pleasure. They were dragged away pleading innocence, and the counsel for C_1 immediately lodged an appeal.

The wheels of Lower Slobbovian justice moved steadily but slowly, and the retrial was held before the Archduchess only 24 years later. In presenting his case C_1's counsel argued that $S = 92 = 2^2 + 2^3 + 2^4 + 2^6$ with $\beta(92) = 4$ satisfied the ratio $S/\beta(S) = 23$ and hence the evidence convicting his client was faulty, while coiners C_4 and C_5 were probably counterfeiters and had been left unpunished.

Lower Slobbovia's very embarrassed Mathematician Laureate was called to explain. In the light of such a concrete counterexample he had to admit that the mapping $S \to S/\beta(S)$ was not, in fact, always 1–1 and therefore did not always admit a unique inverse. Hence in the light of the above evidence he could only conclude C_3 and C_7 were obviously guilty, C_2, C_6, C_8, and C_9 were clearly innocent, but which of the others, C_1, C_4, and C_5, were guilty was not clear. Unable to give a

294   A CENTURY OF MATHEMATICS

decisive conclusion he was invited to share the counterfeiters' fate and was led away muttering things like, "I forgot about carry digits."

The apprentice mathematician was instantly promoted and told the same fate awaited him should he be unable to produce a foolproof technique. After a little thought he stated that, although the restriction to only three weighings appeared to make the problem difficult, it could still be resolved unambiguously, altering only the third weighing. Instead of calling for 2^{i-1} coins from C_i, you should in fact select b^{i-1}, for any integer $b > N$. Let $\beta_b(S)$ be the sum of the digits of S expressed to the base b. Although $S \to S/\beta_b(S)$ would still not be $1-1$ for all $S \in Z^+$, the only integers S generated in this process would be those whose digits to the base b are 0 or 1. In this case should $S/\beta_b(S) = T/\beta_b(T)$ then $\beta_b(T) \cdot S = \beta_b(S) \cdot T$. The digits of $\beta_b(S) \cdot T$ expressed to base b will be either 0 or $\beta_b(S) \le N < b$, and similarly those for $\beta_b(T) \cdot S$ will be 0 or $\beta_b(T)$. As these integers are equal we must find $\beta_b(S) = \beta_b(T)$ and consequently $S = T$.

The new Mathematician Laureate was applauded and as $N = 9$, the court officials set $b = 10$, and selected one coin from C_1, 10 from C_2, etc. Reweighing T', and calculating $D' = T' - (10^N - 1)Wg/9$ the ratio

$$S/\beta_{10}(S) = 222002.2$$

was determined. "The only integer," concluded the young mathematician, "whose digits base 10 are zeros and ones, that satisfies this ratio is $S = 1110011$, so in fact there are *five* counterfeiters, C_1, C_2, C_5, C_6 and C_7!"

There were howls of protest and disbelief. How could this be compatible with the earlier calculation with $S = 69$ or 92? The young mathematician then pointed out that 111011 as an integer (base 2) was $2^6 + 2^5 + 2^4 + 2^1 + 2^0 = 115$, and $115/\beta_2(115) = 23$, a value not previously recognized. These three integers, 69, 92, and 115, are the only three giving the ratio 23, however. We note that if $\beta_2(23r) = r$, then $23r \ge 2^{r-1}$, which is true only for $r < 9$. Testing for $r = 1, 2, 6, 7$, and 8 we find these do not satisfy the ratio and hence we have uncovered all solutions to $S/\beta_2(S) = 23$.

Also 23 is not the only nonunique ratio of $S/\beta_2(S)$. We can show that 26.5, 27, 37, 38.5, 41, 41.5, 46 and many more each are satisfied by multiple values of S. (For related problems, see [3].)

References

2. Julian Braun, The counterfeiters of Lower Slobbovia, this MONTHLY, 61 (1954) 472–473.
3. B. Manuel, Counterfeit coin problems, Math. Mag., 50 (1977) 90–92.

87(1980), 200–201

CLASSROOM NOTES

EDITED BY DEBORAH TEPPER HAIMO AND FRANKLIN TEPPER HAIMO

Proofs That $\Sigma 1/p$ Diverges

Charles Vanden Eynden

The theorem of the title, in which p runs through the primes, was deduced by Euler in 1737 from the "equation"

$$\sum_{n=1}^{\infty} \frac{1}{n} = \prod \left(1 - \frac{1}{p}\right)^{-1}.$$

Kronecker later fixed up this argument by replacing n and p by n^A and p^A, $A > 1$, and letting $A \to 1$ [1].

It is my purpose in this paper to examine some fairly recent (compared to the age of the theorem, anyway) easy proofs of this theorem, where the word "easy" needs some definition. Of course any theorem may be proved very compactly if enough preliminary knowledge is assumed, and many number theory texts derive Euler's theorem from much stronger results. Here we are concerned with direct proofs appropriate for an undergraduate number theory class.

Our presumed audience will know something about infinite series (otherwise the statement of the theorem will not even make sense), but only from calculus, along with some elements of number theory.

We may not assume, for example, familiarity with the connection between the convergence of the series Σa_n and the infinite product $\prod(1 \pm a_n)$. In fact, although our audience knows what it means for a series to converge absolutely, it does not know that rearranging such a series is justified. This means that some arguments must be complicated by replacing series with finite sums.

The proofs below have been left in more or less their original form, except that the notation is standardized. The letter p always represents a prime, and all lower-case letters except e represent positive integers. For given positive integers a and b with $a \leq b$, let P denote the set of primes p satisfying $a \leq p \leq b$, let M be those integers n all of whose prime divisors are in P, and for a given integer x let M_x be all elements of M not exceeding x. Let $|S|$ denote the number of elements in the set S.

Erdös's Proof. Paul Erdös published the following proof in 1938 [2]. If $\Sigma 1/p$ converges we can choose b so that $\Sigma_{p>b} 1/p < \frac{1}{2}$. Take $a = 1$. Suppose $n \in M_x$, and write $n = k^2 m$, where m is square-free. Since $m = \prod_S p$, where S is some subset of P, m can assume at most $2^{|P|}$ values. Also $k \leq \sqrt{n} \leq \sqrt{x}$. Thus $|M_x| \leq 2^{|P|}\sqrt{x}$.

Now the number of positive integers $\leq x$ divisible by a fixed p does not exceed x/p. Thus $x - |M_x|$, the number of such integers divisible by some prime greater than b, satisfies

$$x - |M_x| \leq \sum_{p > b} \frac{x}{p} < \frac{x}{2}.$$

We see

$$\frac{x}{2} < |M_x| \leq 2^{|P|}\sqrt{x} \, ,$$

or $\sqrt{x} < 2^{|P|+1}$, which is clearly false for x sufficiently large.

Comments. The proof above, which is notable for its lack of series manipulations, is given in the classic book by Hardy and Wright [3], as well as by Calvin Long's text [4].

Bellman's and Moser's Proofs. The details of Richard Bellman's 1943 proof [5] and Leo Moser's 1958 proof [6] will be omitted, since both appeared in this MONTHLY. Bellman assumed an a sufficiently large so that $\sum_P 1/p < 1$ with $b = \infty$, and from this derived the convergence of first $\sum_M 1/n$ and then the harmonic series.

Moser derived from the false assumption that $\sum 1/p$ converges the true conclusion that $\pi(x)/x \to 0$ as $x \to \infty$, where $\pi(x)$ denotes (as usual) the number of primes $\leq x$. A contradiction was then produced from this and the assumption that $\sum_P 1/p < \frac{1}{2}$ for large enough a.

Bellman used the rearrangement of positive series several times in the proof, and Moser used the result that a convergent series is Cesàro summable to the same limit [7], which our hypothetical audience is unlikely to have seen.

Dux's Proof. In 1956 Erich Dux [8] gave a proof that began and ended similarly to Bellman's but also made use of the rearrangement of positive series. Let $a = 1$ and, assuming $\sum 1/p$ converges, choose b so that $\sum_{p > b} 1/p = A < 1$. Define M' to be all $n' > 1$ divisible by primes only exceeding b, and M'' to be all positive integers not in M or M'. (Note that 1 is in M.)

Then, since P is finite,

$$\sum_M \frac{1}{n} = \prod_P \left(1 + \frac{1}{p} + \frac{1}{p^2} + \cdots \right) = \prod_P \left(1 - \frac{1}{p} \right)^{-1} < \infty,$$

and

$$\sum_{M'} \frac{1}{n'} \leq \sum_{p > b} \frac{1}{p} + \left(\sum_{p > b} \frac{1}{p} \right)^2 + \cdots = \frac{A}{1 - A} < \infty;$$

so

$$\sum_{M''} \frac{1}{n''} = \left(-1 + \sum_M \frac{1}{n} \right) \sum_{M'} \frac{1}{n'} < \infty.$$

This contradicts the divergence of the harmonic series.

Clarkson's Proof. This 1966 proof by James A. Clarkson [9] calls to mind Euclid's proof that the number of primes is infinite. If $\sum 1/p$ converges, we can choose a so that $\sum_P 1/p < \frac{1}{2}$ for all b. Let $Q = \prod_{p < a} p$. For fixed r, it is possible to choose b large enough so that all the factors of the numbers $1 + iQ$, $1 \leq i \leq r$, are in P, since if $p < a$ then $p \nmid 1 + iQ$.

Now each term of the sum $\sum_{i=1}^{r} 1/(1 + iQ)$ whose denominator is a product of j primes (not necessarily distinct) occurs at least once in the expansion of

$$\left(\sum_{P} \frac{1}{p} \right)^{j} < 2^{-j}. \tag{1}$$

Thus

$$\sum_{i=1}^{r} \frac{1}{1 + iQ} < \sum_{j \geq 1} 2^{-j} \leq 1.$$

But, since r was arbitrary, this implies that the harmonic series converges.

Comments. The expansion of $(\sum_{P} 1/p)^{j}$ recalls the proofs of Bellman and Dux. Getting the inequality after (1) requires rearrangement of infinite series. This proof (with $b = \infty$) is reproduced in Apostol's *Introduction to Analytic Number Theory* [10]. The last sentence of the proof is most easily justified by comparison with $\sum 1/2Qi$.

Two More Proofs. A very simple identity forms the basis for two more proofs of my own design, namely,

$$\left(1 + \frac{1}{p} \right)\left(1 + \frac{1}{p^2} + \frac{1}{p^4} + \cdots + \frac{1}{p^{2k}} \right) = 1 + \frac{1}{p} + \frac{1}{p^2} + \cdots + \frac{1}{p^{2k+1}}.$$

Taking the product over P and letting $k \to \infty$ yields

$$\prod_{P} \left(1 + \frac{1}{p} \right) \sum_{M} \frac{1}{n^2} = \sum_{M} \frac{1}{n}. \tag{2}$$

Since $\sum 1/n^2$ converges and $\sum 1/n$ diverges, this means,

$$\text{for } a = 1, \ \prod_{P} \left(1 + \frac{1}{p} \right) \to \infty \quad \text{as } b \to \infty. \tag{3}$$

Continuation A. Since for $C > 0$, $e^{C} = 1 + C + C^2/2! + \cdots > 1 + C$, we have

$$\prod_{P} \left(1 + \frac{1}{p} \right) < \prod_{P} e^{1/p} = \exp\left(\sum_{P} \frac{1}{p} \right).$$

This, with (3), shows, that $\sum 1/p$ diverges.

Continuation B. By (3) $\prod_{P}(1 + 1/p) \to \infty$ as $b \to \infty$ for any fixed a, and so the same is true for $\sum_{M} 1/n$ by (2). If $\sum 1/p$ converges, we can choose a so that $\sum_{P} 1/p < \frac{1}{2}$ for all b, then choose b and x large enough so that $\sum_{M_x} 1/n > 2$. Since every n in M_x except 1 is of the form pn for $p \in P$ and $n \in M_x$, we have

$$\sum_{P} \frac{1}{p} \sum_{M_x} \frac{1}{n} \geq \sum_{M_x} \frac{1}{n} - 1.$$

Then

$$\frac{1}{2} > \sum_{P} \frac{1}{p} \geq 1 - \left(\sum_{M_x} \frac{1}{n} \right)^{-1} > 1 - \frac{1}{2},$$

a contradiction.

Comments. The proof using Continuation A is so simple (and the theorem is so old) that it would be foolhardy to call it new, although I have not found the arrangement anywhere. Of course $e^C > 1 + C$ may also be proved without recourse to the series for e^C, but I believe most of our hypothetical audience will remember this expansion. Continuation B has a better chance of being a novelty.

The reader should note a very recent proof by Frank Gilfeather and Gary Meisters [11].

References

1. L. E. Dickson, History of the Theory of Numbers, vol. 1, Chelsea, New York, 1952, p. 413.

2. P. Erdös, Uber die Reihe $\Sigma 1/p$, Mathematica, Zutphen. **B.**, 7 (1938) 1–2.

3. G. H. Hardy and E. M. Wright, The Theory of Numbers, Oxford, 1954, pp. 16–17.

4. Calvin T. Long, Elementary Introduction to Number Theory, 2nd ed., Heath, Lexington, 1972, pp. 63–64.

5. R. Bellman, A note on the divergence of a series, this MONTHLY, 50 (1943) 318–319.

6. Leo Moser, On the series $\Sigma 1/p$, this MONTHLY, 65 (1958) 104–105.

7. R. R. Goldberg, Methods of Real Analysis, Blaisdell, New York, 1964, p. 91.

8. Erich Dux, Ein Kürzer Beweis der Divergenz der unendlichen Reihe $\Sigma_{r=1} 1/p_r$, Elem. Math., 11 (1956) 50–51.

9. James A. Clarkson, On the series of prime reciprocals, Proc. Amer. Math. Soc., 17 (1966) 541.

10. Tom M. Apostol, Introduction to Analytic Number Theory, Springer-Verlag, New York, 1976, pp. 18–19.

11. W. G. Leavitt, The sum of the reciprocals of the primes, Two-Year College Math. J., 10 (1979) 198–199.

87(1980), 394–397

Trends in Undergraduate Mathematics: *The CBMS Survey*, by Professor D. J. Albers, Menlo College, and Professor J. W. Jewett, Oklahoma State University.

The report, "Undergraduate Mathematical Sciences in Universities, Four-Year Colleges, and Two-Year Colleges 1975–76" is available at $4 postpaid from CBMS, 2100 Pennsylvania Avenue, N.W., Washington, D.C. 20037.

Professor Albers reported that, from 1970 to 1975, in two-year colleges:

1. Mathematics enrollments increased by 50%. Two-year colleges now account for 37% of *all* undergraduate mathematical science enrollments.

2. In contrast to 1., full-time faculty size increased by only 22%, but their educational qualifications were up, with 11% holding a doctorate.

3. The faculty is young, with a median age of 40.

4. Enrollments in remedial mathematics increased by 81% and now account for 40% of all enrollments.

5. Hand calculators have gained widespread acceptance.

These were compared and contrasted with general trends in two-year colleges, with special attention to the occupational-technical areas.

84(1977), 409

Black Women in Mathematics in the United States

Patricia C. Kenschaft

Increased attention has been focused on women in mathematics during the past decade, but when I was invited to speak on Black women in mathematics, I could find only two references—a talk by Vivienne Malone Mayes [1] at the Summer Meeting in Kalamazoo in 1975 sponsored by the Association for Women in Mathematics, and the AWM panel I chaired in Atlanta in January, 1978 [2]. Since then I have collected much information, and this article tells about the American Black women holding doctoral degrees in mathematics, all but two of whom I have talked with in the past three years.

The 1970 decennial census revealed more than 1,100 Black women who reported themselves as mathematicians. In that census 244 said that they were college or university teachers, and, of these, 178 said that they had finished at least four years of college; 152 had completed five years or more [3]. I have discovered twenty-one American Black women who earned doctorates in pure or applied mathematics before the end of 1980. I would appreciate hearing of others if they exist.

The first year that a Black woman received a Ph.D. degree in pure mathematics from an American university was 1949. In that year there were two: Marjorie Lee Browne at the University of Michigan and Evelyn Boyd Granville at Yale University. The first American white woman, Winifred Edgerton Merrill, to receive such a degree did so in 1886 from Columbia University [4] and the first Black man, Elbert Cox, in 1925 from Cornell [5].

Georgia Caldwell Smith passed the defense for her Ph.D. degree from the University of Pittsburgh in the summer of 1960 but apparently died before it was conferred upon her in early 1961. Thus Gloria Conyers Hewitt became the third American Black woman to be granted a Ph.D. in mathematics when she received hers in 1962 from the University of Washington. But Argelia Velez-Rodriguez, who became a naturalized American citizen in 1972, received her degree from the University of Havana two years earlier, in 1960. (She was apparently the first Black woman to obtain a doctorate in mathematics at that institution, but she knows of

Adapted from an invited address given at the annual meeting of the Association of Mathematics Teachers of New England in Springfield, Massachusetts, on November 2, 1979.

Patricia Kenschaft received her Ph.D. with a specialty in functional analysis from the University of Pennsylvania in 1973 under the direction of Edward Effros. Since then she has taught at Montclair State College in New Jersey. She is the author or co-author of three textbooks for nontechnical majors published by Worth Publishers, Inc., and is currently preparing a paper on the life of Charlotte Scott, vice president of the AMS in 1906.—*Editors* four who have done so since.) Thyrsa Frazier Svager was awarded her Ph.D. degree from Ohio State University in 1965. In 1966 three American Black women received the degree: Eleanor Green Jones from Syracuse University, Vivienne Malone Mayes from the University of Texas at Austin, and Shirley Mathis McBay from the University of Georgia.

In 1967 Geraldine Darden received a doctorate in mathematics from Syracuse University, and in 1968 Mary Lovenia DeConge received hers in mathematics and French from St. Louis University. Etta Zuber Falconer completed her degree at Emory University in 1969. In 1971 Dolores Richard Spikes won hers at Louisiana State University. In 1974 Elayne Arrington-Idowu received her doctorate from the University of Cincinnati, Rada Higgins McCreadie from Ohio State University, and Evelyn Patterson Scott from Wayne State University.

There then appears to have been a gap until Fern Hunt was awarded a doctorate from New York University in 1978 and Fannie Gee from the University of Pittsburgh in 1979. In 1980 Frances Sullivan received hers from the City University of New York, Suzanne Craig from the University of Southern California at Los Angeles, and Sylvia Trimble Bozeman from Emory University.

During my inquiries, I have discovered about a dozen Black women with doctorates in mathematics education, and I feel sure there are more. These women have been leaders not only in teaching and administration, but also in organizing programs and writing. Two of them, Ethel Turner and Joella Gipson, were the author [6] and co-author [5], respectively, of books that contain a list of Blacks who have received doctorates in mathematics and mathematics education. Reference [5] was especially helpful in the preparation of this article because it contains data of the type that appears on a résumé. Both are somewhat out of date, and neither contains anecdotal information or personal views such as those in this article.

Most Black women engaged in mathematical careers do not have doctorates. Their number is probably larger than the number with doctorates would indicate because the economic disincentives for graduate education tend to be greater for Blacks than for whites. I arbitrarily restricted myself to those with doctorates only because the number was sufficiently small for me to interview and they were the ones most interesting to readers of this MONTHLY.

· · ·

What do these women have in common besides mathematical talent? Certainly they all have emotional stamina. In each case there seems to be someone in the family who believed that she was very special and that it was worth sacrificing for her education. All the women apparently remember at least one supportive teacher along the way—someone who knew mathematics and told the young woman that she could be a mathematician.

One might ask why only twenty-one Black women in our country have thus far been able to earn a doctoral degree in mathematics. Despite a variety of efforts to eliminate it, discrimination continues, often perpetrated unconsciously by people of good will. Some of these women feel that the educational hurdles for Blacks are worse now than a decade ago, especially in the large Northern cities. This means that although large numbers of children are responding with interest to the increased publicity given to the sciences, the lack of quality education at the primary and secondary levels is preventing them from fulfilling their scientific ambitions.

The results of past discrimination remain; these include; but are by no means limited to, poorer schools in predominantly Black neighborhoods, the need of young educated Black people to support younger students in their own families, and extra administrative duties devolving on those from underrepresented groups in the mathematical community. Often Blacks and women feel a responsibility to help others in their own groups, which takes time and energy that would otherwise be used in vigorously pursuing their professional careers.

What can be done? Besides encouraging readers of this MONTHLY to support gifted students regardless of race or sex, I feel I can best begin to answer this question by quoting Dolores Spike's statement at the Atlanta [2] panel.*

"I think there is an excellent opportunity for training in the newly emphasized areas of mathematics, such as applied mathematics, in predominantly Black institutions. There the percentage of Ph.D.'s is very low, so this provides a situation where there are able people who could learn in areas where we need more mathematicians.

"What is needed? We need money. But we need more too. Many of these Black women would love to have more training, but they aren't going to leave their families and go to school. I think we have to be innovative in meeting their problems.

"Maybe we can go to them. Maybe we could send visiting specialists, have on-site programs, prepare materials for self-study, and run seminars for these students. I think that this is the way the professional organizations can help these institutions.

"I believe that the mathematical societies should also encourage the direction of more funds from agencies such as the NSF to the kinds of projects that I have enumerated. Another thing I think that these organizations can do is to provide more information to scholars and to college administrators; part of our problem is that these people do not know the plight of Black women mathematicians. We can tell them, but I think that the news coming from a professional organization would have more impact; I think they would be more likely to receive it and perhaps act on it.

"I think that the mathematical organizations could consider a voluntary accreditation program, because I think this would be particularly beneficial to Black institutions. Such institutions will always be on the short end of the funding in states that have predominantly Black schools, but I have noticed that in those subject areas that have accrediting agencies, even with volunteers, they come out a bit better in the distribution of money.

"Lastly, I would suggest that the mathematical community initiate a massive public media campaign. I think we ought to help the public understand what our problems are and what the possible solutions are."

Mathematical talent is both scarce and precious; our society cannot afford to waste it. Possessors of such talent, whatever their race or sex, need encouragement and help if they are to develop fully their abilities and to enter fields previously denied to them. These Black women are role models for those who follow. They should inspire all of us to look for talent wherever it is found and try to provide the needed support.

---

*This is slightly edited and checked by her.

Many people helped in the preparation of this article by responding to brief telephone calls and letters. Several helped in more time-consuming ways: John Houston of Atlanta University, Roosevelt Gentry and Corlis Powell Johnson, both of Jackson State University, and Lee Lorch of York University in Toronto, Canada. My deepest appreciation, however, is extended to the women about whom this article is written; their generous sharing of time, confidences, and encouragement have made the article possible. Talking with them has been a great privilege that has enriched my life.

88(1981), 592–604

Women Presidents of the MAA clockwise from top left: Lida Barrett, Deborah Haimo, Dorothy Bernstein

Can We Make Mathematics Intelligible?

R. P. Boas

Why is it that we mathematicians have such a hard time making ourselves understood? Many people have negative feelings about mathematics, which they blame, rightly or wrongly, on their teachers [1]. Students complain that they cannot understand their textbooks; they have been doing this ever since I was a student, and presumably for much longer than that. Professionals in other disciplines feel compelled to write their own accounts of the mathematics they had trouble with. However, it was not until after I became editor of this MONTHLY that I quite realized how hard it is for mathematicians to write so as to be understood even by other mathematicians (outside of fellow specialists). The number of manuscripts rejected, not for mathematical deficiencies but for general lack of intelligibility, has been shocking. One of my predecessors had much the same experience 35 years earlier [2].

To put it another way, why do we speak and write about mathematics in ways that interfere so dramatically with what we ostensibly want to accomplish? I wish I knew. However, I can at least point out some principles that are frequently violated by teachers and authors. Perhaps they are violated because they contradict what many of my contemporaries seem to consider to be self-evident truths. (They also have little in common with the MAA report on how to teach mathematics [3].)

Abstract Definitions. Suppose you want to teach the "cat" concept to a very young child. Do you explain that a cat is a relatively small, primarily carnivorous mammal with retractile claws, a distinctive sonic output, etc.? I'll bet not. You probably show the kid a lot of different cats, saying "kitty" each time, until it gets the idea. To put it more generally, generalizations are best made by abstraction from experience. They should come one at a time; too many at once overload the circuits.

There is a test for identifying some of the future professional mathematicians at an early age. These are students who instantly comprehend a sentence beginning "Let X be an ordered quintuple $(a, T, \pi, \sigma, \mathscr{B})$, where ... " They are even more promising if they add, "I never really understood it before." Not all professional mathematicians are like this, of course; but you can hardly succeed in becoming a professional unless you can at least understand this kind of writing.

However, unless you are extraordinarily lucky, most of your audience will not be professional mathematicians, will have no intention of becoming professional mathematicians, and will never become professional mathematicians. To begin

---

The author is Professor Emeritus of Mathematics, Northwestern University. The present issue of this MONTHLY is the last under his editorship.

with, they won't understand anything that starts off with an abstract definition (let alone with a dozen at once), because they don't yet have anything to generalize from. Please don't immediately write me angry letters explaining how important abstraction and generalization are for the development of mathematics: I *know* that. I also am sure that when Banach wrote down the axioms for a Banach space he had a lot of specific spaces in mind as models. Besides, I am discussing only the communication of mathematics, not its creation.

For example, if you are going to explain to an average class how to find the distance from a point to a plane, you should first find the distance from $(2, -3, 1)$ to $x - 2y - 4z + 7 = 0$. After that, the general procedure will be almost obvious. Textbooks used to be written that way. It is a good general principle that, if you have made your presentation twice as concrete as you think you should, you have made it at most half as concrete as you ought to.

Remember that *you* have been associating with mathematicians for years and years. By this time you probably not only think like a mathematician but imagine that everybody thinks like a mathematician. Any nonmathematician can tell you differently.

Analogy. Sometimes your audience will understand a new concept better if you explain that it is similar to a more familiar concept. Sometimes this device is a flop. It depends on how well the audience understands the analogous thing. An integral is a limit of a sum; therefore, since sums are simpler (no limiting processes!), students will understand how integrals behave by analogy with how sums behave. Won't they? In practice, they don't seem to. Integrals are simpler than sums for many people, and there may be some deep reason for this [4].

Vocabulary. Never introduce terminology unnecessarily [5]. If you are going to have to mention a countable intersection of open sets—just once!—there is no justification for defining G_δ's and F_σ's.

I have been assured that nobody can really understand systems of linear equations without all the special terminology of modern linear algebra. If you believe this you must have forgotten that people understood systems of linear equations quite well for many years before the modern terminology had been invented. The terminology allows concise statements; but concision is not the alpha and omega of clear exposition. Modern terminology also lets one say more than could be said in old-fashioned presentations. Nevertheless, at the beginning of the subject a lot of the students' effort has to go into memorizing *words* when it could more advantageously go into learning mathematics. Paying more attention to vocabulary than to content obscures the content. This is what leads some students to think that the real difference between Riemann and Lebesgue integration is that in one case you divide up the x-axis and in the other you divide up the y-axis.

If you think you can invent better words than those that are currently in use, you are undoubtedly right. However, you are rather unlikely to get many people except your own students to accept your terminology; and it is unkind to make it hard for your students to understand anyone else's writing. One Bourbaki per century produces about all the neologisms that the mathematical community can absorb.

In any case, if you *must* create new words, you can at least take the trouble to verify that they are not already in use with different meanings. It has not helped communication that "distribution" now means different things in probability and in functional analysis. On the other hand, if you need to use old but unfashionable words it is a good idea to explain what they mean. A friend of mine was rebuked by a naïve referee for "inventing" bizarre words that had actually been invented by Kepler.

It is especially dangerous to assume either that the audience understands your vocabulary already or that the words mean the same to everybody else that they do to you. I know someone who thinks that everybody from high school on up knows all about Fourier transforms, in spite of considerable evidence to the contrary. Other people think that everybody knows what they mean by Abel's theorem, and therefore never say which of Abel's many theorems they are appealing to.

An even more serious problem comes from what (if it didn't violate my principles) I would call geratologisms: that is, words and phrases that, if not actually obsolete in ordinary discourse, are becoming so. Contemporary prose style is simpler and more direct than the style of the nineteenth century—except in textbooks of mathematics. While I was writing this article I was teaching from a calculus book that begins a problem with, "The strength of a beam varies directly as..." I do not know whether the jargon of variation is still used in high schools, but in any case it isn't learned: only one student in a class of 45 had any idea what the book meant (and he was a foreigner). Blame the students if you will, blame the high schools; for my own part I blame the authors of the textbook for not realizing that contemporary students speak a different language. Another current calculus book says, "Particulate matter concentrations in part per million theoretically decrease by an inverse square law." You couldn't get away with that in *Newsweek* or even in *The New Yorker*, but in a textbook...

Authors of textbooks (lecturers, too) need to remember that they are supposed to be addressing the students, not the teachers. What is a function? The textbook wants you to say something like, "a rule which associates to each real number a uniquely specified real number," which certainly defines a function—but hardly in a way that students will comprehend. The point that "a definition is satisfactory only if the students understand it" was already made by Poincaré [6] in 1909, but teachers of mathematics seem not to have paid much attention to it.

The difficulties of a vocabularly are not peculiar to mathematics; similar difficulties are what makes it so frustrating to try to talk to physicians or lawyers. They too insist on a rich technical language because "it is so much more precise that way." So it is, but the refined terminology is clearer only when rigorous distinctions are absolutely necessary. There is no use in emphasizing refined distinctions until the audience knows enough to see that they are needed.

Symbolism is a special kind of terminology. Mathematics can't get along without it. A good deal of progress has depended on the invention of appropriate symbolism. But let's not become so fascinated by the symbols that we forget what they stand for. Our audience (whether it is listening or reading) is going to be less familiar with the symbolism than we are. Hence it is not a good idea (to take a simple example) to say "Let f belong to L^2" instead of "Let f be a measurable function whose square is integrable," unless you are sure that the audience already understands the symbolism. Moreover, if you are not actually going to use L^2 as a Hilbert space, but want only the properties of its elements as functions, the structure of the space is irrelevant and calling attention to it is a form of showing off—mild, but it *is* showing off. If the audience doesn't know the symbolism, it is mystified; if it does know, it will be wondering when you are going to get to the point.

My advice about new terminology applies with even greater force to new symbolism. Do not create new symbolism, or change the old, unnecessarily; and admit (if necessary) that usage varies and explain the existing equivalences. If your $\Phi(x)$ also appears in the literature as $P(x)$ or $P(x) + \frac{1}{2}$ or $F(x)$, *say so.* Irresponsible improvements in notation have already caused enough trouble. I don't know who first thought of using θ in spherical coordinates to mean azimuth instead of colatitude, as it almost universally did and still does in physics and in advanced mathematics. It's superficially a reasonable convention because it makes θ the same as in plane polar coordinates; however, since r is different anyway, that isn't much help. The result is that students who go beyond calculus have to learn all the formulas over again. Such complications don't bother the true-blue pure mathematicians, those who would just as soon see Newton's second law of motion stated as $\mathbf{v} = (d/d\sigma)(\mathscr{R}\mathbf{q})$, but they do bother many students, besides irritating physical scientists.

Proofs. Only professional mathematicians learn anything from proofs. Other people learn from explanations. I'm not sure that even mathematicians learn much from proofs in fields with which they are not familiar. A great deal can be accomplished with arguments that fall short of being formal proofs. I have known a professor (I hesitate to say "teacher") to spend an entire semester on a proof of Cauchy's integral theorem under very general hypotheses. A collection of special cases and examples would have carried more conviction and left time for more varied and interesting material, besides leaving the audience better equipped to understand, apply, generalize, and teach Cauchy's theorem.

I cannot remember who first remarked that a sweater is what a child puts on when its parent feels cold; but a proof is what students have to listen to when the teacher feels shaky about a theorem. It has been claimed [**7**] that "some of the most important results...are so surprising at first sight that nothing short of a proof can make them credible." There are fewer of these than you think.

Experienced parents realize that when a child says "Why?" it doesn't necessarily want to hear a reason; it just wants more conversation. The same principle applies when a class asks for a proof.

Rigor. This is often confused with generality or completeness. In spite of what reviewers are likely to say, there is nothing unrigorous in stating a special case of a theorem instead of the most general case you know, or a simple sufficient condition rather than a complicated one. For example, I prefer to give beginners Dirichlet's test for the convergence of a Fourier series "piecewise monotonic and bounded" is more comprehensible than "bounded variation"; and, in fact, equally useful after one more theorem (learned later).

The compulsion to tell everything you know is one of the worst enemies of effective communication. We mathematicians would get along better with the Physics Department if, for example, we could bring ourselves to admit that, although their students need some Fourier analysis for quantum mechanics, they don't need a whole semester's worth—two weeks is nearer the mark.

Being more thorough than necessary is closely allied to **pedantry**, which (my dictionary says) is "excessive emphasis of trivial details."

Here's an example. Suppose students are looking for a local minimum of a differentiable function f, and they find critical points at $x = 2$, $x = 5$, and nowhere else. Suppose also that they do not want to use (or are told not to use) the second derivative. Some textbooks will tell them to check $f(2 + h)$ and $f(2 - h)$ for all small h. Students naturally prefer to check $f(3)$ and $f(1)$. The pedantic teacher says, "No"; the honest teacher admits that any point up to the next critical point will do.

Enthusiasm. Teachers are often urged to show enthusiasm for their subjects. Did you ever have to listen to a really enthusiastic specialist holding forth on something that you did not know and did not want to know anything about, say the bronze coinage of Poldavia in the twelfth century or "the doctrine of the enclitic *De*" [8]? Well, then.

Skills. A great deal of the mathematics that many mathematicians support themselves by teaching consists of subjects like elementary algebra or calculus or numerical analysis—skills, in short. It is not always easy to tell whether a student has acquired a skill or, as we like to put it, "really" learned a subject. The difficulty is much like that of deciding whether apes can use language in a linguistically interesting way or whether they have just become very clever at pushing buttons and waving their hands [9]. Mathematical skills are like any other kind. If you are learning to play the piano, you usually start by practicing under supervision; you don't begin with theoretical lectures on acoustical vibrations and the internal structure of the instrument. Similarly for mathematical skills. We often read or hear arguments about the relative merits of lectures and discussions, as if these were the only two ways to conduct a class. Having students practice under supervision is another and very effective way. Unfortunately it is both untraditional and expensive.

Even research in mathematics is, to a considerable extent, a teachable skill. A student of G. H. Hardy's once described to me how it was done. If you were a student of Hardy's, he gave you a problem that he was sure you could solve. You solved it. Then he asked you to generalize it in a specific way. You did that. Then he suggested another generalization, and so on. After a certain number of iterations, you were finding (and solving) your own problems. You didn't necessarily learn to be a second Gauss that way, but you could learn to do useful work.

Lectures. These are great for arousing the emotions. As a means of instruction, they ought to have become obsolete when the printing press was invented. We had a second chance when the Xerox machine was invented, but we seem to have muffed it. If you *have* to lecture, you can at least hand out copies of what you said (or wish you had said). I know mathematicians who contend that only through their lectures can they communicate their personal attitudes toward their subjects. This may be true at an advanced level, for pre-professional students. Otherwise I wonder whether these mathematicians' personalities are really worth learning about, and (if so) whether the students couldn't learn them better some other way (over coffee in the cafeteria, for example.)

One of the great mysteries is: How can people manage to extract useful information from incomprehensible nonsense? In fact, we can and do. Read, for example, in Morris Kline's book [10] about the history of the teaching of calculus. Perhaps this talent that we have can explain the popularity of lectures. One incomprehensible lecture is not enough, but a whole course may be effective in a way that one incomprehensible book never can. I still content that a comprehensible book is even better.

Conclusion. I used to advise neophyte teachers: "Think of what your teachers did that you particularly disliked—and don't do it." This was good advice as far as it went, but it didn't go far enough. My tentative answer to the question in my title is, "Yes; but don't be guided by introspection." You cannot expect to communicate effectively (whether in the classroom or in writing) unless and until you understand your audience. This is not an easy lesson to learn.

References

1. See, for example, Sydney J. Harris, column for February 9, 1980, Chicago Sun-Times and elsewhere.

2. L. R. Ford, Retrospect, this MONTHLY, 53 (1946) 582–585.

3. College Mathematics: Suggestions on How to Teach It, Mathematical Association of America, 1972.

4. D. R. Stoutemyer, Symbolic computation comes of age, SIAM News, 12, no. 6 (December 1979) 1, 2, 9.

5. The same point has been made by P. R. Halmos in How to Write Mathematics, L'Enseignement mathématique, (2) 16 (1970) 123–152.

6. H. Poincaré, Science et méthode, 1909, Book II, Chapter 2.

7. H. and B. S. Jeffreys, Methods of Mathematical Physics, 2nd ed., Cambridge University Press, 1950, p. v.

8. Robert Browning, "A Grammarian's Funeral," in The Complete Poetic and Dramatic Works of Robert Browning, Houghton Mifflin, Boston and New York, 1895, pp. 279–280.

9. For example, E. S. Savage-Rumbaugh, D. M. Rumbaugh, and S. Boysen, Do apes use language? Amer. Scientist, 68 (1989) 49–61.

10. Morris Kline, Mathematics: The Loss of Certainty, Oxford University Press, New York, 1980.

88(1981), 727–731

RETROACTIVE EDITORIAL POLICY

R. P. BOAS

"Let me make it clear to you
This is what we'll never do."

It really doesn't matter if you don't know how to spell:
You'll find that many readers understand you just as well;
And once the spelling's gone to pot, why then I rather guess
It doesn't really matter if the syntax is a mess;
    But not in *my* journal.

You really shouldn't worry that it's all been said before,
Since checking out the sources can be an awful bore.
And don't be very troubled if you seem to plagiarize
If you copy comething really good, you just might win a prize;
    But not in *my* journal.

We often note that authors, even those whose work is strong,
They sometimes go too far and say a thing or two that's wrong.
You needn't worry very much about a stray mistake:
If you can fool the referee, what difference does it make?
    But not in *my* journal.

Oh, inspiration's wonderful but second thoughts are best,
And don't you think we must have made those silly rules in jest?
You seekers for perfection who've made an utter goof,
You'll have a chance to fix it up: rewrite in galley proof;
    But not in *my* journal.

89(1982), 32

The Disk with the College Education

Herbert S. Wilf

The title is somewhat exaggerated, but the calculators-or-no-calculators dilemma that haunt the teaching of elementary school mathematics is heading in the direction of college mathematics and this article is intended as a distant early-warning signal.

I have in my home a small personal computer. About 500,000 small personal computers have been sold in this country, of which a healthy fraction are owned by individuals. I use mine primarily for word processing (this article was written on it), for writing programs that do various mathematical jobs related to my teaching or to my research, for playing games, for keeping class rolls, etc.

A new program has recently been made available for my little computer, one whose talents seem worthy of comment here because it knows calculus; in fact, as you read these words, some of your students may be doing their homework with it.

The program is called muMATH; it was written by the Soft Warehouse, and is distributed in the United States by Microsoft Consumer Products of Bellevue, Washington. It costs about $75 and is supplied on a 5-inch floppy disk with an (inadequate) instruction manual.

The program on the disk does numerical calculation to high precision, or symbolic manipulation of expressions. The numerical calculation, which is less important as far as this article is concerned, is in rational arithmetic and is done with 611-digit accuracy. Thus, for example, when the program is loaded, the question

$$?30!;$$

yields the instant answer

$$@265252859812191105863630848000000$$

The question

$$?1 + 1/2 + 1/3 + 1/4 + 1/5 + 1/6 + 1/7;$$

elicits

$$@363/140$$

and so forth.

But these are fairly standard calculator-type questions. The first glimmer that a nontrivial intelligence lives on the disk comes with the request for $\sqrt{12}$,

$$?12\uparrow(1/2);$$

(the up arrow means "to the power"), whence the response

$$@2*3\uparrow(1/2)$$

At least the disk has been to junior high school. The next few samples show it in grades 9–12:

$$?(X + 2*Y)\uparrow 3;$$

$$@12*X*Y\uparrow 2 + 6*X\uparrow 2*Y + X\uparrow 3 + 8*Y\uparrow 3$$

$$?\mathrm{COS}(5*Y);$$

$$@-20*\mathrm{COS(Y)}\uparrow 3 + 16*\mathrm{COS(Y)}\uparrow 5 + 5*\mathrm{COS(Y)}$$

(Tschebycheff polynomials, anyone?).

The disk, however, has graduated from high school. Here it is in a freshman calculus course. To differentiate $x \sin x$ with respect to x just ask

$$?\mathrm{DIF}(X*\mathrm{SIN}(X), X);$$

to obtain

$$@X*\mathrm{COS}(X) + \mathrm{SIN}(X)$$

. . .

Programs that do symbolic manipulation of mathematical expressions are not new. The MACSYMA program of MIT has been doing it for years. What is new is the sudden mass availability of a program with these capabilities, and the promise that more of the same is in prospect.

This year for the first time there are widely available pocket computers: objects about the size of calculators that have thousands of bytes of memory and speak BASIC. These pocket computers are not yet quite powerful enough to handle a sophisticated program like muMATH. In a few years, though, they probably will be.

As teachers of mathematics, our responses might range all the way from a declaration that "no computers are allowed in exams or to help with homework" to the if-you-can't-lick-'em-then-join-em approach (teach the students how to use their clever little computers).

Will we allow students to bring them into exams? Use them to do homework? How will the content of calculus courses be affected? Will we take the advice that we have been dispensing to teachers in the primary grades: that they should teach more of concepts and less of mechanics. What happens when $29.95 pocket computers can do all of the above and solve standard forms of differential equations, do multiple integrals, vector analysis, and what-have-you?

Excuse me if I don't have answers, I wanted only to raise the questions and beat a hasty retreat.

89(1982), 4–7

'Cos some of us don't know the way
To say the name of André Weil,
It leaves the rest of us the while
Confusing him with Hermann Weyl.

RICHARD K. GUY

The Meaning of the Conjecture $P \neq NP$ For Mathematical Logic

Jan Mycielski

I think that the conjecture $P \neq NP$ is not as widely taught in courses of mathematical logic as it should be, in view of its capital importance for the foundation of mathematics. Therefore I am writing this note in the hope that all logicians will always include it in the introductory courses of their subject although it does not appear yet in the appropriate books. The original paper of S. Cook [1], where the conjecture was formulated, was indeed written from the point of view of logic but it became the domain of computer scientists (see [2]), particularly because of a paper of Karp [3] where the combinatorial or computational aspects of the conjecture were developed in a very suggestive way.

Let us assume that the teacher has already presented the concept of a first order theory and the concept of a Turing machine (neither the concepts of a decidable theory nor that of a recursive function are needed). Then he may proceed as follows:

By a *normal theory* we shall mean a theory which is formalized with a finite alphabet in first order logic with equality and is axiomatizable by a finite set of axioms and axiom schemata in which one can prove $\exists xy[x \neq y]$. (In [4] it is proved that every theory which is recursively axiomatizable and contains a minimal amount of arithmetic or set theory is normal.) By a proof in a normal theory we mean a Hilbert style proof from the axioms.

Let Σ be a finite alphabet of Σ^* the set of all words, i.e., finite sequences of elements of Σ. For any $\xi \in \Sigma^*$, $|\xi|$ denotes the length of ξ. Now we introduce a more abstract concept of a theory which we will call a $\tau\pi$-theory.

A $\tau\pi$-*theory* is a set of pairs $T \subseteq \Sigma^* \times \Sigma^*$ such that there exists a polynomial $P(x, y)$ and a Turing machine M such that, for any $(\tau, \pi) \in \Sigma^* \times \Sigma^*$, M can decide in time $\leq P(|\tau|, |\pi|)$ if $(\tau, \pi) \in T$.

If $(\tau, \pi) \in T$, then τ is called a *theorem* of T and π is called a *proof* of τ in T.

Every normal theory defines a $\tau\pi$-theory since the time necessary to check the correctness of a Hilbert style proof in a normal theory can be estimated from above by a polynomial in the length of that proof.

Now, a $\tau\pi$-theory T will be called *amenable* (to *automatization*) iff there exists another polynomial $P_0(x, y)$ and another Turing machine M_0 such that, given any word $\tau \in \Sigma^*$ and any positive integer n, M_0 can decide in time $\leq P_0(|\tau|, n)$ if there exists a $\pi \in \Sigma^*$ with $|\pi| \leq n$ and such that $(\tau, \pi) \in T$. (Notice that if we replaced the condition $\leq P_0(|\tau|, n)$ by the condition $\leq P_0(|\tau|, c^n)$ where $c = \text{card}\Sigma$, then the concept would trivialize since every $\tau\pi$-theory would be amenable. In fact, given a time $P_0(|\tau|, c^n)$, the machine can form all sequences of symbols of length n and find out if any of them is a proof of τ.)

It is clear that, after Gödel's 1931 discovery that all sufficiently strong theories are undecidable, the next question which presents itself is to ask if the $\tau\pi$-theories (corresponding to normal theories) are amenable. But we had to wait until 1970 (the paper of Cook [1]) for a clear statement of that question. It can be asked in many equivalent ways and the proofs of their equivalence are very ingenious (see [2]). Now we will state three such ways which are most striking to the logician.

PROPOSITION. *The following three statements are equivalent to each other:*

    (i) *There exists a $\tau\pi$-theory which is not amenable.*
    (ii) *Every $\tau\pi$-theory which is defined by a normal theory is not amenable.*
    (iii) $P \neq NP$.

This proposition follows immediately from the theorem of Cook that the set of satisfiable formulas of propositional calculus is NP-complete. (We refer the reader to [1], [2], [3].) It is not known if $P \neq NP$; this is the conjecture of Cook. I believe that this is the most outstanding problem of contemporary mathematics (I rank it higher than the Riemann-ζ conjecture). The above proposition is the best way of explaining the conjecture's capital importance in foundations of mathematics. (Its great importance in theoretical computer science is well known (see [2]).)

References

1. S. A. Cook, The Complexity of Theorem-Proving Procedures, Proc. 3rd Annual ACM Sympos. on Theory of Computing, 1970, pp. 151–158.

2. M. R. Garey and D. S. Johnson, Computers and Intractability, a Guide to the Theory of NP-Completeness. Freeman, San Francisco, 1979.

3. R. M. Karp, Reducibility among combinatorial problems, in the collection Complexity of Computer Computations, edited by J. W. Thatcher, Plenum Press, New York-London, 1972.

4. R. L. Vaught, Axiomatizability by a schema, J. Symbolic Logic, 32 (1967) 473–479.

90(1983), 129–130

When I consider how my grants are spent,
Ere half my days in these dark halls and wide;
My teaching talent, which I'd love to hide,
Lodged with me useless, though my soul more bent
To serve therewith my Provost, and present
My teaching ratings, lest he returning chide;
"Do Deans exact much teaching, grants denied?"
I fondly ask: the Provost, to prevent
That murmur, soon replies, "Deans do not need
Either that teaching or those grants. Who best
Does his committee work, serves best. His state
Is Deanly: thousands at his bidding speed
And fly to Washington, disdaining rest;
They also serve who only loaf and wait."

—Edwin Hewitt

The Editor's Corner: The New Mersenne Conjecture

P. T. Bateman, J. L. Selfridge and S. S. Wagstaff, Jr.

It is well known that Mersenne stated in his *Cogitata* [4] that, of the fifty-five primes $p \leqslant 257$, $2^p - 1$ is itself prime only for the eleven values

$$p = 2, 3, 5, 7, 13, 17, 19, 31, 67, 127, \text{ and } 257.$$

It is also well known that his list had five errors: $p = 67$ and 257 should have been removed from the list while $p = 61, 89$, and 107 should have been added to it.

Several authors [1, 2, 3] have speculated about how Mersenne formed his list. It is easy to notice that all numbers on his (incorrect) list lie within 3 of some power of 2. However, Mersenne certainly knew that $2^{11} - 1$ is composite and hence that not all primes $p = 2^k \pm 3$ produce prime $M_p = 2^p - 1$. The next prime of this form not on Mersenne's list is $p = 29$. He surely knew that M_{29} is composite, as it has the small divisor 233. Also 263 divides $2^{131} - 1$. Mersenne's list is explained by the rule

M_p is prime if and only if p is a prime of one of the forms $2^k \pm 1$ or $2^{2k} \pm 3$ (1)

except for the omission of $p = 61$. In fact Mersenne stated in [5, Chap. 21, p. 182] a rule very similar to (1). (The verb "differs"—not "exceeds," as some have guessed—is omitted from this sentence, but Mersenne supplied it in a corrigendum on the back of page 235.) Drake [2] quotes this sentence from [5], locates the missing verb and argues that (1) was in fact Mersenne's rule. He suggests that 61 was missing from [4] either because of a typographical error or because Mersenne mistakenly believed that M_{61} is composite. When copying a list, like "..., 61, 67, ...", containing two adjacent similar items, it is a common error to omit the first of these (here "61").

Now the question presents itself: Is there a neat way to distinguish the Mersenne hits like 31, 61, 127 from the Mersenne misses like $67, 257, \ldots$ and $89, 107, \ldots$? When $(2^{127} + 1)/3$ was proved prime, we began looking at the other $(2^p + 1)/3$. We noticed that they were prime for the hits and composite for the misses! Is this accidental? Will "a little more computing" find a counterexample?

We replace (1) by this new, related conjecture that when both sides of (1) are true, $(2^p + 1)/3$ is prime, and when (1) is false, $(2^p + 1)/3$ is composite. Restating this conjecture we get the

NEW MERSENNE CONJECTURE. *If two of the following statements about an odd positive integer p are true, then the third one is also true.*

(a) $p = 2^k \pm 1$ or $p = 4^k \pm 3$.
(b) M_p is prime.
(c) $(2^p + 1)/3$ is prime.

It is not necessary to assume that p is prime, for if p is composite (or 1), then statements (b) and (c) are both false and the conjecture holds.

It is easy to find examples of primes p for which all three statements are true ($p = 3, 5, 7, 13, 17, 19, 31, 61, 127$) or all three are false ($p = 29, 37, 41, 47, \ldots$) or exactly one is true ($p = 67, 257, 1021, \ldots$ for only (a) true; $p = 89, 107, 521, \ldots$ for only (b) true; and $p = 11, 23, 43, 79, \ldots$ for only (c) true). However, the New Mersenne Conjecture is true for all p less than 100000, which is the current limit of the search for Mersenne primes. It is valid also for all p between 10^5 and 10^6 for which at least one of the three statements is known to hold. We expect that the three statements are true simultaneously only for the nine primes mentioned above.

The Table above summarizes what is known about our conjecture. It lists all odd primes p satisfying at least one of these three conditions:

(1) $p < 1000000$ and $p = 2^k \pm 1$ or $p = 4^k \pm 3$.
(2) $p < 100000$ and $2^p - 1$ is prime.
(3) $p < 4000$ and $(2^p + 1)/3$ is prime.

When a number is asserted to be composite, a factor is given if one is known. The factors of M_{131071} and M_{524287} were found by Robinson [6]. The 1065-digit number $(2^{3539} + 1)/3$ passed a probabilistic primality test, but we did not give a complete proof that it is prime.

It is a simple consequence of quadratic reciprocity that if $p \equiv 1 \pmod 4$, then the factors of $2^p - 1$ are congruent to 1 or $6p + 1 \pmod{8p}$, and if $p \equiv 3 \pmod 4$, then the factors of $2^p - 1$ are congruent to 1 or $2p + 1 \pmod{8p}$. This observation is the starting point for a heuristic argument [7] which concludes that the number of p less than y for which M_p is prime is about $e^\gamma \log_2 y \approx 1.78 \log_2 y$, where γ is Euler's constant.

Likewise, one can show that if $p \equiv 1 \pmod 4$, then the factors of $(2^p + 1)/3$ are congruent to 1 or $2p + 1 \pmod{8p}$, and if $p \equiv 3 \pmod 4$, then the factors of $(2^p + 1)/3$ are congruent to 1 or $6p + 1 \pmod{8p}$. A heuristic argument like the one mentioned above concludes that the number of p less than y for which $(2^p + 1)/3$ is prime is also about $e^\gamma \log_2 y$.

This total number of natural numbers less than y with one of the forms $2^k \pm 1$ or $4^k \pm 3$ is about $3 \log_2 y$. Hence, the number of primes less than y with one of these forms is $O(\log y)$.

In view of the foregoing heuristics and the fact that there are about $y/\log y$ primes less than y, the probability that any one of the three statements holds for a randomly chosen prime p less than y is $O(y^{-1} \log^2 y)$. If the three statements were independent random events, then the expected number of primes p greater than L for which at least two of the statements hold is about $C \int_L^\infty y^{-2} \log^4 y \, dy$, which is finite. Substituting $L = 100000$ gives an upper bound on the expected number of failures of the New Mersenne Conjecture. Assuming a reasonable value for C (about 9) we find that the expected number of failures is less than 1. This is one

TABLE for "The New Mersenne Conjecture"

p	$p = 2^k \pm 1$ or $4^k \pm 3$?	$2^p - 1$ prime?	$(2^p + 1)/3$ prime?
3	yes (-1)	yes	yes
5	yes $(+1)$	yes	yes
7	yes $(-1$ or $+3)$	yes	yes
11	no	no: 23	yes
13	yes (-3)	yes	yes
17	yes $(+1)$	yes	yes
19	yes $(+3)$	yes	yes
23	no	no: 47	yes
31	yes (-1)	yes	yes
43	no	no: 431	yes
61	yes (-3)	yes	yes
67	yes $(+3)$	no: 193707721	no: 7327657
79	no	no: 2867	yes
89	no	yes	no: 179
101	no	no: 7432339208719	yes
107	no	yes	no: 643
127	yes (-1)	yes	yes
167	no	no: 2349023	yes
191	no	no: 383	yes
199	no	no: 164504919713	yes
257	yes $(+1)$	no: 535006138814359	no: 37239639534523
313	no	no: 10960009	yes
347	no	no: 14143189112952632419639	yes
521	no	yes	no: 510203
607	no	yes	no: 115331
701	no	no: 796337	yes
1021	yes (-3)	no: 40841	no: 10211
1279	no	yes	no: 706009
1709	no	no: 379399	yes
2203	no	yes	no: 13219
2281	no	yes	no: 22811
2617	no	no: 78511	yes
3217	no	yes	no: 7489177
3539	no	no: 7079	yes (prp)
4093	yes (-3)	no	no
4099	yes $(+3)$	no: 73783	no: 2164273
4253	no	yes	no: 118071787
4423	no	yes	no
8191	yes (-1)	no: 338193759479	no
9689	no	yes	no: 19379
9941	no	yes	no
11213	no	yes	no
16381	yes (-3)	no	no: 163811
19937	no	yes	no
21701	no	yes	no: 43403
23209	no	yes	no: 4688219
44497	no	yes	no: 2135857
65537	yes $(+1)$	no	no
65539	yes $(+3)$	no	no: 58599599603
86243	no	yes	no
110503	no	yes	no
131071	yes (-1)	no: 231733529	no: 2883563
132049	no	yes	no
216091	no	yes	no
262147	yes $(+3)$	no: 268179002471	no: 4194353
524287	yes (-1)	no: 62914441	no

reason why we believe that the conjecture is true. Another reason is that it holds for all p less than 100000 as well as those larger p for which it has been tested.

We are grateful to Duncan A. Buell and Jeff Young for testing the primality of $(2^p + 1)/3$ for several $p > 50000$, using a Cray 2 computer.

References

1. R. C. Archibald, Mersenne's numbers, *Scripta Math.*, 3 (1935), 113.

2. Stillman Drake, The rule behind 'Mersenne's numbers', *Physis-Riv. Internaz. Storia Sci.*, 13 (1971) 421–424. MR 58#26870.

3. Malcolm R. Heyworth, A conjecture on Mersenne's conjecture, *New Zealand Math. Mag.*, 19 (1982) 147–151. MR 85a:11002.

4. M. Mersenne, Cogitata Physico Mathematica, Parisiis, 1644, Praefatio Generalis No. 19.

5. _____, Novarum Observationum Physico-Mathematicarum, Tomus III, Parisiis, 1647.

6. Raphael M. Robinson, Some factorizations of numbers of the form $2^n \pm 1$, *Math. Tables Aids Comput.* 11 (1957) 265–268, MR 20 #832.

7. S. S. Wagstaff, Jr., Divisors of Mersenne numbers, *Math. Comp.*, 40 (1983) 385–397, MR 84j: 10052.

96(1989), 125–128

The Latest Mersenne Prime

Will they ever stop coming? David Slowinski and Paul Gage, of Cray Research, recently announced the discovery of the latest (and largest) Mersenne prime. The 32nd known Mersenne prime is $2^{756839} - 1$, a number with 227,831 digits. The number was shown prime using a program written by Slowinski and Gage on a Cray-2 computer at Harwell Laboratory in Didcot, England.

Proving a random number of this size is prime would be impossible. (Trial division, for example, would be futile—there are about 10^{113910} primes to divide.) For Mersenne primes, there is the famous Lucas-Lehmer test: $M_p = 2^p - 1$ is prime if and only if M_p divides U_p where $\{U_n\}$ is the sequence of numbers starting with $U_2 = 4$ and defined recursively by $U_n = U_{n-1}^2 - 2$. Raising numbers to powers—even such large powers—is possible with some clever work. Squaring a number with over 200,000 digits is not easy, however. Slowinski and Gage used an algorithm of Schonhage and Strassen that employs the Fast Fourier Transform (in a clever implementation by Dennis Kuba, also of Cray Research). Checking the M_{756839} for primality the first time still required many hours of computer time; rechecking it on a machine with 16 processors required 20 minutes.

Before this, the largest known Mersenne prime was M_{216092}; the next before that is M_{110503} (discovered by Colquitt and Welsh, *Mathematics of Computation*, 56:194, April 1991, pp. 867–870). Are there others in between? No one is sure. The computer at Harwell discovered the new prime after checking only 85 exponents. Slowinski, an old hand at finding Mersenne Primes, says, "We were incredibly lucky." Slowinski seems to have more than his share.

99(1992), 360

Index of Names

Abel, Niels Henrik 42, 110, 273
Adam 177
Adams, C. R. 153, 249
Agnew, R. P. 182
Ahlfors, Lars 250
Aiken 267
Alaoglu, Leonidas 153
Albers, D. J. 298
Albert, A. A. 113
Alder, Henry L. 250
Alexander 64, 187
Alexanderson, G. L. 235
Aley, Robert J. 80
Allendoerfer, C. B. 113
Anderegg, F. 14
Andree, R. V. 192
Ankeny, N. C. 245
Apostol, T. 297
Archibald, R. C. 41
Archimedes 98
Arnold, B. H. 153
Arrington-Idowu, Elayne 300
Artin, Emil 113, 140, 253
Atiyah, Sir Michael 249

Babbage, Charles 267
Bach, J. S. 41
Bacher, R. F. 117
Bachmann 86
Baker, C. L. 242
Ball, W. W. Rouse 19
Ballantine, J. P. 70
Bamberger 86
Bandy, J. M. 30
Barrett, Lida 302
Bass, H. 249
Bateman, P. T. 313
Beatley, Ralph 107
Becher, Franklin A. 7
Beethoven, L. v. 55
Bell, Eric Temple 72, 75, 93
Bell, Clifford 235
Bellman, Richard 296–297
Bers, L. 250
Berstein, Dorothy 302

Besicovitch 271
Bethe, H. A. 117
Beurling 250
Bing, R. H. 238
Birkhoff, Garrett 98, 113, 251, 283
Birkhoff, George David 98, 143, 252–253
Bliss, Gilbert Ames 54, 58, 60
Boas, R. P., Jr. 97, 108, 229, 245, 276, 282, 303, 308
Bôcher 161
Bohnenblust 148
Bolyai, János 41, 43, 55
Boole 48
Borel 143
Bott, Raoul 215, 249
Bourbaki 304
Bozeman, Sylvia Trimble 300
Brigg, Henry 85
Broad, C. D. 46
Brouwer, L. E. J. 166
Brown, B. H. 95
Brown, E. W. 159
Browne, Marjorie Lee 299
Browning, Robert 93, 172
Bruns 135–136
Buchanan, H. E. 175
Buell, Duncan A. 316
Bundy, McGeorge 252
Burt, R. C. 90
Bush, V. 87
Byerly, W. E. 48
Byrne, Col. W. E. 183

Cairns, W. D. 25, 118, 174, 184
Cajori, Florian 2, 11, 28, 47, 70
Cantor, Georg 7
Carleman, T. 195
Carmichael, R. D. 13, 61, 63–64
Cartan, E. J. 110
Cartan 248
Cauchy, Augustin-Louis 137
Cayley, Arthur 11
Chapman, Sydney 271
Chauvenet, William 50, 60
Chern, S-S. 248, 250